W9-ARZ-054

MODERN ALGEBRA: AN INTRODUCTION

MODERN ALGEBRA
AN INTRODUCTION

John R. Durbin
The University of Texas at Austin

John Wiley & Sons, New York • Chichester • Brisbane • Toronto

Cover Illustration by Angie Lee

Library of Congress Cataloging in Publication Data:

Durbin, John R.
Modern algebra.

Includes indexes.

1. Algebra, Abstract. I. Title.
QA162.D87 512′.02 78-15778
ISBN 0-471-02158-X

Printed in the United States of America

10 9 8 7 6 5 4

To Jane

Preface

This book is an introduction to modern (abstract) algebra for undergraduates who may not have had a previous course in linear algebra. The first third of the book presents the core of the subject; the remainder of the book is designed to be as flexible as possible, and covers the traditional topics as well as some topics that have not been traditional.

I have responded to the increasing and entirely reasonable tendency of students to ask "What is it good for?" in two ways: First, by including an informal orientation in the Introduction. Second, by including more than is usual on the applications of algebra.

The case for modern algebra rests, however, on more than the applications that can be presented in an introductory textbook. The basic ideas and ways of thought of algebra permeate nearly every part of mathematics, and the subject has a coherence that is well worth the attention of every serious undergraduate mathematics student—not only future secondary school teachers and graduate students, but also computer science students and others who will use discrete mathematics. Moreover, no subject is better suited than modern algebra to cultivate the ability to handle abstract ideas; that ability must not be undervalued in the present-day rush to immediate relevance. Therefore, although I have included a number of applications, I have been concerned first of all with introducing the most important kinds of algebraic structures as objects naturally deserving attention, and with trying to help students improve their ability to handle abstract ideas.

Some may regret, as I do, that various topics have not been developed further in the book. It simply seemed that for most audiences the depth found here would be sufficient, and the book should not be unnecessarily long. Students studying algebra in graduate school will obviously go further, as will those who are interested in one or more of the applications.

The most basic ideas of modern algebra—group, ring, integral domain, field, and isomorphism—are introduced in Chapters I to V. In one quarter there should be time to cover those chapters and also at least one additional topic—perhaps part or all of Chapter VI, which draws on the earlier material to analyze each of the familiar number systems. A one-semester course could add Chapter VII, on group homomorphisms, and also one or more additional chapters—possibly applications of groups in combinatorics and symmetry, more about rings, the chapter on algebraic coding, or lattices and Boolean algebras. In all, if each section is studied in detail there should be enough material for a full-year course.

Most of the sections of the book can be covered in one lecture each; that is especially true of the sections in Chapters I to VII. The pace of a course will obviously depend on students' backgrounds and the number and types of problems assigned. Some of the problems, especially later in the book, contain fairly substantial extensions of the material in the text. Notice that elementary facts about sets, logic, proofs, and mathematical induction are collected in the appendixes.

Nearly all of the book can be understood without a knowledge of linear algebra. The notable exceptions are Section 37 (the Euclidean group), part of Section 49 (finite fields), and Chapter XVI (algebraic coding); Appendix D contains a concise review of the linear algebra that is used in those parts of the book. Appendix D also presents the standard examples of groups and rings of matrices and linear transformations, and thereby provides a ready reference for instructors who choose to include those examples as an integral part of the course.

Historical remarks are inserted throughout the book, and references and remarks are given in notes at the end of most chapters. Any reference number (such as [1]) refers to the list at the end of the chapter in which the number appears.

Here are several possibilities for courses based on the book. Prerequisites can be determined from the table showing the interdependence of chapters, which follows this preface.

Introduction to the basic ideas (group, ring, integral domain, field, isomorphism): Chapters I to V.

Additional material on the familiar number systems and group homomorphisms: Chapters VI and VII.

Applications of groups: Chapters VIII and IX.

Additional material on rings: Chapters X to XII.

Fields and polynomials, and their historical applications: Chapters X to XV.

Applications of discrete mathematics (such as for computer sciences):

- Chapter VIII (Sections 31, 32)—combinatorics
- Chapter X—integers
- Chapter XI (Sections 40–42)—polynomials
- Chapter XII (Sections 44–46)—background for finite fields
- Chapter XIII—finite fields
- Chapter XVI—algebraic coding
- Chapter XVII—lattices and Boolean algebras

It is a pleasure to acknowledge the advice I have received from Professors J. E. Adney, Jr., Michigan State University; Gene Levy, University of Oklahoma; and Merry McDonald, Northwest Missouri State University. I am also grateful for the support and cooperation given by Andrew E. Ford, Jr., of John Wiley. The book is dedicated to my wife, Jane; my deepest thanks must go to her.

John R. Durbin

Interdependence of Chapters

I. Mappings and Operations

Chapter	I	II	III	IV	V	VI	VII	VIII	IX	X	XI	XII	XIII	XIV	XV	XVI
II. Introduction to Groups	∘															
III. Equivalence and Congruence	∘	∘														
IV. Groups	∘	∘	∘													
V. Introduction to Rings	∘	∘	∘	∘												
VI. The Familiar Number Systems	∘	∘	∘	∘	∘											
VII. Group Homomorphisms	∘	∘	∘	∘												
VIII. Applications of Permutation Groups	∘	∘	∘	∘			∘									
IX. Symmetry[†]	∘	∘	∘	∘												
X. Factorization of Integers	∘	∘	∘													
XI. Polynomials	∘	∘	∘	∘	∘	∘				∘						
XII. Quotient Rings	∘	∘	∘	∘	∘	∘	∘			∘	∘					
XIII. Field Extensions	∘	∘	∘	∘	∘	∘	∘			∘	∘	∘				
XIV. Polynomial Equations	∘	∘	∘	∘	∘	∘	∘			∘	∘	∘	∘			
XV. Geometric Constructions	∘	∘	∘	∘	∘	∘	∘			∘	∘	∘	∘			
XVI. Algebraic Coding[†]	∘	∘	∘	∘	∘											
XVII. Lattices and Boolean Algebras	∘	∘	∘	∘	∘											

Note: A dot (∘) in a square indicates that the chapter with its number above the square is a prerequisite for the chapter with its number to the right of the square. Example: The prerequisites for Chapter VIII are Chapters I, II, III, IV, and VII. In some cases only part of a chapter is a prerequisite.

[†]Section 37 (in Chapter IX) and Chapter XVI also require linear algebra, which is reviewed in Appendix D.

Contents

INTRODUCTION 1

Chapter I

Mappings and Operations

1. Mappings 11
2. Composition. Invertible Mappings 17
3. Operations 20
4. Composition as an Operation 25

Chapter II

Introduction to Groups

5. Definition and Examples 30
6. Permutations 35
7. Subgroups 40
8. Groups and Symmetry 45

Chapter III

Equivalence and Congruence

9. Equivalence Relations 51
10. Congruence. The Division Algorithm 55
11. Integers Modulo *n* 59

Chapter IV

Groups

12. Elementary Properties 64
13. Cosets. Direct Products 69
14. Lagrange's Theorem 73
15. Isomorphism 76
16. More on Isomorphism 80
17. Cayley's Theorem 85

Chapter V

Introduction to Rings

18. Definition and Examples 89
19. Integral Domains. Subrings 95
20. Fields 98
21. Isomorphism. Characteristic 102

Chapter VI

The Familiar Number Systems

22. Ordered Integral Domains 108
23. The Integers 111
24. Fields of Quotients. The Field of Rational Numbers 113
25. Ordered Fields. The Field of Real Numbers 117
26. The Field of Complex Numbers 121
27. Complex Roots of Unity 127

Chapter VII

Group Homomorphisms

28. Homomorphisms of Groups. Kernels 133
29. Quotient Groups 138
30. The Fundamental Homomorphism Theorem 141

Chapter VIII

Applications of Permutation Groups

31. Groups Acting on Sets 148
32. Burnside-Pólya Counting 153
33. Sylow's Theorem 159

Chapter IX

Symmetry

34. Finite Symmetry Groups 164
35. Infinite Two-Dimensional Symmetry Groups 173
36. On Crystallographic Groups 178
37. The Euclidean Group 185

Chapter X

Factorization of Integers

38. Greatest Common Divisors. The Euclidean Algorithm 190
39. The Fundamental Theorem of Arithmetic 195

Chapter XI

Polynomials

40. Definition and Elementary Properties 198
Appendix to Section 40 201
41. The Division Algorithm 204
42. Factorization of Polynomials 208
43. Unique Factorization Domains 213

Chapter XII

Quotient Rings

44. Homomorphisms of Rings. Ideals 219
45. Quotient Rings 223

46. Quotient Rings of $F[x]$ 226
47. Factorization and Ideals 230

Chapter XIII

Field Extensions

48. Adjoining Roots 236
49. Finite Fields 240

Chapter XIV

Polynomial Equations

50. Roots of a Polynomial 246
51. Rational Roots. Conjugate Roots 248
52. An Introduction to Galois Theory 250

Chapter XV

Geometric Constructions

53. Three Famous Problems 254
54. Constructible Numbers 259
55. Impossible Constructions 260

Chapter XVI

Algebraic Coding

56. Introduction 263
57. Linear Codes 267
58. Standard Decoding 271
59. Error Probability 275

Chapter XVII

Lattices and Boolean Algebras

60. Partially Ordered Sets 280
61. Lattices 285

62. Boolean Algebras 289
63. Finite Boolean Algebras 295
64. Switching 299

Appendix A

Sets 304

Appendix B

Proofs 307

Appendix C

Mathematical Induction 312

Appendix D

Linear Algebra 315
Photo Credit List 321
Index 323

MODERN ALGEBRA: AN INTRODUCTION

Figure 1

Arabic (13th century).

San Francesco in Assisi (13th century).

San Francesco in Assisi (13th century).

INTRODUCTION

Modern algebra—like any other branch of mathematics—can be mastered only by working up carefully from the most basic ideas and examples. But that takes time, and some of the goals will not be clear until you reach them. This section is meant to help sustain you along the way. You may read this all at first, or begin with Chapter I and then return here at your leisure. The purpose is simply to convey some feeling for how modern algebra developed and for the kinds of problems it can help solve.

Please note that this section is not a survey of all of modern algebra. There is no discussion of the applications of algebra in modern theoretical physics, for instance, or of some of the deeper applications of algebra in mathematics itself; in fact, such applications are not covered even in the text. The examples here have been chosen because they can be understood without special background. Even at that, by its very purpose this section must occasionally be vague; we can worry about details and proofs when we get to the text itself.

Symmetry. Symmetrical designs like those in Figure 1 have been used for decoration throughout history. Each of these designs is built up from a basic irreducible component; such components for the designs in Figure 1 are shown in Figure 2. In each case the plane can be filled without overlap if the basic component is repeated by appropriate combinations of rotation (twisting), translation (sliding), and reflection (such as interchanging left and right).

Although the artistic possibilities for the components are unlimited, there is another sense in which the possibilities are *not* unlimited. Notice that the lower two examples in Figure 1 will look the same if the page is

Figure 2

turned upside down, but the top example will not. Because of this we could say that the top example has a different symmetry type from the other two examples. Continuing, notice that the middle example lacks the strict left-right symmetry of the top and bottom examples. (For instance, in the irreducible component for the middle example the ring passes under the white band on the left, but over the white band on the right.) We can conclude, then, that the three examples somehow represent three different symmetry types—different combinations of rotations, translations, and reflections are needed to build up the three different designs from their basic components. If we were to look at more designs, we would find examples of still other symmetry types, and we would also find many examples that could be distinguished from one another but not on the basis of symmetry type alone. At some stage we would feel the need for a more precise definition of "symmetry type." There is such a definition, and it turns out that in terms of that definition there are exactly 17 different symmetry types of plane-filling designs. Figure 35.9 shows one example of each type. Although each of the 17 types occurs in decorations from ancient civilizations, it was only in the nineteenth century that these possibilities were fully understood. The key to making "symmetry type" precise, and also to determining the number of different symmetry types, is the idea of a *group*.

Groups. The idea of a group is one of the focal points of modern algebra. Like all significant mathematical ideas, this idea is general and abstract, and it is interesting and important because of the cumulative interest and importance of its many special cases. Roughly, a group is a set of elements that can be combined through some operation such as addition or multiplication, subject to some definite rules like those that govern ordinary addition of numbers. The elements may be something other than numbers, however, and the operation something other than the usual operations of arithmetic. For instance, the elements of the groups used to study symme-

Figure 3

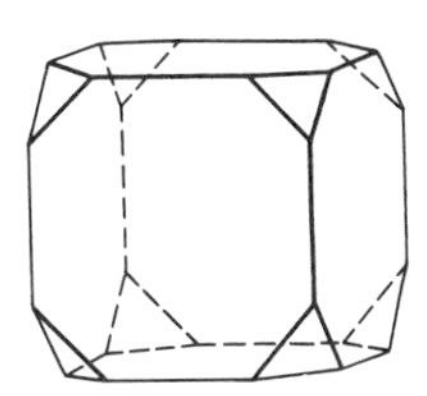

try are things like rotations, translations, and reflections. The precise definition of *group* will be given in Section 5.

Crystallography. Think again about symmetry type, but now move from two dimensions up to three. One easy example here is given by moving a cube repeatedly along the directions perpendicular to its faces. Other examples can be very complicated, and the general problem of finding all symmetry types was not easy. Like the problem in two dimensions, however, it was also settled in the nineteenth century: in contrast to the 17 different symmetry types of plane-filling designs, there are 230 different types of symmetry for figures that fill three-dimensional space. Again, groups provide the key. And with this use of groups we have arrived at an application to science, for ideas used to solve this three-dimensional problem are just what are needed to classify crystals according to symmetry type. The symmetry type of a crystal is a measure of the regular pattern in which the component atoms or molecules arrange themselves. Although this pattern is an internal property of the crystal, which may require x-ray techniques for analysis, symmetry is often evident from the external or surface features of the crystal. Figure 3 illustrates this with a picture of galena crystals: the repeated occurrence of the shape on the left, in the picture on the right, is a consequence of the internal symmetry of galena. (Galena is the chief ore mineral of lead, and has properties that make it useful in electronics.) Classification by symmetry type is at the heart of crystallography, and is used in parts of physics, chemistry, and mineralogy.

Figure 4

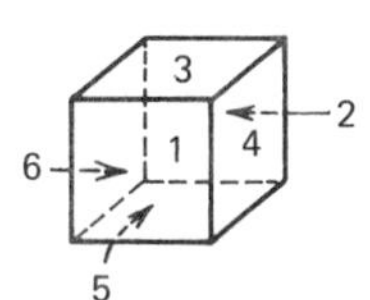

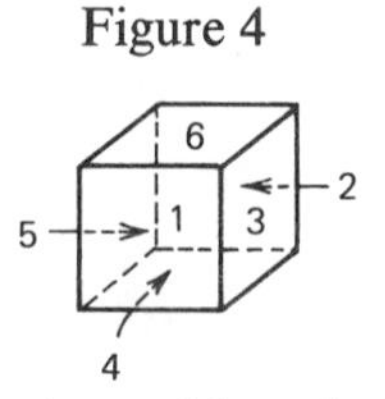

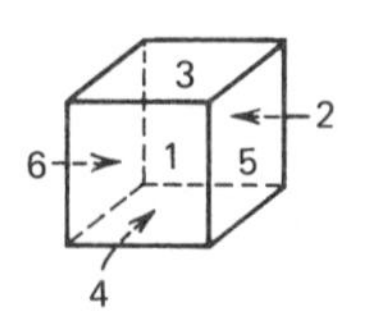

In each case, 1 is on the front and 2 is on the back.

Combinatorics. Although one of the preceding connections of groups is with design, and the other is with science, they both have to do with symmetry. Here is another example. The numbers 1 through 6 can be placed on the faces of a die (cube) in 720 different ways. But only 30 of these ways are distinguishable—if the numbers are put on more than 30 dice, then at least two of the dice can be made to look the same through rotation. In Figure 4, for example, the middle arrangement differs from the left-hand arrangement only by a rotation, but no rotation of the middle die would make it look like the right-hand die. The problem of counting the number of distinguishable arrangements belongs to the domain of combinatorics, and if you are good at systematic counting you can solve it without groups. But the problem can also be solved with an appropriate group, and this provides a way of viewing the problem that is almost indispensable for more complicated problems. In each case a group is used to account systematically for the symmetry in the problem (such as the symmetry of a cube).

Algebraic equations. The examples thus far have had to do with geometrical symmetry, but groups were first studied for a different reason. Much of the early history of modern algebra was tied closely to questions about equations. In beginning algebra we learn that each linear (first degree) equation $ax+b=0$ $(a\neq 0)$ has a unique solution $x=-b/a$, and each quadratic (second degree) equation $ax^2+bx+c=0$ $(a\neq 0)$ has solutions

$$x=\frac{-b\pm\sqrt{b^2-4ac}}{2a}. \tag{1}$$

Methods for solving these equations were known by the sixteenth century. For example, particular types of quadratics had been handled by the ancient Egyptians, Babylonians, and Greeks, and by the Hindus and Arabs in the Middle Ages. But what about equations of degree higher than two? For instance, is there a general procedure or formula, like (1), for writing the solutions of a cubic (third degree) equation $ax^3+bx^2+cx+d=0$ in terms of the coefficients a, b, c, and d? Italian algebraists discovered in the sixteenth century that the answer is Yes, not only for cubics but also for quartic (fourth degree) equations.

These solutions for cubics and quartics are fairly complicated, and their detailed form is not important here. What is important is that all of this leads to the following more general question: Can the solutions of each algebraic equation

$$a_n x^n + a_{n-1} x^{n-1} + \cdots + a_0 = 0 \tag{2}$$

be derived from the coefficients $a_n, a_{n-1}, \ldots, a_0$ by addition, subtraction, multiplication, division, and extraction of roots, each applied only finitely many times—or briefly, as we now say, is (2) *solvable by radicals*? By early in the nineteenth century mathematicians knew that the answer was No: for each $n>4$ there are equations of degree n that are not solvable by radicals. However, *some* equations of each degree $n>4$ are solvable by radicals, and so there is the new problem of how to determine whether a given equation is or is not solvable in this way. With this problem we are brought back to groups: with each equation (2) we can associate a group, and the French mathematician Evariste Galois (1811–1832) discovered that properties of this group reveal whether the equation is solvable by radicals. The group associated with an equation measures an abstruse kind of symmetry involving the solutions of the equation. Thus the abstract idea of a group can be used to analyze both geometrical symmetry and solvability by radicals. Groups arise in other contexts, as well, but we cannot pursue them all here.

Rings and fields. The theory of groups was not the only part of algebra to be stimulated by questions about equations. One question that arises when first studying quadratic equations has to do with square roots of negative numbers: What can we say about the solutions in (1) if $b^2-4ac<0$? Nowadays this creates little problem, since we know about complex numbers, and we generally feel just as comfortable with them as we do with integers and real numbers. But this is true only because earlier mathematicians worked out a clear understanding of the properties of all the familiar number systems. Two more ideas from modern algebra—*ring* and *field*—were essential for this. Roughly, a ring is a set with operations like addition, subtraction, and multiplication; and a field is a ring in which division is also possible. Precise definitions of ring and field will be given at the appropriate places in the text, where we shall also see how these ideas can be used to characterize the familiar number systems. There will be no surprises when we relate rings and fields to number systems, but the following application of fields to geometry should be more unexpected.

Figure 5

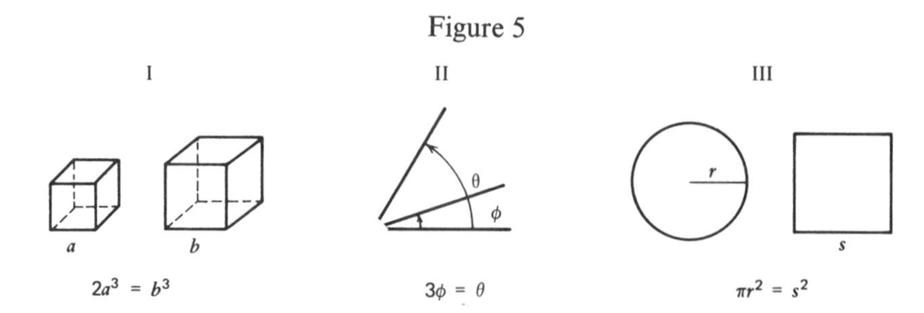

Geometric constructions. Among the geometric construction problems left unsolved by the ancient Greeks, three became especially famous. Each involved the construction of one geometrical segment from another, using only unmarked straightedge and compass (Figure 5):

- **I.** Construct the edge of a cube having twice the volume of a given cube.
- **II.** Show that every angle can be trisected.
- **III.** Construct the side of a square having the same area as a circle of given radius.

These problems remained unsolved for more than 2000 years. Then, in the nineteenth century, it was proved that the constructions are impossible. What makes this interesting for us is that although the constructions were to be geometric, the proofs of their impossibility involve algebra. And the key algebraic concepts needed are the same as those used to analyze solvability by radicals; these include facts about fields that go far beyond what is necessary to characterize the familiar number systems.

Number theory. The motivation for studying some of the deeper properties of rings came from a source totally different from the applications already mentioned. *Pythagorean triples* are triples (x,y,z) of positive integers such that $x^2+y^2=z^2$. That is, they are the triples of integers that can occur as lengths of sides of right triangles (relative to an appropriate unit of length). Examples are (3, 4, 5), (5, 12, 13), (8, 15, 17), and (199, 19800, 19801). The Greek mathematician Diophantus derived a method for determining all such triples around A.D. 250. (The Babylonians had determined many Pythagorean triples by much the same method around 1500 B.C.) In reading about this problem in Diophantus' book *Arithmetica*, the French mathematician Pierre de Fermat (1601–1665) was led to introduce one of the most famous problems in mathematics: Are there nonzero integers x,y,z such that

$$x^n + y^n = z^n,$$

for any integer $n > 2$? Actually, Fermat claimed that there are no such integers, and this claim eventually came to be known as *Fermat's Last Theorem*. But Fermat did not give a proof of his claim, and the problem of constructing a proof has defied some of the best mathematicians of the past 300 years. (It is now known that there are no solutions for $n < 100{,}000$, but that hardly exhausts the possibilities.) One of the most important lessons from the history of mathematics, however, is that the gains from unsuccessful attempts to solve problems can be as great as the gains from successful ones. Regardless of whether the problem of Fermat's Last Theorem is ever solved, the attempts to solve it—and several other problems in number theory—have had a profound effect on number theory and algebra. In fact, these problems led to the creation of a new branch of mathematics, *algebraic number theory*, which lies on the boundary between number theory and algebra. Much of the early work in the theory of rings, which is now thought of as part of algebra, also belongs to algebraic number theory. This work was used in uncovering most of what is known about Fermat's Last Theorem and a number of other difficult problems in number theory, and it has also been used in branches of mathematics totally unrelated to number theory.

Order. The three basic kinds of systems we have discussed—groups, rings, and fields—are examples of what are known as algebraic structures. Each such structure involves one or more operations like addition or multiplication of numbers. Some algebraic structures also involve a notion of order, like $\subseteq$ for sets and $\leq$ for numbers. For example, order must be taken into account in studying the familiar number systems. One formal idea that grew from questions about order is that of a *lattice*. Lattices can be represented by diagrams like those in Figure 6: the example on the left shows the subsets of $\{x, y, z\}$, with a sequence of segments connecting one set to another above it if the first set is contained in the second; the example on the right shows the positive factors of 30, with a sequence of segments connecting one integer to another above it if the first integer is a

Figure 6

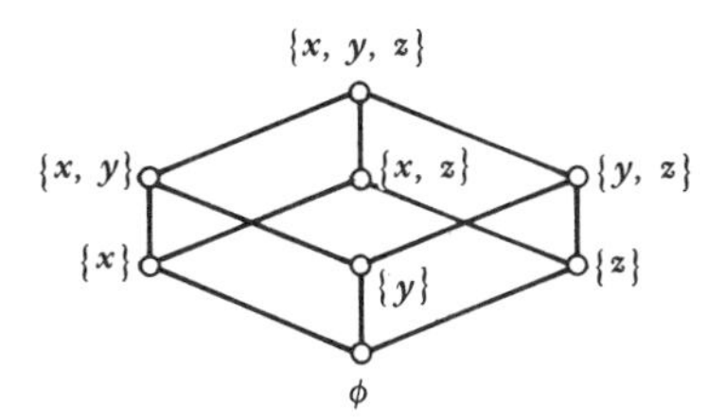

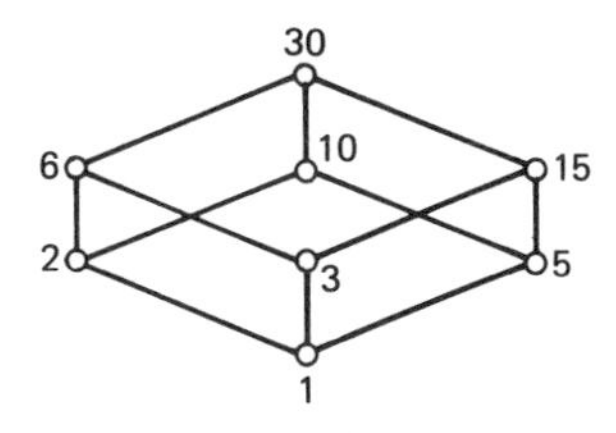

factor of the second. The similarity of these two diagrams suggests one of the purposes of lattice theory, just as the similarity of certain symmetric figures suggests one of the purposes of group theory. Lattice theory is concerned with analyzing the notion of order (subject to some definite rules), and with describing in abstract terms just what is behind the similarity of diagrams like those in Figure 6. Of course there is more to this study of order than diagrams. Lattices were first studied as natural generalizations of *Boolean algebras*, which were themselves introduced in the mid-nineteenth century by the British mathematician George Boole (1815–1864) for the purpose of giving an algebraic analysis of formal logic. The first significant use of lattices outside of this connection with logic was in ring theory and algebraic number theory; this interdependence of different branches of algebra is certainly not uncommon in modern mathematics—in fact, it is one of its characteristic features.

Computer-related algebra. A number of applications of modern algebra have grown with the advent of electronic computers and communication systems. These applications make use of many of the general ideas first introduced to handle much older problems. For example, one such application involves the use of Boolean algebras to study the design of computers and switching circuits. Another application is to algebraic coding (Chapter XVI), which uses, among other things, finite fields—these are systems that have only finitely many elements but are otherwise much like the system of real numbers. Applications that use tools from modern algebra and combinatorics belong to the general area of discrete applied mathematics; this can be contrasted with classical applied mathematics, which uses tools from calculus and its extensions.

General remarks. Each topic discussed in this section will be touched on in the book, but they cannot all be treated thoroughly. It would take more than one volume to do that, and in any event there is even more to algebra than the topics introduced in this section might suggest. One of the methods used by the American Mathematical Society to classify current research divides mathematics into eight broad areas: algebra and the theory of numbers, analysis, applied mathematics, geometry, logic and foundations, statistics and probability, topology, and miscellaneous. Although the major branches represented in this list are in many ways interdependent, it is nonetheless true that each branch tends to have its own special outlook and its own special methods and techniques. The goal of this book is to go as far as possible in getting across the outlook and methods and techniques of algebra or, more precisely, that part of algebra devoted to the study of algebraic structures.

Most of the chapters end with notes that list other books, including some where more historical background can be found. Here are 10 general references that are concerned with history; the notes at the end of Chapter XIV give a short list of more advanced general references on modern algebra.

Aleksandrov, A. D. et. al., eds., *Mathematics: Its Content, Methods, and Meaning*, 3 vols., The MIT Press, Cambridge, Mass., 1969. (English translation of the 1956 edition.) This gives an overview of modern mathematics written by well-known Russian mathematicians.

Bell, E. T., *Development of Mathematics*, 2nd ed., McGraw-Hill, New York, 1945.

______, *Men of Mathematics*, Simon and Schuster, New York, 1957. These two books by E. T. Bell are especially lively, though slightly romanticized.

Bourbaki, N., *Éléments d'Historie des Mathématiques*, 2nd ed., Hermann, 1969. A concise history for those who can read French.

Boyer, C. B., *A History of Mathematics*, John Wiley, New York, 1968. An excellent comprehensive survey.

Encyclopedia Britannica, 1974. This contains a number of useful survey articles. See especially those on algebra and the history of mathematics.

Gaffney, M. P., and L. A. Steen, *Annotated Bibliography of Expository Writing in the Mathematical Sciences*, The Mathematical Association of America, 1976.

Kline, M., *Mathematical Thought from Ancient to Modern Times*, Oxford University Press, London, 1972. Another excellent comprehensive history; more complete on modern topics than Boyer's book.

May, K. O., *Bibliography and Research Manual of the History of Mathematics*, University of Toronto Press, Toronto, 1973.

Nový, L., *Origins of Modern Algebra*, Noordhoff, Leyden, The Netherlands, 1973. A detailed account of algebra in the important period from 1770 to 1870.

Remarks on proofs. A first course in modern algebra often has another goal in addition to algebra: that of introducing the axiomatic method and the construction of careful proofs. Some of the "nuts and bolts" of proofs are reviewed in Appendix B. In addition, here are two general pieces of advice.

First, do not become discouraged if constructing proofs seems difficult. It is difficult for nearly everyone. Don't be deceived by the form in which mathematicians generally display their finished products. Behind a polished proof of any significance there has often been a great deal of struggle and frustration and ruined paper. There is no other way to discover good mathematics, and for most of us there is no other way to learn it.

Second, after you have constructed a proof or solved a problem, it is good to remember that no one else can be expected to know what is in your mind or what you have discarded. They have only what you tell them or what you write. For communicating proofs it would be hard to find sounder advice than that given by Quintilian, 1900 years ago:

> One should not aim at being possible to understand, but at being impossible to misunderstand.

THE GREEK ALPHABET

Α	α	alpha
Β	β	beta
Γ	γ	gamma
Δ	δ	delta
Ε	ϵ	epsilon
Ζ	ζ	zeta
Η	η	eta
Θ	θ	theta
Ι	ι	iota
Κ	κ	kappa
Λ	λ	lambda
Μ	μ	mu
Ν	ν	nu
Ξ	ξ	xi
Ο	o	omicron
Π	π	pi
Ρ	ρ	rho
Σ	σ	sigma
Τ	τ	tau
Υ	υ	upsilon
Φ	ϕ	phi
Χ	χ	chi
Ψ	ψ	psi
Ω	ω	omega

CHAPTER I

Mappings and Operations

The most fundamental concept in modern algebra is that of an operation on a set. Addition and the other operations in the familiar number systems are examples, but we shall see that the general concept of operation is much broader than that. Before looking at operations, however, we shall devote several sections to some basic terminology and facts about mappings, which will be just as important for us as operations. The words *function* and *mapping* are synonymous; therefore, at least some of this material on mappings will be familiar from calculus and elsewhere. Our context for mappings will be more general than that in calculus, however, and one of the reasons we shall use *mapping* rather than *function* is to emphasize this generality. Notice that elementary facts about sets are collected in Appendix A; they will be used without explicit reference, and probably should be reviewed at the start. You are also urged to read Appendix B, which reviews some elements of logic and offers suggestions regarding proofs.

SECTION 1. MAPPINGS

Mappings are important throughout mathematics. In calculus, for instance, we study mappings (functions) that assign real numbers to real numbers. The mapping given by $f(x)=x^2$, for example, assigns to each real number x the real number x^2. The mapping given by $f(x)=\sin x$ assigns to each real number x the real number $\sin x$. The set $\mathbb{R}$ of real numbers plays two roles in these examples: first, $x \in \mathbb{R}$, and second, $f(x) \in \mathbb{R}$. In general these roles can be played by sets other than $\mathbb{R}$. Thus in the definition of mapping, which follows, S and T can denote any sets whatsoever.

Definition. A *mapping* from a set S to a set T is a relationship (rule, correspondence) that assigns to each element of S a uniquely determined element of T. The set S is called the *domain* of the mapping, and the set T is called the *codomain*.

Mappings will generally be denoted by Greek letters, and, to indicate that α is a mapping from S to T, we shall write $\alpha: S \to T$ or $S \xrightarrow{\alpha} T$. If x is an element of S, then $\alpha(x)$ will denote the unique element of T that is assigned to x; the element $\alpha(x)$ is called the *image* of x under the mapping α. Sometimes there will be a formula for $\alpha(x)$, as in the examples $f(x) = x^2$ and $f(x) = \sin x$. But that certainly need not be the case.

Example 1.1. Let $S = \{x, y, z\}$ and $T = \{1,2,3\}$. Then α defined by $\alpha(x) = 2$, $\alpha(y) = 1$, $\alpha(z) = 3$ is a mapping from S to T.

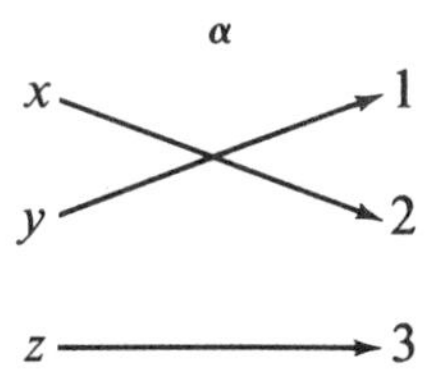

Another mapping, $\beta: S \to T$, is given by $\beta(x) = 1$, $\beta(y) = 3$, and $\beta(z) = 1$.

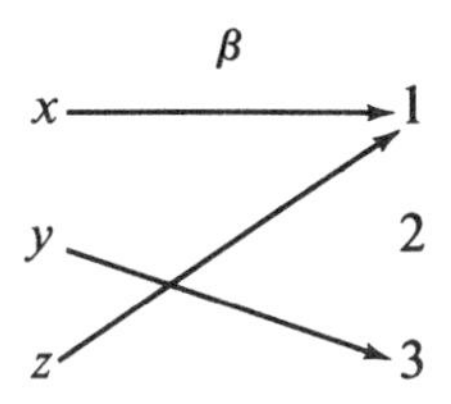

Example 1.2. If S is any set, we shall use ι (iota) to denote the *identity mapping* from S to S; this is defined by $\iota(x) = x$ for each $x \in S$. If it is necessary to indicate which set S is being considered, ι_S can be written instead of ι.

It is sometimes convenient to write $x \xrightarrow{\alpha} y$ or $x \mapsto y$ to indicate that y is the image of x under a mapping.

Example 1.3. The rule $(s,t) \mapsto s + t$ for each ordered pair of real numbers s and t defines a mapping from the set of ordered pairs of real numbers to the set of real numbers. [That (s,t) is *ordered* means that it is to be distinguished from (t,s); the necessity for this can be seen if addition is replaced by subtraction: $t - s \neq s - t$ unless $s = t$.] Mappings of this kind, assigning single elements of a set to pairs of elements from the same set, will be discussed at length in Section 3.

If $\alpha: S \to T$, and A is a subset of S, then $\alpha(A)$ will denote the set of elements of T that are images of elements of A under the mapping α. In set-builder notation (described in Appendix A),

$$\alpha(A) = \{y \mid y = \alpha(x) \text{ for some } x \in A\}.$$

Example 1.4. With α and β as in Example 1.1,

$$\alpha(\{x,z\}) = \{2,3\} \qquad \beta(\{x,z\}) = \{1\}.$$

If $\alpha: S \to T$, then $\alpha(S)$ will be called the *image* of α. If $\alpha: S \to T$ and $\alpha(S) = T$, then α is said to be *onto*. Thus α is onto if for each $y \in T$ there is at least one $x \in S$ such that $\alpha(x) = y$. (This use of the preposition *onto* as an adjective may be regrettable, but it is solidly established.)

Example 1.5. The mapping α in Example 1.1 is onto. The mapping β in Example 1.1 is not onto; its image is $\{1,3\}$, which is a proper subset of $\{1,2,3\}$, the codomain of β.

With $f(x) = x^2$ and $f(x) = \sin x$ thought of as mappings from $\mathbb{R}$ to $\mathbb{R}$, neither is onto. The image of $f(x) = x^2$ is the set of nonnegative real numbers, and the image of $f(x) = \sin x$ is the set of real numbers between -1 and 1, inclusive.

A mapping $\alpha: S \to T$ is said to be *one-to-one* if

$$x_1 \neq x_2 \quad \text{implies} \quad \alpha(x_1) \neq \alpha(x_2) \qquad (x_1, x_2 \in S);$$

that is, if unequal elements in the domain have unequal images in the codomain.

Example 1.6. The mapping α in Example 1.1 is one-to-one. The mapping β in Example 1.1 is not one-to-one, because $x \neq z$ but $\beta(x) = \beta(z)$. The mapping $f(x) = x^2$, with domain $\mathbb{R}$, is not one-to-one, because $f(x) = x^2 = f(-x)$ although $x \neq -x$ when $x \neq 0$. The mapping $f(x) = \sin x$, with domain $\mathbb{R}$, is not one-to-one, because $f(x) = \sin x = \sin(x + 2n\pi) = f(x + 2n\pi)$ for each $x \in \mathbb{R}$ and each integer n.

The contrapositive (Appendix B) of

$$x_1 \neq x_2 \quad \text{implies} \quad \alpha(x_1) \neq \alpha(x_2)$$

is

$$\alpha(x_1) = \alpha(x_2) \qquad \text{implies} \qquad x_1 = x_2.$$

Therefore, since a statement and its contrapositive are logically equivalent, we see that

$$\alpha : S \to T \text{ is one-to-one}$$

iff[†]

$$\alpha(x_1) = \alpha(x_2) \qquad \text{implies} \qquad x_1 = x_2 \quad (x_1, x_2 \in S).$$

It is sometimes easier to work with this latter condition than with that given in the definition. With $\alpha : \mathbb{R} \to \mathbb{R}$ defined by $\alpha(x) = x - 1$, for instance, we have that $\alpha(x_1) = \alpha(x_2)$ implies $x_1 - 1 = x_2 - 1$, which implies $x_1 = x_2$; therefore, α is one-to-one.

Notice that the identity mapping on any set is both onto and one-to-one. Notice also that a mapping of a finite set to itself is onto iff it is one-to-one. In contrast, there are mappings of any infinite set to itself that are one-to-one but not onto, and also mappings that are onto but not one-to-one.

Example 1.7. Define mappings α and β from the set of natural numbers, $\{1, 2, 3, \ldots\}$, to itself, by

$$\alpha(n) = 2n$$

and

$$\beta(n) = \begin{cases} (n+1)/2 & \text{if } n \text{ is odd} \\ n/2 & \text{if } n \text{ is even.} \end{cases}$$

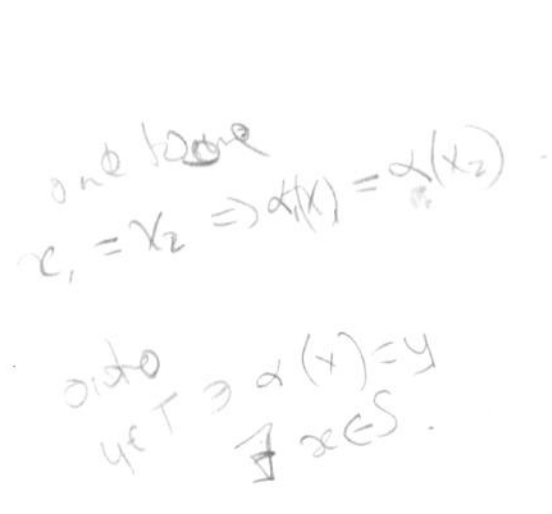

1 2 3 4 5 6 7 8 ...

α

1 2 3 4 5 6 7 8 ...

1 2 3 4 5 6 7 8 ...

β

1 2 3 4 5 6 7 8 ...

Then α is one-to-one but not onto; and β is onto but not one-to-one. The existence of such mappings is precisely what distinguishes infinite sets from finite sets. In fact, a set S can be defined as *infinite* if there exists a mapping from S to S that is one-to-one but not onto.

[†]We follow the practice, now widely accepted by mathematicians, of using "iff" to denote "if and only if."

Suppose that α and β are the mappings, each with the set of real numbers as domain, defined by $\alpha(x)=(x+1)^2$ and $\beta(x)=x^2+2x+1$. Because $(x+1)^2=x^2+2x+1$ for all real numbers x, it is reasonable to think of the mappings α and β as being *equal*. In general, any two mappings α and β are said to be *equal* if their domains are equal and if $\alpha(x)=\beta(x)$ for every x in that common domain.

We close this section with one remark on notation, and another on terminology.

The symbols $\mathbb{N}$, $\mathbb{Z}$, $\mathbb{Q}$, $\mathbb{R}$, and $\mathbb{C}$ will be reserved to denote the following sets:

- $\mathbb{N}$ the set of all natural numbers, $\{1,2,3,\ldots\}$
- $\mathbb{Z}$ the set of all integers, $\{\ldots, -2,-1,0,1,2,\cdots\}$
- $\mathbb{Q}$ the set of all rational numbers, that is, real numbers that can be expressed in the form a/b, with $a,b\in\mathbb{Z}$ and $b\neq 0$
- $\mathbb{R}$ the set of all real numbers
- $\mathbb{C}$ the set of all complex numbers

Familiarity with basic properties of $\mathbb{N}$, $\mathbb{Z}$, $\mathbb{Q}$, and $\mathbb{R}$ will be assumed throughout; all of these sets will be studied in Chapter VI.

Although we shall not use the following terminology, it should be mentioned because you may see it elsewhere. Sometimes, a one-to-one mapping is called an *injection*, an onto mapping is called a *surjection*, and a mapping that is both one-to-one and onto is called a *bijection*. Also, what we are calling the codomain of a mapping is sometimes called the *range* of the mapping. Regrettably, *range* is also used for what we are calling the image of a mapping; this ambiguity over *range* is one reason for avoiding its use.

PROBLEMS

1.1. Let $S=\{w,x,y,z\}$ and $T=\{1,2,3,4\}$, and define $\alpha:S\to T$ and $\beta:S\to T$ by $\alpha(w)=2$, $\alpha(x)=4$, $\alpha(y)=1$, $\alpha(z)=2$ and $\beta(w)=4$, $\beta(x)=2$, $\beta(y)=3$, $\beta(z)=1$.
(a) Is α one-to-one? Is β one-to-one? Is α onto? Is β onto?
(b) Let $A=\{w,y\}$ and $B=\{x,y,z\}$. Determine each of the following subsets of T: $\alpha(A)$, $\beta(B)$, $\alpha(A\cap B)$, $\beta(A\cup B)$.

1.2. Let α, β, and γ be mappings from $\mathbb{Z}$ to $\mathbb{Z}$ defined by $\alpha(n)=2n$, $\beta(n)=n+1$, and $\gamma(n)=n^3$ for each $n\in\mathbb{Z}$.
(a) Which of the three mappings are onto?
(b) Which of the three mappings are one-to-one?
(c) Determine $\alpha(\mathbb{N})$, $\beta(\mathbb{N})$, and $\gamma(\mathbb{N})$.

1.3. Assume S and T as in Example 1.1.
(a) How many mappings are there from S to T?
(b) How many mappings are there from S onto T?
(c) How many one-to-one mappings are there from S to T?

1.4. Assume that S and T are sets and $\alpha: S \to T$ and $\beta: S \to T$. Complete each of the following statements. (The discussion of quantifiers in Appendix B may be helpful here.)

(a) α is not onto iff for some $y \in T \ldots$.

(b) α is not one-to-one iff $\ldots$.

(c) $\alpha \neq \beta$ iff $\ldots$.

1.5. Assume that S and T are finite sets containing m and n elements, respectively.

(a) How many mappings are there from S to T?

(b) How many one-to-one mappings are there from S to T? (Consider two cases: $m > n$ and $m \leq n$.)

1.6. A mapping $f: \mathbb{R} \to \mathbb{R}$ is onto iff each horizontal line (line parallel to the x-axis) intersects the graph of f at least once.

(a) Formulate a similar condition for $f: \mathbb{R} \to \mathbb{R}$ to be one-to-one.

(b) Formulate a similar condition for $f: \mathbb{R} \to \mathbb{R}$ to be both one-to-one and onto.

1.7. Each f below defines a mapping from $\mathbb{R}$ to $\mathbb{R}$. Determine which of these mappings are onto and which are one-to-one. (Problem 1.6 may be helpful.) Also determine $f(P)$ in each case, for P the set of positive real numbers.

(a) $f(x) = 2x$ (b) $f(x) = x - 4$

(c) $f(x) = x^3$ (d) $f(x) = x^2 + x$

(e) $f(x) = e^x$ (f) $f(x) = \tan x$

1.8. For each ordered pair (a, b) of integers define a mapping $\alpha_{a,b}: \mathbb{Z} \to \mathbb{Z}$ by $\alpha_{a,b}(n) = an + b$.

(a) For which pairs (a, b) is $\alpha_{a,b}$ onto?

(b) For which pairs (a, b) is $\alpha_{a,b}$ one-to-one?

1.9. With β as defined in Example 1.7, for each $n \in \mathbb{N}$ the equation $\beta(x) = n$ has exactly two solutions. (The solutions of $\beta(x) = 2$ are $x = 3$ and $x = 4$, for example.)

(a) Define a mapping $\gamma: \mathbb{N} \to \mathbb{N}$ such that for each $n \in \mathbb{N}$ the equation $\gamma(x) = n$ has exactly three solutions.

(b) Define a mapping $\gamma: \mathbb{N} \to \mathbb{N}$ such that for each $n \in \mathbb{N}$ the equation $\gamma(x) = n$ has exactly n solutions. (It suffices to describe γ in words.)

(c) Define a mapping $\gamma: \mathbb{N} \to \mathbb{N}$ such that for each $n \in \mathbb{N}$ the equation $\gamma(x) = n$ has infinitely many solutions.

1.10. Prove that there is a mapping from a set to itself that is one-to-one but not onto iff there is a mapping from the set to itself that is onto but not one-to-one. (Compare Example 1.7.)

1.11. Let A denote the set of odd natural numbers, B the set of even natural numbers, and C the set of natural numbers that are multiples of 4. With α and β as in Example 1.7, determine each of the following sets.

(a) $\alpha(A)$, $\alpha(B)$, and $\alpha(C)$

(b) $\beta(A)$, $\beta(B)$, and $\beta(C)$

1.12. Prove that if $\alpha: S \to T$, and A and B are subsets of S, then $\alpha(A \cup B) = \alpha(A) \cup \alpha(B)$.

1.13. (a) Prove that if $\alpha: S \to T$, and A and B are subsets of S, then $\alpha(A \cap B) \subseteq \alpha(A) \cap \alpha(B)$.

(b) Give an example (specific S, T, A, B, and α) to show that equality need not hold in part (a). (For the simplest examples S will have two elements.)

1.14. Prove that a mapping $\alpha: S \rightarrow T$ is one-to-one iff $\alpha(A \cap B) = \alpha(A) \cap \alpha(B)$ for every pair of subsets A and B of S. (Compare Problem 1.13.)

SECTION 2. COMPOSITION. INVERTIBLE MAPPINGS

Assume that $\alpha: S \rightarrow T$ and $\beta: T \rightarrow U$. Then $\alpha(x) \in T$ for each $x \in S$, and so it makes sense to write $\beta(\alpha(x))$, which is an element of U:

$$x \overset{\alpha}{\mapsto} \alpha(x) \overset{\beta}{\mapsto} \beta(\alpha(x)).$$

Application of α followed by β in this way is seen to yield a mapping from S to U. This mapping is called the *composition* of α and β; it is denoted by $\beta \circ \alpha$. Thus, by definition,

$$(\beta \circ \alpha)(x) = \beta(\alpha(x))$$

for each $x \in S$. Note carefully: *in $\beta \circ \alpha$, it is α, the mapping on the right, that is applied first.*

Example 2.1. Let $S = \{x, y, z\}$, $T = \{1, 2, 3\}$, and $U = \{a, b, c\}$. Define $\alpha: S \rightarrow T$ by $\alpha(x) = 2$, $\alpha(y) = 1$, $\alpha(z) = 3$. Define $\beta: T \rightarrow U$ by $\beta(1) = b$, $\beta(2) = c$, $\beta(3) = a$. Then $(\beta \circ \alpha)(x) = c$, $(\beta \circ \alpha)(y) = b$, and $(\beta \circ \alpha)(z) = a$.

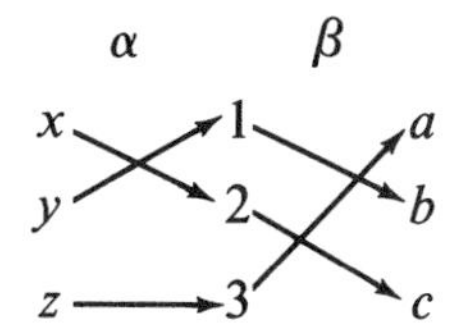

Example 2.2. Let α and β denote mappings, each with the set of real numbers as both domain and codomain, defined by

$$\alpha(x) = x^2 + 2 \quad \text{and} \quad \beta(x) = x - 1.$$

Then

$$\begin{aligned}(\alpha \circ \beta)(x) &= \alpha(\beta(x)) \\ &= \alpha(x-1) \\ &= (x-1)^2 + 2 \\ &= x^2 - 2x + 3\end{aligned}$$

while

$$\begin{aligned}(\beta \circ \alpha)(x) &= \beta(\alpha(x)) \\ &= \beta(x^2+2) \\ &= (x^2+2) - 1 \\ &= x^2 + 1.\end{aligned}$$

In particular, for example, $(\alpha\circ\beta)(0)=3$ but $(\beta\circ\alpha)(0)=1$. Thus $\beta\circ\alpha$ and $\alpha\circ\beta$ need not be equal, even if both are defined.

It will be important to know which compositions are onto and which are one-to-one. The following theorem provides the answer.

Theorem 2.1. *Assume that* $\alpha: S\to T$ *and* $\beta: T\to U$.

(*a*) *If* α *and* β *are onto, then* $\beta\circ\alpha$ *is onto.*
(*b*) *If* $\beta\circ\alpha$ *is onto, then* β *is onto.*
(*c*) *If* α *and* β *are one-to-one, then* $\beta\circ\alpha$ *is one-to-one.*
(*d*) *If* $\beta\circ\alpha$ *is one-to-one, then* α *is one-to-one.*

PROOF. (a) Assume that both α and β are onto. To prove that $\beta\circ\alpha$ is onto, we must establish that if $z\in U$ then there is an element $x\in S$ such that $(\beta\circ\alpha)(x)=z$. Because β is onto, we know there is an element $y\in T$ such that $\beta(y)=z$. But α is onto, so there is an element $x\in S$ such that $\alpha(x)=y$. Putting these facts together, we have $(\beta\circ\alpha)(x)=\beta(\alpha(x))=\beta(y)=z$, as required.

(b) Assume that $\beta\circ\alpha$ is onto, and that $z\in U$. Then there is $x\in S$ such that $(\beta\circ\alpha)(x)=z$. But then $\beta(\alpha(x))=z$ with $\alpha(x)\in T$; hence β is onto.

(c) Assume that both α and β are one-to-one, and that $(\beta\circ\alpha)(x_1)=(\beta\circ\alpha)(x_2)$ $(x_1,x_2\in S)$. Then $\beta(\alpha(x_1))=\beta(\alpha(x_2))$, so that $\alpha(x_1)=\alpha(x_2)$ since β is one-to-one; therefore $x_1=x_2$ since α is one-to-one. This proves that $\beta\circ\alpha$ is one-to-one.

(d) Assume that $\beta\circ\alpha$ is one-to-one. If $x_1,x_2\in S$ and $\alpha(x_1)=\alpha(x_2)$, then $\beta(\alpha(x_1))=\beta(\alpha(x_2))$; that is, $(\beta\circ\alpha)(x_1)=(\beta\circ\alpha)(x_2)$. This implies $x_1=x_2$ because $\beta\circ\alpha$ is one-to-one. Therefore, α is one-to-one. □

A mapping $\beta: T\to S$ is an *inverse* of $\alpha: S\to T$ if both $\beta\circ\alpha=\iota_S$ and $\alpha\circ\beta=\iota_T$. A mapping is said to be *invertible* if it has an inverse. If a mapping is invertible, its inverse is unique (this will be Problem 4.9).

Example 2.3. The mapping α in Example 2.1 is invertible. Its inverse is the mapping γ defined by $\gamma(1)=y$, $\gamma(2)=x$, $\gamma(3)=z$. For example, $(\gamma\circ\alpha)(x)=\gamma(2)=x=\iota_S(x)$ and $(\alpha\circ\gamma)(1)=\alpha(y)=1=\iota_T(1)$. (There are also four other equations to be checked, of course.) In terms of the diagram in Example 2.1, γ is gotten by reversing the direction of the arrows under α.

Example 2.4. The mapping $\alpha: \mathbb{R}\to\mathbb{R}$ defined by $\alpha(x)=x^2$ is not invertible. It fails on two counts, by the following theorem.

Theorem 2.2. *A mapping is invertible iff it is both one-to-one and onto.*

PROOF. First assume that $\alpha: S \to T$ is invertible, with inverse β. Then $\beta \circ \alpha$, being the identity mapping on S, is one-to-one; therefore α must be one-to-one by Theorem 2.1. On the other hand, $\alpha \circ \beta$, being the identity mapping on T, is onto; therefore α must be onto by Theorem 2.1. This proves that if α is invertible, then it is both one-to-one and onto.

Now assume that $\alpha: S \to T$ is both one-to-one and onto. We shall show that α is invertible by describing an inverse. Assume $t \in T$. Then, because α is onto, there is at least one element $s \in S$ such that $\alpha(s) = t$; but α is also one-to-one, so that this element s must be unique; let $\beta(t) = s$. This can be done for each element $t \in T$, and in this way we obtain a mapping $\beta: T \to S$. Moreover, from the way in which β is constructed it can be seen that $\beta \circ \alpha = \iota_S$ and $\alpha \circ \beta = \iota_T$, so that β is an inverse of α. Thus α is invertible. □

Warning: Some authors write mappings on the right, rather than on the left. Our $\alpha(x)$ becomes, for them, $x\alpha$. Then in $\beta \circ \alpha$ it is β, the mapping on the left, that is applied first, because $x(\beta \circ \alpha) = (x\beta)\alpha$. Although we shall consistently write mappings on the left, it is important when reading other sources to take note of which convention is being followed.

PROBLEMS

2.1. Let α, β, and γ be mappings from $\mathbb{Z}$ to $\mathbb{Z}$ defined by $\alpha(n) = 2n$, $\beta(n) = n + 1$ and $\gamma(n) = n^2$. Write a formula for each of the following compositions. Also determine the image in each case.

(a) $\alpha \circ \alpha$ (b) $\beta \circ \alpha$ (c) $\gamma \circ \alpha$
(d) $\alpha \circ \beta$ (e) $\beta \circ \beta$ (f) $\gamma \circ \beta$
(g) $\alpha \circ \gamma$ (h) $\beta \circ \gamma$ (i) $\gamma \circ \gamma$

2.2. Prove that if $\alpha: S \to T$, then $\alpha \circ \iota_S = \alpha$ and $\iota_T \circ \alpha = \alpha$.

2.3. Describe the inverse of the mapping β in Example 2.1.

2.4. Each f below defines an invertible mapping from $\mathbb{R}$ to $\mathbb{R}$. Write a formula for the inverse (denote it g) in each case.

(a) $f(x) = 2x$ (b) $f(x) = x - 4$
(c) $f(x) = 2x - 4$ (d) $f(x) = x^3$

2.5. Prove that the inverse of an invertible mapping is invertible.

2.6. Give an example of sets S, T, and U and mappings $\alpha: S \to T$ and $\beta: T \to U$ such that $\beta \circ \alpha$ is onto but α is not onto. [Compare Theorem 2.1(b).]

2.7. Give an example of sets S, T, and U and mappings $\alpha: S \to T$ and $\beta: T \to U$ such that $\beta \circ \alpha$ is one-to-one but β is not one-to-one. [Compare Theorem 2.1(d).]

2.8. Prove that if $\alpha: S \to T$, $\beta: T \to U$, $\gamma: T \to U$, α is onto, and $\beta \circ \alpha = \gamma \circ \alpha$, then $\beta = \gamma$.

2.9. Prove that if $\beta: S \to T$, $\gamma: S \to T$, $\alpha: T \to U$, α is one-to-one, and $\alpha \circ \beta = \alpha \circ \gamma$, then $\beta = \gamma$.

2.10. Give an example to show that the condition "α is onto" cannot be omitted from Problem 2.8.

2.11. Give an example to show that the condition "α is one-to-one" cannot be omitted from Problem 2.9.

2.12. Complete the following statement: A mapping $\alpha: S \to T$ is not invertible iff α is not one-to-one....

2.13. For α a mapping, decide whether each of the following is true or false. (Appendix B compares *if* and *only if*.)
(a) α is invertible if α is one-to-one.
(b) α is invertible only if α is one-to-one.
(c) α is invertible if α is onto.
(d) α is invertible only if α is onto.

2.14. For α and β as in Example 1.7, determine both $\beta \circ \alpha$ and $\alpha \circ \beta$. Is either α or β invertible?

2.15. Consider f and g, mappings from $\mathbb{R}$ to $\mathbb{R}$, defined by $f(x) = \sin x$ and $g(x) = 2x$. Is $f \circ g$ equal to $g \circ f$?

2.16. Which of the functions sine, cosine, and tangent, as mappings from $\mathbb{R}$ to $\mathbb{R}$, are invertible?

2.17. Assume that $\alpha: S \to T$ and $\beta: T \to U$. Use Theorems 2.1 and 2.2 to prove each of the following statements.
(a) If α and β are invertible, then $\beta \circ \alpha$ is invertible.
(b) If $\beta \circ \alpha$ is invertible, then β is onto and α is one-to-one.

SECTION 3. OPERATIONS

If one integer is added to another, the result is an integer. If one integer is subtracted from another, the result is also an integer. These examples—addition and subtraction of integers—are special cases of what are known as *operations*. In each case there is a set (here the integers), and a relationship that assigns to each ordered pair (a,b) of elements of that set another element of the same set: $a+b$ in one case, $a-b$ in the other. The general definition of operation is as follows.

Definition. An *operation* on a set S is a relationship (rule, correspondence) that assigns to each ordered pair of elements of S a uniquely determined element of S.

Thus an operation is a special kind of mapping: First, $S \times S$, the *Cartesian product* of S with S, is the set of all ordered pairs (a,b) with $a \in S$ and $b \in S$ (Appendix A). Then an operation on S is simply a mapping from $S \times S$ to S. In the case of addition as an operation on the integers, $(a,b) \mapsto a+b$.

Example 3.1. On the set of positive integers, multiplication is an operation: $(m,n)\mapsto mn$, where mn has the usual meaning, m times n. Division is not an operation on the set of positive integers, because $m \div n$ is not necessarily a positive integer ($1 \div 2 = \frac{1}{2}$, for instance).

The last example illustrates a point worth stressing. In order to have an operation on a set S, it is essential that if $a,b \in S$, then the image of the ordered pair (a,b) be in S. This property of an operation is referred to as *closure*, or we say that S is *closed* with respect to the operation.

Example 3.2. The mapping $(m,n)\mapsto m^n$ defines an operation on the set of positive integers. Notice that $(3,2)\mapsto 3^2 = 9$, while $(2,3)\mapsto 2^3 = 8$, so that, just as with subtraction, the order makes a difference.

If there is an established symbol to denote the image of a pair under an operation, as in the case of $a+b$ for addition of numbers, then that symbol is used. Otherwise some other symbol is adopted, such as $(a,b)\mapsto a*b$ or just $(a,b)\mapsto ab$, for instance, where it must be specified what $a*b$ or ab is to mean in each case.

Example 3.3. Let S denote any nonempty set, and let $M(S)$ denote the set of all mappings from S to S. Suppose that $\alpha \in M(S)$ and $\beta \in M(S)$. Then $\alpha: S \to S$, $\beta: S \to S$, and $\beta \circ \alpha: S \to S$, so that $\beta \circ \alpha \in M(S)$. Thus $\circ$ is an operation on $M(S)$. We shall return to this in Section 4.

Example 3.4. If S is a finite set, then we can specify an operation on S by means of a table, similar to the addition and multiplication tables used in beginning arithmetic. We first form a square, and then list the elements of S across the top and also down the left-hand side. For an operation $*$, we put $a*b$ at the intersection of the (horizontal) row with a at the left and the (vertical) column with b at the top. For Table 3.1, $u*v = w$, $v*u = v$, and so forth. Any way of filling in the nine spaces in the square, with entries chosen from the set $\{u,v,w\}$, will define an operation on $\{u,v,w\}$. Changing one or more of the nine entries will give a different operation. (If the

Table 3.1

$*$	u	v	w
u	u	w	w
v	v	v	v
w	w	u	v

nine entries are left unchanged, but $*$ is changed to some other symbol, the result would not be considered a different operation.) Tables defining operations in this way are called *Cayley* tables, after the British mathematician Arthur Cayley (1821–1895).

Example 3.5. Matrix multiplication is an operation on the set of all 2×2 matrices with real numbers as entries:

$$\begin{bmatrix} a & b \\ c & d \end{bmatrix}\begin{bmatrix} w & x \\ y & z \end{bmatrix} = \begin{bmatrix} aw+by & ax+bz \\ cw+dy & cx+dz \end{bmatrix}.$$

More generally, both matrix multiplication and matrix addition are operations on the set of all $n\times n$ matrices with real numbers as entries, for any positive integer n. Such examples are discussed more fully in Appendix D.

What we are calling *operations* are often called *binary operations*, to emphasize that they are mappings of ordered *pairs*, rather than mappings of single elements or ordered triples or such. [Examples of operations that are not binary are $a\mapsto -a$ and $(a,b,c)\mapsto a(b+c)$, where $a,b,c\in\mathbb{R}$.] We shall have no occasion to discuss explicitly any operations other than binary operations, so we shall not carry along the extra word *binary*.

The notion of operation is so fundamental in algebra that algebra could almost be defined as the study of operations (with binary operations being the most important). But this would be something like defining mathematics as the study of sets and mappings—there is no question of the importance of these concepts, but at the same time they are too general to be of real interest. In calculus, for example, it is not *all* functions $f:\mathbb{R}\to\mathbb{R}$ that are of interest, but only functions that have some property such as continuity or differentiability. In the same way, in algebra the operations that are of interest usually possess certain special properties. We now introduce some of the most important of these properties.

Definition. An operation $*$ on a set S is said to be *associative* if it satisfies the condition

$$a*(b*c) = (a*b)*c \qquad \text{associative law}$$

for all $a,b,c\in S$.

For example, addition of real numbers is associative: $a+(b+c)=(a+b)+c$ always. Subtraction of real numbers, however, is not associative: $2-(3-4)=2-(-1)=3$ but $(2-3)-4=(-1)-4=-5$. Notice that if the equation in the associative law fails for even one triple (a,b,c), then the operation is not associative. (See the discussion in Appendix B on the negation of statements with quantifiers.)

Multiplication of real numbers is associative: $a(bc)=(ab)c$. But the operation defined in Example 3.2 is not associative: denote it by $*$, so that $m*n=m^n$; then for example, $2*(3*2)=2*(3^2)=2*9=2^9=512$ but $(2*3)*2 = 2^3*2=8*2=8^2=64$.

To motivate the next definition, think of the following properties of the integers 0 and 1: $m+0=0+m=m$ and $m\cdot 1=1\cdot m=m$, for every integer m.

Definition. An element e in a set S is an *identity* (or *identity element*) for an operation $*$ on S if

$$e*a = a*e = a \qquad \text{for each } a\in S.$$

Thus 0 is an identity for addition of integers, and 1 is an identity for multiplication of integers. Note that the definition requires *both* $e*a=a$ *and* $a*e=a$, for each $a\in S$. (See Problems 3.2 and 3.4.) A similar remark applies to the next definition.

Definition. Assume that $*$ is an operation on S, with identity e, and that $a\in S$. An element b in S is an *inverse* of a relative to $*$ if

$$a*b = b*a = e.$$

Example 3.6. Relative to addition as an operation on the integers, each integer has an inverse, its negative: $a+(-a)=(-a)+a=0$ for each integer a. It is important to notice that the inverse of an element must be in the set under consideration. Relative to addition as an operation on the set of nonnegative integers, no element other than 0 has an inverse: the negative of a positive integer is negative.

Example 3.7. Relative to multiplication, each real number different from 0 has an inverse, its reciprocal: $a\cdot(1/a)=(1/a)\cdot a=1$. Multiplication is also an operation on the set of integers (with identity 1), but only 1 and -1 have inverses in this case.

We have seen that it is possible to have an operation $*$ and elements a and b such that $a*b\neq b*a$ (subtraction of integers, or Example 3.2, for instance). Operations for which this cannot happen are numerous enough and important enough to deserve a special name.

Definition. An operation $*$ on a set S is said to be *commutative* if

$$a*b = b*a \qquad \text{commutative law}$$

for all $a,b\in S$.

Addition and multiplication of integers are commutative. Other examples will occur in the problems and elsewhere.

PROBLEMS

3.1. Which of the following define operations on the set of integers? Of those that do, which are associative? Which are commutative? Which have identity elements?

(a) $m*n=mn+1$ (b) $m*n=(m+n)/2$
(c) $m*n=m$ (d) $m*n=mn^2$
(e) $m*n=m^2+n^2$ (f) $m*n=2^{mn}$
(g) $m*n=3$ (h) $m*n=\sqrt{mn}$

3.2. Verify that the operation in Example 3.2 has no identity element.

3.3. There is an identity for the operation in Example 3.3. What is it?

3.4. (a) Verify that the operation in Example 3.4 has no identity element.
(b) Change one entry in the table in Example 3.4 so that u becomes an identity element.

3.5. Does $(m,n)\to m^n$ define an operation on the set of all integers? (Compare Example 3.2.)

3.6. If $*$ is an operation on S, T is a subset of S, and T is closed with respect to $*$, then two of the following three statements are necessarily true but one may be false. Which two are true?
(a) If $*$ is associative on S, then $*$ is associative on T.
(b) If there is an identity element for $*$ on S, then there is an identity element for $*$ on T.
(c) If $*$ is commutative on S, then $*$ is commutative on T.

3.7 Assume that $*$ is an operation on S. Complete each of the following statements.
(a) $*$ is not associative iff $a*(b*c)\neq(a*b)*c\ldots.$
(b) $*$ is not commutative iff $a*b\neq b*a\ldots.$
(c) $e\in S$ is not an identity element for $*$ iff... .
(d) There is no identity element for $*$ iff for each $e\in S$ there is an element $a\in S$ such that... .

3.8. (a) Complete the following Cayley table in such a way that u becomes an identity element. In how many ways can this be done?

*	u	v
u		
v		

(b) Can the table be completed in such a way that u and v both become identity elements? Why or why not?
(c) Prove: An operation $*$ on a set S (any S) can have at most one identity element.

3.9. How many different operations are there on a 1-element set? 2-element set? 3-element set? n-element set? (See the remarks in Example 3.4.)

3.10. Complete the following table in a way that makes $*$ commutative.

$*$	a	b	c	d
a	a	b		d
b		c		
c	c	d	a	b
d		a		c

3.11. How many different commutative operations are there on a 1-element set? 2-element set? 3-element set? n-element set? (The Cayley table for a commutative operation must have a special kind of symmetry.)

3.12. (a) Determine the smallest subset A of $\mathbb{Z}$ such that $2 \in A$ and A is closed with respect to addition.

(b) Determine the smallest subset B of $\mathbb{Q}$ such that $2 \in B$ and B is closed with respect to addition and division.

3.13. Complete the following table in such a way that $*$ is commutative and has an identity element, and each element has an inverse. (There is only one correct solution. First explain why y must be the identity element.)

$*$	w	x	y	z
w	y			x
x	z	w		
y				
z				w

SECTION 4. COMPOSITION AS AN OPERATION

In Example 3.3 we saw that if S is any nonempty set, then composition is an operation on $M(S)$, the set of all mappings from S to S. It will be worthwhile to look more closely at this operation, for its importance is matched only by that of addition and the other operations on the familiar number systems. The most general properties are summarized in the following theorem.

Theorem 4.1. *Let S denote any nonempty set.*

(*a*) *Composition is an associative operation on $M(S)$, with identity element ι_S.*

(*b*) *Composition is an associative operation on the set of all invertible mappings in $M(S)$, with identity ι_S.*

PROOF. Associativity means that $\gamma\circ(\beta\circ\alpha)=(\gamma\circ\beta)\circ\alpha$ for all $\alpha,\beta,\gamma\in M(S)$. By the definition of equality for mappings, this means that

$$[\gamma\circ(\beta\circ\alpha)](x)=[(\gamma\circ\beta)\circ\alpha](x)$$

for each $x\in S$. To verify this, we can write

$$\begin{aligned}[\gamma\circ(\beta\circ\alpha)](x) &= \gamma((\beta\circ\alpha)(x))\\ &= \gamma(\beta(\alpha(x)))\\ &= (\gamma\circ\beta)(\alpha(x))\\ &= [(\gamma\circ\beta)\circ\alpha](x).\end{aligned}$$

(As in all proofs, it is important to understand the justification for each step. Here is a good test: Could you explain it to someone else?)

It is easy to verify that $\iota_S\circ\alpha=\alpha\circ\iota_S=\alpha$ for each $\alpha\in M(S)$, which proves that ι_S is an identity element. (Compare Problem 2.2 with $S=T$.)

Now move to part (b) of the theorem. Notice that the question of whether $\gamma\in M(S)$ is an inverse of $\alpha\in M(S)$ means the same whether taken in the sense of Section 2 (preceding Example 2.3), or in the sense of Section 3 (preceding Example 3.6): $\gamma\circ\alpha=\alpha\circ\gamma=\iota_S$. To prove that composition is an operation on the set of all invertible mappings in $M(S)$, assume that $\alpha,\beta\in M(S)$ and that both α and β are invertible. Then α and β are one-to-one and onto by Theorem 2.2. Therefore $\beta\circ\alpha$ is both one-to-one and onto by Theorem 2.1, parts (c) and (a). But this implies that $\beta\circ\alpha$ is invertible, again by Theorem 2.2.

Since composition is associative as an operation on all of $M(S)$, it is certainly associative when restricted to the invertible elements of $M(S)$. The proof is now finished by the observation that ι_S is invertible. □

Example 2.2 shows that composition as an operation on $M(\mathbb{R})$ is not commutative. Problem 4.2 gives a more general statement about commutativity.

Notice that composition is an operation on a subset of $M(S)$ iff that subset is closed with respect to the operation. Many important operations involve composition on special sets of invertible mappings. We conclude this section by giving two examples of this; more examples will come later.

Example 4.1. Let p denote a fixed point in a plane, and let G denote the set of all rotations in the plane about the point p. Composition is an operation on G: If α_1 and α_2 are rotations about p, then $\alpha_2\circ\alpha_1$ is the rotation obtained by first performing α_1 and then α_2 (Figure 4.1). For example, if α_1 denotes clockwise rotation through 70°, and α_2 clockwise rotation through 345°, then $\alpha_2\circ\alpha_1$ is clockwise rotation through 415° or, equivalently, 55°, since we can ignore multiples of 360°. This operation is associative by Theorem 4.1.

Figure 4.1

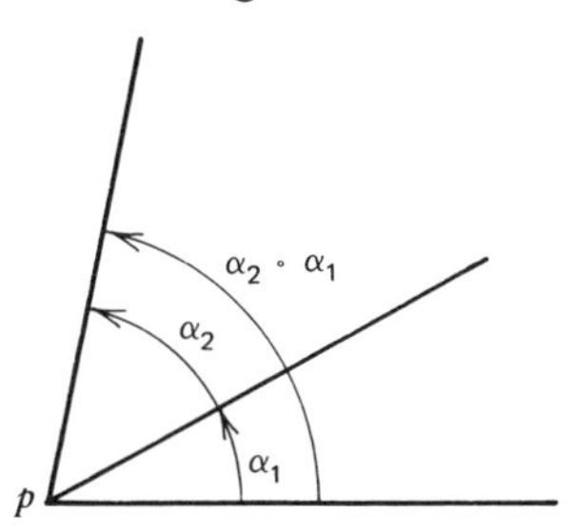

An identity element is rotation through 0°. And each rotation has an inverse: rotation of the same magnitude in the opposite direction. Finally, as an operation on G, composition is commutative.

Example 4.2. For each ordered pair (a,b) of real numbers with $a \neq 0$, let $\alpha_{a,b} : \mathbb{R} \to \mathbb{R}$ be defined by $\alpha_{a,b}(x) = ax + b$. Let A denote the set of all such mappings. Then composition is an operation on A. To verify this, let (a,b) and (c,d) be ordered pairs of real numbers with $a \neq 0$ and $c \neq 0$. Then $ac \neq 0$, and

$$\begin{aligned}(\alpha_{a,b} \circ \alpha_{c,d})(x) &= \alpha_{a,b}(\alpha_{c,d}(x)) \\ &= \alpha_{a,b}(cx+d) \\ &= a(cx+d)+b \\ &= acx + ad + b \\ &= \alpha_{ac,ad+b}(x).\end{aligned}$$

Thus $\alpha_{a,b} \circ \alpha_{c,d} = \alpha_{ac,ad+b}$. Notice that A is a subset of $M(\mathbb{R})$, and that composition is, again, associative. Further properties of this example are brought out in the problems.

Each mapping $\alpha_{a,b}$ can be interpreted geometrically by considering what it does to the points on a real line. If $a > 1$, for instance, then $\alpha_{a,0}$ magnifies the distance of each point from the origin by a factor of a: $\alpha_{a,0}(x) = ax$. Also, if $b > 0$, then $\alpha_{1,b}$ translates each point b units to the right (assuming the real line directed to the right): $\alpha_{1,b}(x) = x + b$. Thus $\alpha_{a,b} = \alpha_{1,b} \circ \alpha_{a,0}$ consists of the first action followed by the second (Figure 4.2). Problem 4.6 asks what happens in other cases.

Figure 4.2

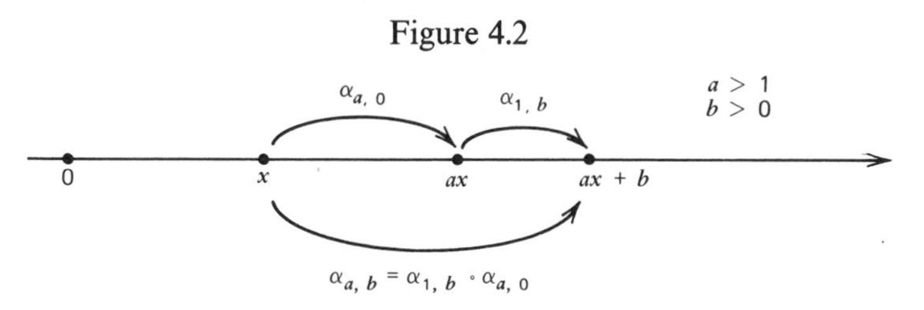

PROBLEMS

4.1. With $S=\{a,b\}$, the set $M(S)$ contains four elements; denote these by π,ρ,σ,τ, defined as follows:

$$\pi(a)=a \qquad \rho(a)=a \qquad \sigma(a)=b \qquad \tau(a)=b$$

$$\pi(b)=a \qquad \rho(b)=b \qquad \sigma(b)=a \qquad \tau(b)=b$$

(a) Construct the Cayley table for composition ($\circ$) as an operation on $M(S)=\{\pi,\rho,\sigma,\tau\}$. (As a start, $\rho\circ\tau=\tau$ and $\sigma\circ\tau=\pi$.)
(b) What is the identity element?
(c) Is $\circ$ commutative as an operation on $M(S)$?
(d) Which elements of $M(S)$ are invertible?
(e) Is $\circ$ commutative as an operation on the set of invertible elements in $M(S)$?

4.2. Verify that if S contains more than one element, then composition is not a commutative operation on $M(S)$. (Try "constant" mappings, like π and τ in Problem 4.1.)

4.3 Consider Example 4.1, and let ρ_1,ρ_2,ρ_3, and ρ_4 denote clockwise rotation through 0, 90, 180, and 270°, respectively. Composition is an operation on $\{\rho_1,\rho_2,\rho_3,\rho_4\}$ ($\rho_4\circ\rho_2=\rho_1$, for instance). Construct the corresponding Cayley table. Is there an identity element? Does each element have an inverse?

4.4 (a) Consider Example 4.1, and let α denote clockwise rotation through $\pi/2$ radians. Let H denote the smallest subset of G such that $\alpha\in H$ and H is closed with respect to $\circ$. Determine H.
(b) Same as (a) with $\pi/2$ replaced by $\pi/6$.
(c) Same as (a) with $\pi/2$ replaced by π/k ($k\in\mathbb{N}$). How many distinct elements (rotations) does H contain? (Treat rotations as indistinct if they differ by an integral multiple of 2π.)

4.5. Consider the operation $\circ$ on the set A in Example 4.2.
(a) Verify that $\alpha_{1,0}$ is an identity element.
(b) Prove that each $\alpha_{a,b}\in A$ is invertible by verifying that it is one-to-one and onto. (Remember that $a\neq 0$.)
(c) Determine (c,d) so that $\alpha_{c,d}$ is an inverse of $\alpha_{a,b}$.

4.6. Give a geometric interpretation of $\alpha_{a,b}$ in Example 4.2 under each of the following conditions.
(a) $0<a<1$ and $b=0$.
(b) $a<0$ and $b=0$.
(c) $a=1$ and $b<0$.

4.7. Let B and C denote the following subsets of A in Example 4.2:

$$B = \{\alpha_{a,0}|a\in\mathbb{R} \text{ and } a\neq 0\}$$
$$C = \{\alpha_{1,b}|b\in\mathbb{R}\}.$$

(a) Verify that $\circ$ is an operation on B. Is it associative? Commutative? Is there an identity element?
(b) Verify that $\circ$ is an operation on C. Is it associative? Commutative? Is there an identity element?

(c) Verify that each mapping in A is the composition of a mapping in B and a mapping in C.

4.8. Consider Example 4.2 and let D denote the smallest subset of A such that $\alpha_{1,1} \in D$ and D is closed with respect to $\circ$. Determine D.

4.9. Prove that each invertible mapping has a unique inverse. (Assume that $\alpha: S \to T$, $\beta: T \to S$, $\gamma: T \to S$, $\beta \circ \alpha = \gamma \circ \alpha = \iota_S$, and $\alpha \circ \beta = \alpha \circ \gamma = \iota_T$. Show that $\beta = \gamma$ by computing $\beta \circ \alpha \circ \gamma$ in two different ways.)

4.10. Let S denote the set of all real numbers except 0 and 1. Consider the six mappings from S to S defined as follows:

$$\alpha_1(x) = x \qquad \alpha_2(x) = \frac{1}{x} \qquad \alpha_3(x) = 1 - x$$

$$\alpha_4(x) = 1 - \frac{1}{x} \qquad \alpha_5(x) = \frac{1}{1-x} \qquad \alpha_6(x) = \frac{x}{x-1}.$$

(a) Verify that composition is an operation on $\{\alpha_1, \ldots, \alpha_6\}$ by constructing a Cayley table. [As a start, $(\alpha_6 \circ \alpha_2)(x) = \alpha_6(1/x) = (1/x)/[(1/x) - 1] = 1/(1 - x) = \alpha_5(x)$, so $\alpha_6 \circ \alpha_2 = \alpha_5$.]

(b) There is an identity element. What is it?

(c) Show that each of the six elements has an inverse.

(d) Is $\circ$ commutative as an operation on $\{\alpha_1, \ldots, \alpha_6\}$?

(e) The operation $\circ$ on $\{\alpha_1, \ldots, \alpha_6\}$ is associative. Why? (No calculations are necessary.)

4.11 Prove that if $\alpha: S \to T$, $\beta: T \to U$, and $\gamma: U \to V$ are any mappings, then $\gamma \circ (\beta \circ \alpha) = (\gamma \circ \beta) \circ \alpha$. (The associativity in Theorem 4.1 is the special case of this with $S = T = U = V$.)

4.12. Is composition an operation on the set of all continuous mappings from $\mathbb{R}$ to $\mathbb{R}$? (The answer requires either a theorem or an example from calculus.)

4.13. Prove that if V is a vector space, then composition is an operation on the set of all linear transformations from V to V. (This requires some knowledge of linear algebra. The basic ideas are reviewed in Appendix D.)

CHAPTER II

Introduction to Groups

We are now ready for one of the central ideas of modern mathematics, that of a group. In the first chapter we met a number of examples of sets with operations, and we observed that such operations may or may not possess any of several special properties such as associativity or the existence of an identity element. We shall see that a group is a set together with an operation such that certain specified properties are required to hold. These properties have been singled out because they arise naturally in many important special cases. By studying groups we can arrive at a clearer understanding of each of those special cases.

This chapter gives an introduction to groups through examples and a connection with symmetry. Later chapters will treat the general theory and applications of groups.

SECTION 5. DEFINITION AND EXAMPLES

It takes patience to appreciate the diverse ways in which groups arise, but one of these ways is so familiar that we can use it to ease our way into the basic definition. To this end, recall the following three things about the set of integers with respect to addition. First, addition is associative. Second, 0 is an identity element. And third, relative to 0, each integer has an inverse (its negative). Much more can be said about the integers, of course, but these are the properties that are important at the moment; for they show that the integers with addition form a group, in the sense of the following definition.

Definition. A *group* is a set G together with an operation $*$ on G such that each of the following axioms is satisfied:

associativity

$a*(b*c)=(a*b)*c$ for all $a,b,c \in G$

existence of identity element

there is an element $e \in G$ such that $a*e=e*a=a$ for each $a \in G$, and

existence of inverse elements

for each $a \in G$ there is an element $b \in G$ such that $a*b=b*a=e$.

Notice that a group consists of a *pair* of things, a set *and* an operation on that set. Very often, a group is referred to by naming only the underlying set, but that is safe only if it is clear what operation is intended. Whenever we refer to "the group of integers" the operation is meant to be addition.

Example 5.1. The set of even integers together with addition is a group. Addition is an operation because the sum of two even integers is an even integer. The associative law is true for all integers, so it is certainly true for even integers. The identity element is 0 (an even integer), and the inverse of an even integer x is $-x$ (again an even integer).

Example 5.2. The set of positive integers with addition is not a group, because there is no identity element. Even if we considered the positive integers along with 0 we would not get a group, because no element other than 0 would have an inverse.

Example 5.3. The set $\{0\}$ together with addition is a group. Notice that because a group must contain an identity element, the set underlying a group must always contain at least one element.

Example 5.4. The set of positive rational numbers with multiplication is a group. If r/s and u/v are positive (where none of the integers r, s, u, or v is 0), then $(r/s)(u/v)=ru/sv$ is also positive. The associative law we take to be a generally known fact from arithmetic. (We shall have more to say about this in Chapter VI.) The identity element is 1 and the inverse of r/s is s/r ($r\neq 0$, $s\neq 0$).

Example 5.5. Tables 5.1 and 5.2 define operations on the set $\{a,b,c\}$ that yield groups. In Table 5.1 ($*$), a is an identity element and the inverses of

Table 5.1

$*$	a	b	c
a	a	b	c
b	b	c	a
c	c	a	b

Table 5.2

#	a	b	c
a	c	a	b
b	a	b	c
c	b	c	a

a, b, and c are a, c, and b, respectively. In Table 5.2 (#), b is an identity element and the inverses of a, b, and c are c, b, and a, respectively. The verification of associativity is Problem 5.8. This example illustrates why, in general, we should specify the operation, not just the set, when talking about a group.

Example 5.6. If S is any nonempty set, then the set of all invertible mappings in $M(S)$ is a group with composition as the operation. This is merely a restatement of Theorem 4.1(b). We shall return to groups of this type in the next section.

Example 5.7. Let p denote a fixed point in a plane, and let G denote the set of all rotations of the plane about the point p. In Example 4.1 we observed that composition is an operation on this set G, and we also verified everything needed to show that this gives a group.

Example 5.8. Let A denote the set of all mappings $\alpha_{a,b}: \mathbb{R} \to \mathbb{R}$, as defined in Example 4.2. Recall that $a, b \in \mathbb{R}$, $a \neq 0$, and $\alpha_{a,b}(x) = ax + b$ for each $x \in \mathbb{R}$. With composition of mappings as the operation, this yields a group. The rule for composition was worked out in Example 4.2. The identity element is $\alpha_{1,0}$ [Problem 4.5(a)], and the inverse of $\alpha_{a,b}$ is $\alpha_{a^{-1}, -a^{-1}b}$ (Problem 5.9).

Examination of the groups given thus far will reveal that in each case there is only one identity element. The definition requires that there be one; the point now is that there can be no other. Similarly, each element in each group has only one inverse element. Here is a formal statement and proof of these two facts.

Theorem 5.1. *Assume that G together with $*$ is a group.*
(*a*) *The identity element of G is unique. That is, if e and f are elements of G such that*

$$e*a = a*e = a \quad \text{for each } a \in G$$

and

$$f*a = a*f = a \quad \text{for each } a \in G,$$

then $e=f$.

(*b*) *Each element in a group has a unique inverse. That is, if a, x, and y are elements of G, e is the identity element of G, and*

$$a*x = x*a = e$$

and

$$a*y = y*a = e,$$

then $x=y$.

PROOF. (a) Assume that e and f are as stated. Then $e*a=a$ for *each* $a \in G$, and so, in particular, $e*f=f$. Similarly, using $a=e$ in $a*f=a$, we have $e*f=e$. Thus $f=e*f=e$, and so $e=f$, as claimed.

(b) With a, x, and y as stated, write

$$\begin{aligned} x &= x*e && (e \text{ is the identity}) \\ &= x*(a*y) && (a*y=e) \\ &= (x*a)*y && (\text{associativity}) \\ &= e*y && (x*a=e) \\ &= y && (e \text{ is the identity}). \quad \square \end{aligned}$$

Because of Theorem 5.1, it makes sense to speak of *the* identity element of a group, and *the* inverse of a group element. It is customary to use a^{-1} for the inverse of a group element a when there is no conflicting natural notation (such as $-a$ in the case of the integers with addition). Thus $a*a^{-1}=a^{-1}*a=e$. This is in accordance with the notation for inverses relative to multiplication of real numbers, and with notation for other exponents in groups, to be introduced in Section 12. If several groups are being considered at once, e_G can be used in place of e to denote the identity of a group G.

The number of elements in the set underlying a group is called the *order* of the group; we denote this $|G|$. A group is said to be *finite* or *infinite* depending on whether its order is finite or infinite. The group in Example 5.3 has order 1, and those in Example 5.5 have order 3. We shall compute the orders of the groups in Example 5.6, for S finite, in the next section. All other groups considered thus far are infinite.

PROBLEMS

5.1. Determine which of the following sets of numbers form groups with respect to the given operations. For those that do, give the identity element and the inverse of each element. For those that do not, give a reason.
(a) $\{1\}$, multiplication.
(b) $\{0\}$, multiplication.
(c) All nonzero rational numbers, multiplication.
(d) All rational numbers, addition.
(e) All rational numbers, multiplication.
(f) $\{-1, 1\}$, multiplication.
(g) $\{-1, 0, 1\}$, addition.
(h) All integers, multiplication.
(i) $\{n \mid n = 10k \text{ for some } k \in \mathbb{Z}\}$, addition.
(j) All nonzero rational numbers, division.
(k) All integers, subtraction.

5.2. (a) Verify that $\{2^m \mid m \in \mathbb{Z}\}$ is a group with respect to multiplication.
(b) Verify that $\{2^m 3^n \mid m, n \in \mathbb{Z}\}$ is a group with respect to multiplication.

5.3. Verify that $M(2, \mathbb{Z})$, the set of all 2×2 matrices with integers as entries, forms a group with respect to matrix addition,

$$\begin{bmatrix} a & b \\ c & d \end{bmatrix} + \begin{bmatrix} w & x \\ y & z \end{bmatrix} = \begin{bmatrix} a+w & b+x \\ c+y & d+z \end{bmatrix}.$$

5.4. If $|S| > 1$, then $M(S)$ is not a group with respect to composition. Why?

5.5. Let F denote the set of all functions $f: \mathbb{R} \to \mathbb{R}$. For $f, g \in F$, define $f+g$ by $(f+g)(x) = f(x) + g(x)$ for all $x \in \mathbb{R}$. Then $f+g \in F$. Verify that with this operation F is a group.

5.6. Let $H = \{f \mid f: \mathbb{R} \to \mathbb{R} \text{ and } f(x) \neq 0 \text{ for all } x \in \mathbb{R}\}$. For $f, g \in H$, define fg by $(fg)(x) = f(x)g(x)$ for all $x \in \mathbb{R}$. Then $fg \in H$. Verify that with this operation H is a group. How does this group differ from the group of invertible mappings in $M(\mathbb{R})$ (Example 5.6)?

5.7. (For those who have studied linear algebra.) Verify that the set of all invertible (nonsingular) 2×2 matrices with real numbers as entries forms a group with respect to matrix multiplication.

5.8. Verify the associative law for the operation $*$ in Example 5.5. (Notice that each time the identity is involved there is really no problem.)

5.9. For Example 5.8, verify the claim that the inverse of $\alpha_{a,b}$ is $\alpha_{a^{-1}, -a^{-1}b}$.

5.10. If $\{a, b\}$ with operation $*$ is to be a group, with a the identity element, then what must the Cayley table be?

5.11. If $\{x,y,z\}$ with operation $*$ is to be a group, with x the identity element, then what must the Cayley table be?

5.12. Prove that if G is a group, $a \in G$, and $a*b=b$ for some $b \in G$, then a must be the identity element of G.

5.13. There are four assumptions in Theorem 5.1(a):

$$e*a = a \text{ for each } a \in G, \qquad f*a = a \text{ for each } a \in G,$$

$$a*e = a \text{ for each } a \in G, \qquad a*f = a \text{ for each } a \in G.$$

The proof of Theorem 5.1(a) uses only two of these assumptions. Which two?

5.14. Formulate a question like that in Problem 5.13 with Theorem 5.1(b) in place of Theorem 5.1(a). Answer it.

SECTION 6. PERMUTATIONS

A *permutation* of a nonempty set S is a one-to-one mapping from S onto S. Because a mapping from S to S is one-to-one and onto iff it is invertible, the permutations of S are the same as the invertible elements in $M(S)$. We observed in Example 5.6 that with composition as the operation these invertible elements form a group; such groups are of sufficient interest for us to state this as a theorem.

Theorem 6.1. *The set of all permutations of a nonempty set S is a group with respect to composition. This group is called the* symmetric group *on S, and will be denoted* Sym(S).

Any group whose elements are permutations, with composition as the operation, is called a *permutation group*; or, if we want to specify the underlying set S, a *permutation group on S*. Therefore Sym(S) is a permutation group on S. But, in general, a permutation group on S need not contain *all* permutations of S. Examples are the group of rotations of a plane about a fixed point p (Example 5.7), and the group of mappings $\alpha_{a,b}$: $\mathbb{R} \to \mathbb{R}$ $(a \neq 0)$ (Example 5.8). We shall see that permutation groups have a number of important applications. At the moment we concentrate on some elementary facts about the groups Sym(S), especially for S finite.

When S is the set $\{1,2,\ldots,n\}$, consisting of the first n positive integers, the group Sym(S) is commonly denoted S_n. An element α of S_n can be conveniently represented as follows. First write $1,2,\ldots,n$, and then below each number k write its image $\alpha(k)$. Thus

$$\begin{pmatrix} 1 & 2 & 3 & 4 \\ 2 & 4 & 1 & 3 \end{pmatrix}$$

represents the permutation in S_4 defined by $1 \mapsto 2$, $2 \mapsto 4$, $3 \mapsto 1$, $4 \mapsto 3$. The identity element of S_n is

$$\begin{pmatrix} 1 & 2 & \cdots & n \\ 1 & 2 & \cdots & n \end{pmatrix}.$$

The inverse of an element is obtained by reading from bottom entry to top entry rather than from top to bottom: if 1 appears beneath 3 in α, then 3 will appear beneath 1 in α^{-1}. Thus

$$\begin{pmatrix} 1 & 2 & 3 & 4 \\ 2 & 4 & 1 & 3 \end{pmatrix}^{-1} = \begin{pmatrix} 1 & 2 & 3 & 4 \\ 3 & 1 & 4 & 2 \end{pmatrix}.$$

In composing permutations we always follow the same convention we use in composing any other mappings: read *from right to left*. Thus

$$\begin{pmatrix} 1 & 2 & 3 & 4 \\ 2 & 4 & 1 & 3 \end{pmatrix} \circ \begin{pmatrix} 1 & 2 & 3 & 4 \\ 3 & 4 & 1 & 2 \end{pmatrix} = \begin{pmatrix} 1 & 2 & 3 & 4 \\ 1 & 3 & 2 & 4 \end{pmatrix}$$

but

$$\begin{pmatrix} 1 & 2 & 3 & 4 \\ 3 & 4 & 1 & 2 \end{pmatrix} \circ \begin{pmatrix} 1 & 2 & 3 & 4 \\ 2 & 4 & 1 & 3 \end{pmatrix} = \begin{pmatrix} 1 & 2 & 3 & 4 \\ 4 & 2 & 3 & 1 \end{pmatrix}.$$

Warning: Some authors compose permutations from left to right. One must check in each case to see what convention is being followed.

Example 6.1 The elements of S_3 are

$$\pi_1 = \begin{pmatrix} 1 & 2 & 3 \\ 1 & 2 & 3 \end{pmatrix} \quad \pi_2 = \begin{pmatrix} 1 & 2 & 3 \\ 2 & 3 & 1 \end{pmatrix} \quad \pi_3 = \begin{pmatrix} 1 & 2 & 3 \\ 3 & 1 & 2 \end{pmatrix}$$

$$\pi_4 = \begin{pmatrix} 1 & 2 & 3 \\ 2 & 1 & 3 \end{pmatrix} \quad \pi_5 = \begin{pmatrix} 1 & 2 & 3 \\ 3 & 2 & 1 \end{pmatrix} \quad \pi_6 = \begin{pmatrix} 1 & 2 & 3 \\ 1 & 3 & 2 \end{pmatrix}.$$

The labels $\pi_1, \ldots, \pi_6$ are arbitrary, but will be as defined here whenever we refer to S_3. Table 6.1 is the Cayley table for this group.

Table 6.1

$\circ$	π_1	π_2	π_3	π_4	π_5	π_6
π_1	π_1	π_2	π_3	π_4	π_5	π_6
π_2	π_2	π_3	π_1	π_5	π_6	π_4
π_3	π_3	π_1	π_2	π_6	π_4	π_5
π_4	π_4	π_6	π_5	π_1	π_3	π_2
π_5	π_5	π_4	π_6	π_2	π_1	π_3
π_6	π_6	π_5	π_4	π_3	π_2	π_1

A reminder of two of our conventions: Across from π_3 and below π_5 is $\pi_3 \circ \pi_5$, which is the permutation obtained by applying first π_5 and then π_3.

The group S_1 has order 1; S_2 has order 2; and, as we have just seen, S_3 has order 6. To give a general formula for the order of S_n, we first recall that if n is a positive integer then $n!$ (read *n factorial*) denotes $1 \cdot 2 \cdot \cdots \cdot n$, the product of all positive integers up to and including n. Thus $1! = 1$, $2! = 2$, $3! = 6$, $5! = 120$, $20! = 2{,}432{,}902{,}008{,}176{,}640{,}000$.

Theorem 6.2. *The order S_n is $n!$.*

PROOF. The problem of computing the number of elements in S_n is the same as that of computing the number of different ways the integers $1, 2, \ldots, n$ can be placed in the n blanks indicated (using each integer just once):

$$\begin{pmatrix} 1 & 2 & \cdots & n \\ \_ & \_ & \cdots & \_ \end{pmatrix}.$$

If we begin filling these blanks from the left, there are n possibilities for the first blank. Once that choice has been made, there remain $n-1$ possibilities for the second blank; then $n-2$ possibilities for the third blank; and so on. The theorem follows by repeated application of this basic counting principle: If one thing can be done in r different ways, and after that a second thing can be done in s different ways, then the two things can be done together in rs different ways. □

A group is said to be *Abelian* if its operation is commutative ($a*b = b*a$ for all a, b). This is in honor of the Norwegian mathematician N. H. Abel (1802–1829), whose contributions will be discussed in Chapter XIV. *Non-Abelian* means not Abelian. Here is one more fact about the groups S_n.

Theorem 6.3. *If $n \geq 3$, then S_n is non-Abelian.*

PROOF. If α and β in S_n are defined by

$$\alpha = \begin{pmatrix} 1 & 2 & 3 & \cdots \\ 1 & 3 & 2 & \cdots \end{pmatrix} \quad \text{and} \quad \beta = \begin{pmatrix} 1 & 2 & 3 & \cdots \\ 3 & 2 & 1 & \cdots \end{pmatrix},$$

with each number after 3 mapped to itself in each case, then

$$\beta \circ \alpha = \begin{pmatrix} 1 & 2 & 3 & \cdots \\ 3 & 1 & 2 & \cdots \end{pmatrix} \quad \text{but} \quad \alpha \circ \beta = \begin{pmatrix} 1 & 2 & 3 & \cdots \\ 2 & 3 & 1 & \cdots \end{pmatrix}.$$

Thus $\beta \circ \alpha \neq \alpha \circ \beta$, and the group is non-Abelian. □

It is easily seen that S_1 and S_2 are Abelian. Indeed, if S is *any* set containing only 1 or 2 elements, then Sym(S) is Abelian. On the other hand, if S contains more than 2 elements, then Sym(S) is non-Abelian (Problem 6.5).

Elements of S_n are frequently written using *cycle notation*. If S is a set, and $a_1, a_2, \ldots, a_k \in S$, then $(a_1 a_2 \cdots a_k)$ denotes the permutation of S for which

$$a_1 \mapsto a_2, \quad a_2 \mapsto a_3, \quad \ldots, \quad a_{k-1} \mapsto a_k, \quad a_k \mapsto a_1,$$

and

$$x \mapsto x \quad \text{for all other } x \in S.$$

Such a permutation is called a *cycle*, or a *k-cycle*. If $a \in S$, then (a) is the identity permutation of S. Cycles are composed just as any other permutations (except that the symbol $\circ$ is usually omitted). We shall occasionally refer to a composition of cycles, or of other permutations, as a *product*; this is common practice. If necessary for clarity, $(a_1, a_2, \ldots, a_k)$ can be written in place of $(a_1 a_2 \cdots a_k)$.

Example 6.2. In S_5,

$$(1\ 2\ 4) = \begin{pmatrix} 1 & 2 & 3 & 4 & 5 \\ 2 & 4 & 3 & 1 & 5 \end{pmatrix}$$

$$(1\ 2\ 4)(3\ 5) = \begin{pmatrix} 1 & 2 & 3 & 4 & 5 \\ 2 & 4 & 3 & 1 & 5 \end{pmatrix} \circ \begin{pmatrix} 1 & 2 & 3 & 4 & 5 \\ 1 & 2 & 5 & 4 & 3 \end{pmatrix} = \begin{pmatrix} 1 & 2 & 3 & 4 & 5 \\ 2 & 4 & 5 & 1 & 3 \end{pmatrix}$$

$$(1\ 2\ 4)(3\ 4) = (1\ 2\ 4\ 3)$$

$$(3\ 4)(1\ 2\ 4) = (1\ 2\ 3\ 4)$$

$$(1\ 2\ 4)(3)(5) = (1\ 2\ 4)$$

$$(1\ 2\ 3\ 4) = (2\ 3\ 4\ 1) = (3\ 4\ 1\ 2) = (4\ 1\ 2\ 3).$$

Notice that $(1\ 2\ 4)$ can denote an element of S_n for any $n \geq 4$. For instance, in S_4 it is the same as $(1\ 2\ 4)(3)$, and in S_5 it is the same as $(1\ 2\ 4)(3)(5)$. This ambiguity seldom causes trouble.

Cycles $(a_1 a_2 \cdots a_m)$ and $(b_1 b_2 \cdots b_n)$ are *disjoint* if $a_i \neq b_j$ for all i,j. Disjoint cycles commute; that is, if α and β represent disjoint cycles, then $\alpha\beta = \beta\alpha$ (Problem 6.16). *Any permutation of a finite set is either a cycle or can be written as a product of pairwise disjoint cycles; and, except for the order in which the cycles are written, and the inclusion or omission of 1-cycles, this can be done in only one way.* We shall omit the proof of that statement, but it is illustrated in the following example. In each case, we simply start the first cycle with 1, continue till we get back to 1, and then close the first cycle; then start the second cycle with the smallest number not in the first cycle, continue till we get back to that number, and then close the second cycle; and so on, never repeating a number that has already appeared. Other facts about cycles are given in the problems.

Example 6.3. In each of the following equations the cycles on the right are pairwise disjoint.

$$\begin{pmatrix} 1 & 2 & 3 & 4 & 5 \\ 3 & 4 & 1 & 5 & 2 \end{pmatrix} = (1\ \ 3)(2\ \ 4\ \ 5) = (2\ \ 4\ \ 5)(1\ \ 3)$$

$$(1\ \ 4\ \ 5)(2\ \ 3\ \ 5) = (1\ \ 4\ \ 5\ \ 2\ \ 3)$$

$$(2\ \ 4)(1\ \ 3\ \ 2)(2\ \ 5\ \ 4) = (1\ \ 3\ \ 4)(2\ \ 5) = (2\ \ 5)(1\ \ 3\ \ 4)$$

$$(1\ \ 6)(1\ \ 5)(1\ \ 4)(1\ \ 3)(1\ \ 2) = (1\ \ 2\ \ 3\ \ 4\ \ 5\ \ 6)$$

$$(1\ \ 2\ \ 3\ \ 4)^{-1} = (4\ \ 3\ \ 2\ \ 1) = (1\ \ 4\ \ 3\ \ 2)$$

$$(1\ \ 5\ \ 4\ \ 6\ \ 3\ \ 2)(4\ \ 3\ \ 6)(2\ \ 5) = (1\ \ 5)(2\ \ 4)(3)(6) = (1\ \ 5)(2\ \ 4).$$

PROBLEMS

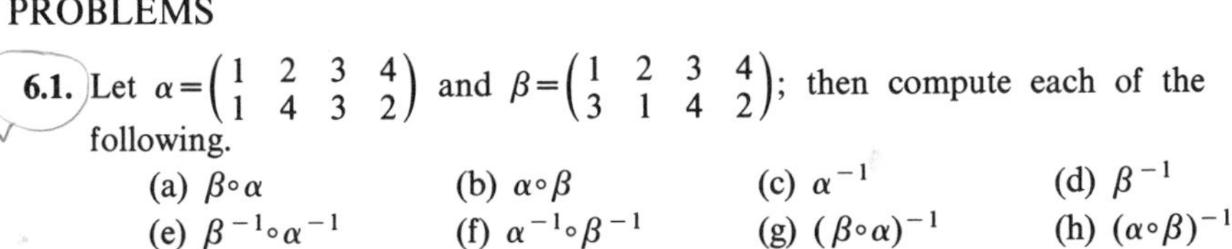

6.1. Let $\alpha = \begin{pmatrix} 1 & 2 & 3 & 4 \\ 1 & 4 & 3 & 2 \end{pmatrix}$ and $\beta = \begin{pmatrix} 1 & 2 & 3 & 4 \\ 3 & 1 & 4 & 2 \end{pmatrix}$; then compute each of the following.

(a) $\beta \circ \alpha$ (b) $\alpha \circ \beta$ (c) α^{-1} (d) β^{-1}

(e) $\beta^{-1} \circ \alpha^{-1}$ (f) $\alpha^{-1} \circ \beta^{-1}$ (g) $(\beta \circ \alpha)^{-1}$ (h) $(\alpha \circ \beta)^{-1}$

6.2. Repeat Problem 6.1 using $\alpha = \begin{pmatrix} 1 & 2 & 3 & 4 \\ 3 & 4 & 1 & 2 \end{pmatrix}$ and $\beta = \begin{pmatrix} 1 & 2 & 3 & 4 \\ 4 & 3 & 1 & 2 \end{pmatrix}$.

6.3. (a) Write out the elements of S_4.

(b) Which elements of S_4 are their own inverse?

6.4. Verify that in S_3 (Example 6.1) $\pi_1^{-1} = \pi_1$, $\pi_2^{-1} = \pi_3$, $\pi_3^{-1} = \pi_2$, $\pi_4^{-1} = \pi_4$, $\pi_5^{-1} = \pi_5$, and $\pi_6^{-1} = \pi_6$.

6.5. Prove that if S contains at least three elements, then Sym(S) is non-Abelian. (The main idea is already in the proof of Theorem 6.3.)

6.6. If $|S| = 2$, then composition, as an operation on $M(S)$, is not commutative (Problem 4.2); but Sym(S) is Abelian. Explain.

6.7. Verify that the group in Example 5.8 is non-Abelian.

6.8. Consider the problem of completing the following Cayley table.

*	u	v	w	x	y	z
u						
v						
w						
x						
y						
z						

(a) In how many ways can the table be completed (so that $*$ is an operation)?
(b) Same as (a) but with u required to be an identity element.
(c) Make use of the Cayley table for S_3 (Example 6.1) to complete this table in such a way as to obtain a group. (Consider replacement of π_1 by u, π_2 by v, and so on.)

6.9. Complete: If G (with operation $*$) is a group, then G is non-Abelian iff $a*b \neq b*a \ldots$.

6.10. (a) How many elements of S_3 map 3 to 3?
(b) How many elements of S_n map n to n?

6.11. Write the elements of S_3 (Example 6.1) using cycle notation.

6.12. Show that every element of S_n is a 2-cycle or can be written as a product of 2-cycles. [*Suggestion*: $(a_1a_2 \cdots a_k) = (a_1a_k) \cdots (a_1a_3)(a_1a_2)$.]

6.13. (a) Write $(a_1a_2 \cdots a_k)^{-1}$ in cycle notation (without the symbol for inverse).
(b) For which values of k will every k-cycle be its own inverse?

6.14. Rewrite the proof of Theorem 6.3 using cycle notation.

6.15. Write each of the following as a single cycle or a product of disjoint cycles.

(a) $\begin{pmatrix} 1 & 2 & 3 & 4 & 5 & 6 \\ 3 & 5 & 6 & 4 & 2 & 1 \end{pmatrix}$ (b) $(1\ \ 2)(1\ \ 3)(1\ \ 4)$

(c) $(3\ \ 1)(4\ \ 2)(5\ \ 2\ \ 3)$ (d) $(1\ \ 4\ \ 5)(1\ \ 2\ \ 3\ \ 5)(1\ \ 3)$

6.16. Explain why $\alpha\beta = \beta\alpha$ if α and β represent disjoint cycles.

SECTION 7. SUBGROUPS

The set of even integers is a subset of the set of all integers, and both sets are groups with respect to addition. Thus the even integers form a subgroup of the group of all integers, according to the following definition.

Definition. A subset H of a group G is a *subgroup* of G if H is itself a group with respect to the operation on G.

Example 7.1. (a) The group of integers with addition is a subgroup of the group of real numbers with addition.
(b) With multiplication, $\{1, -1\}$ is a subgroup of the group of nonzero real numbers.
(c) Any group is a subgroup of itself.
(d) If e is the identity of a group G, then $\{e\}$ is a subgroup of G.

It follows at once from the definition of subgroup that if G is a group with operation $*$, H is a subgroup of G, and $a, b \in H$, then $a*b \in H$. That is, the set H must be closed with respect to the operation $*$. Of the other conditions for H to be a group, the associative law is satisfied automatically: if $a*(b*c) = (a*b)*c$ is true for all elements of G, then it is certainly true for all elements in any subset of G. Notice that if f is the identity of H

and e the identity of G, then $f = e$, because $f*f = f$ and the only solution in G to the equation $x*x = x$ is e (Problem 7.6).

The following theorem will make it easier to determine which subsets of a group are subgroups.

Theorem 7.1. *Let G be a group with operation $*$, and let H be a subset of G. Then H is a subgroup of G iff*

(a) H is nonempty,

(b) if $a \in H$ and $b \in H$, then $a*b \in H$, and

(c) if $a \in H$, then $a^{-1} \in H$.

PROOF. Assume H to be a subgroup. Then, being a group, it must contain at least an identity element and thus be nonempty, confirming condition (a). The necessity of closure for a group, condition (b), has already been pointed out. Now consider condition (c). If $a \in H$, then a must have an inverse in the set H. If b denotes that inverse element, then $a*b = e$ (the identity of both H and G); thus necessarily $b = a^{-1}$ (the inverse of a in G) because of the uniqueness of inverses [Theorem 5.1(b)].

Assume now that H is a subset satisfying (a), (b), and (c); we must verify that it is a subgroup with respect to $*$. Property (b) ensures that $*$ is an operation on H. We have already observed that the associative law is satisfied for elements of H since it is satisfied for all elements of G. To show that H contains e, the identity of G, let x denote any element of H [there is such an element by (a)]. By condition (c), $x^{-1} \in H$, and then by condition (b), $x*x^{-1} = e \in H$. Thus H is a subgroup. □

Problem 7.16 contains a variation on Theorem 7.1, showing how (b) and (c) can be combined into a single condition. If H is known to be a finite set, then condition (c) of Theorem 7.1 can be omitted altogether (Problem 12.24).

Example 7.2. Table 7.1 is the Cayley table for a subgroup of S_3, with the notation that of Example 6.1.

Table 7.1

$\circ$	π_1	π_2	π_3
π_1	π_1	π_2	π_3
π_2	π_2	π_3	π_1
π_3	π_3	π_1	π_2

Certainly $\{\pi_1,\pi_2,\pi_3\}$ is nonempty. Closure is fulfilled because nothing appears in the table except π_1, π_2, and π_3. Also, condition (c) of Theorem 7.1 is satisfied because $\pi_1^{-1}=\pi_1$, $\pi_2^{-1}=\pi_3$, and $\pi_3^{-1}=\pi_2$, and all are in $\{\pi_1,\pi_2,\pi_3\}$.

Example 7.3. If k is an integer, then the set of all integral multiples of k satisfies the conditions of Theorem 7.1 and is therefore a subgroup of $\mathbb{Z}$. (In this case, the inverse of an element is the negative of the element.) Let G and H denote such subgroups consisting of all multiples of 3 and all multiples of 4, respectively. Then $G\cap H$ is the set of all multiples of 12, which is also a subgroup of $\mathbb{Z}$. This is a particular case of the following theorem.

Theorem 7.2. *If $\mathcal{C}$ denotes any collection of subgroups of a group G, then the intersection of all groups in $\mathcal{C}$ is also a subgroup of G.*

PROOF. Let H denote the intersection in question; we shall verify that H satisfies the three conditions in Theorem 7.1. Each subgroup in $\mathcal{C}$ contains e, the identity of G, so that $e\in H$ and H is nonempty. If $a,b\in H$, then a and b belong to each subgroup in $\mathcal{C}$ and thus $a*b$ belongs to each subgroup in $\mathcal{C}$ so that $a*b\in H$. Finally, if $a\in H$, then a^{-1} belongs to each subgroup in $\mathcal{C}$ because a does, and thus $a^{-1}\in H$. □

Theorem 7.2 permits us to associate with each subset of a group a unique smallest subgroup containing that subset. As an example of what this means, think of the set $\{10\}$ as a subset of the group of integers. Any subgroup containing 10 must also contain $10+10=20$, then $20+10=30$, and so on: it must contain $10n$ for each positive integer n, just because of closure. But the subgroup must also contain 0 (the identity) and the negative (inverse) of each of its elements. Thus it must contain all integral multiples of 10: $\{\ldots,-20,-10,0,10,20,\ldots\}$. This is a subgroup, and it is the smallest subgroup containing $\{10\}$.

Theorem 7.3. *Let S be any subset of a group G, and let $\langle S\rangle$ denote the intersection of all subgroups of G that contain S. Then $\langle S\rangle$ is the unique smallest subgroup of G that contains S, in the sense that*

(*a*) *$\langle S\rangle$ contains S,*
(*b*) *$\langle S\rangle$ is a subgroup, and*
(*c*) *if H is any subgroup of G that contains S, then H contains $\langle S\rangle$.*

Proof. First notice that there is always at least one subgroup of G containing S, namely G itself. With $\langle S\rangle$ as defined, $\langle S\rangle$ certainly contains S: the intersection of any collection of subsets each containing S will contain S, whether the subsets are subgroups or not. Next, $\langle S\rangle$ is a subgroup by Theorem 7.2. Finally, condition (c) is simply a property of the intersection of sets: if H is a subgroup containing S, then H is a member of the collection of subgroups whose intersection is $\langle S\rangle$, and thus H contains $\langle S\rangle$.

To justify use of the term *unique* in the theorem, assume that $[S]$ is also a subgroup of G satisfying conditions (a), (b), and (c) (with $[S]$ in place of $\langle S\rangle$). By making use of both lists of statements (a), (b), and (c), we can deduce both that $[S]$ contains $\langle S\rangle$ and that $\langle S\rangle$ contains $[S]$ (Problem 7.5). Thus $[S]=\langle S\rangle$. □

We say that S *generates* the subgroup $\langle S\rangle$, and that $\langle S\rangle$ is *generated* by S. If a subgroup is generated by a subset containing only a single element, then the subgroup is said to be *cyclic*. Thus the subgroup in the illustration preceding Theorem 7.3 is cyclic, namely $\langle\{10\}\rangle$, or, as we usually write, $\langle 10\rangle$. The group of integers is cyclic, generated by either $\{1\}$ or $\{-1\}$. The group of even integers is $\langle 2\rangle=\langle -2\rangle$.

Example 7.4. The subgroup $\langle 9,12\rangle$ of the integers must contain $12+(-9)=3$. Therefore $\langle 9,12\rangle$ must contain all multiples of 3. That is, $\langle 9,12\rangle\supseteq\langle 3\rangle$. But also $\{9,12\}\subseteq\langle 3\rangle$, and therefore $\langle 9,12\rangle=\langle 3\rangle$.

Example 7.5. The subgroup of S_3 in Example 7.2 is cyclic, generated by $\pi_2=(1\ \ 2\ \ 3)$. It is also generated by $\pi_3=(1\ \ 3\ \ 2)$. The group S_3 itself is not cyclic, but $S_3=\langle\pi_2,\pi_4\rangle$, for example. (See Problem 7.7.)

To determine just which elements are in a subgroup $\langle S\rangle$, we must in general make repeated use of Theorem 7.1, beginning with S and obtaining larger and larger sets until we arrive at a subgroup. At the first step we adjoin to S all elements $a*b$ for $a,b\in S$, and also all elements a^{-1} for $a\in S$. This is then repeated with S replaced by the (possibly larger) set consisting of S together with the elements adjoined at the first step. And so on. In this way it can be seen that $\langle S\rangle$ must contain all elements $a_1*a_2*\cdots*a_k$, where k is a positive integer and each of $a_1,a_2,\ldots,a_k$ is either an element of S or the inverse of an element of S. In fact, if S is nonempty then $\langle S\rangle$ will consist precisely of the set of all such elements $a_1*a_2*\cdots*a_k$ (Problem 7.17). As Example 7.4 shows, however, in special cases $\langle S\rangle$ can be determined more directly than this.

PROBLEMS

7.1. Verify each of the following equalities for subgroups of $\mathbb{Z}$.
(a) $\langle 2,3\rangle=\langle 1\rangle$
(b) $\langle -4,8\rangle=\langle 4\rangle$
(c) $\langle -8,10\rangle=\langle 2\rangle$
(d) $\langle 6,10,15\rangle=\langle 1\rangle$

7.2. Determine a necessary and sufficient condition for $\langle m\rangle\subseteq\langle n\rangle$, if $m,n\in\mathbb{Z}$.

7.3. If $\varnothing$ denotes the empty set, then what is $\langle\varnothing\rangle$ (in any group G)?

7.4. For which subsets S of a group G is $S=\langle S\rangle$?

7.5. Write a careful proof of the uniqueness of $\langle S\rangle$ in Theorem 7.3. (That is, fill in the details in the last paragraph of the proof.)

7.6. Prove that if G is a group with identity e, $x\in G$, and $x*x=x$, then $x=e$. (*Suggestion*: How would you prove this in $\mathbb{Z}$ with $+$ in place of $*$, and 0 in place of e?)

7.7. Determine the elements in each of the cyclic subgroups of S_3.

7.8. (a) Determine the elements in the subgroup $\langle(1\ \ 2\ \ 3\ \ 4)\rangle$ of S_4.
(b) Determine the elements in the subgroup $\langle(1\ \ 2\ \ 3\ \ 4\ \ 5)\rangle$ of S_5.
(c) What is the order of the subgroup $\langle(1\ \ 2\cdots n)\rangle$ of S_n?

7.9. (a) Let α denote the clockwise rotation of the plane through 90° about a fixed point p ($\alpha\in G$ in Example 5.7). Determine $\langle\alpha\rangle$.
(b) Repeat part (a) with 40° in place of 90°.
(c) What is the order of the subgroup $\langle\alpha\rangle$ if α denotes clockwise rotation through $(360/k)°$ $(k\in\mathbb{N})$?
(d) Repeat part (c) with $(360m/n)°$ in place of $(360/k)°$ $(m,n\in\mathbb{N})$?

7.10. Verify that each of the following is a subgroup of the group in Example 5.8. Characterize each subgroup in geometric terms (see Example 4.2).
(a) $\{\alpha_{a,0}\mid a\in\mathbb{R},\ a\neq 0\}$
(b) $\{\alpha_{1,b}\mid b\in\mathbb{R}\}$

7.11. Which of the following are subgroups of the group in Example 5.8?
(a) $\{\alpha_{a,0}\mid a\in\mathbb{Q},\ a\neq 0\}$
(b) $\{\alpha_{a,0}\mid a\in\mathbb{Z},\ a\neq 0\}$
(c) $\{\alpha_{1,b}\mid b\in\mathbb{Z}\}$
(d) $\{\alpha_{1,b}\mid b\in\mathbb{N}\}$

7.12. In each of the following parts, determine $n\in\mathbb{N}$ so that equality holds. Justify each answer.
(a) $\langle 5,-2\rangle=\langle n\rangle$
(b) $\langle 14,-7\rangle=\langle n\rangle$
(c) $\langle -10,14,35\rangle=\langle n\rangle$
(d) $\langle 21,-15\rangle=\langle n\rangle$

7.13. Complete: A group G is not cyclic iff… .

7.14. Prove that the group of rationals (operation addition) is not cyclic.

7.15. The mappings $\{\alpha_1,\ldots,\alpha_6\}$ in Problem 4.10 form a group of order 6 with respect to composition. Determine the elements in each cyclic subgroup of this group.

7.16. Prove that if G is a group with operation $*$, and H is a subset of G, then H is a subgroup of G iff
(a) H is nonempty, and
(b) if $a\in H$ and $b\in H$, then $a*b^{-1}\in H$.

7.17. Prove that if G is a group with operation $*$, and S is a nonempty subset of G, then $\langle S \rangle$ is the set of all $a_1 * a_2 * \cdots * a_k$, where k is a positive integer and each of $a_1, a_2, \ldots, a_k$ is either an element of S or the inverse of an element of S. [*Suggestion*: Show that the set described satisfies the conditions (a), (b), and (c) in Theorem 7.3, which characterize $\langle S \rangle$.]

SECTION 8. GROUPS AND SYMMETRY

Much of the importance of groups comes from their connection with symmetry. Just as numbers can be used to measure size (once a unit of measurement has been chosen), groups can be used to measure symmetry. With each figure we associate a group, and this group characterizes the symmetry of the figure. This application of groups extends from geometry to crystallography, and will be introduced in this section and then discussed more fully in Chapter IX. Another connection with symmetry—more abstract and not geometrical—arises in the study of algebraic equations. A group is associated with each equation, and this group characterizes a type of symmetry involving the solutions of the equation; questions about the solvability of an equation can be answered by studying the group associated with the equation. This application will be discussed in Chapter XIV.

To begin, we need a theorem about groups of permutations. Assume that G is a permutation group on S, and that T is a subset of S. Let

$$G_T = \{\alpha \in G \mid \alpha(t) = t \text{ for each } t \in T\}.$$

We say that the elements of G_T *leave T elementwise invariant*.

Example 8.1. Let $S = \{1,2,3,4\}$, $G = \mathrm{Sym}(S) = S_4$, and $T = \{1,2\}$. Then $G_T = \{(1)(2)(3)(4), (1)(2)(34)\}$. This is a subgroup of S_4.

Just as in the example, G_T is always a subgroup. Before proving that, however, we introduce another subset of G closely related to G_T. If α is a permutation of S, and T is a subset of S, then $\alpha(T)$ is the set of all elements $\alpha(t)$ for $t \in T$. Let

$$G_{(T)} = \{\alpha \in G \mid \alpha(T) = T\}.$$

Thus, if $\alpha \in G_{(T)}$, then α may permute the elements of T among themselves, but it sends no element of T outside T. Notice that $G_T \subseteq G_{(T)}$, whatever G and T are. We say that the elements of $G_{(T)}$ *leave T invariant*.

Example 8.2. With S, G, and T as in the previous example,

$$G_{(T)} = \{(1)(2)(3)(4), (1)(2)(34), (12)(3)(4), (12)(34)\}.$$

This also is a subgroup of S_4.

Theorem 8.1. *If G is a permutation group on S, and T is a subset of S, then G_T and $G_{(T)}$ are subgroups of G.*

PROOF. Apply Theorem 7.1, first to G_T. The identity, ι, is in G_T so that G_T is nonempty. If $\alpha, \beta \in G_T$, then $(\alpha \circ \beta)(t) = \alpha(\beta(t)) = \alpha(t) = t$ for every $t \in T$ so that $\alpha \circ \beta \in G_T$. Finally, if $\alpha \in G_T$, then $\alpha(t) = t$ for every $t \in T$; therefore, applying α^{-1} to both sides, $\alpha^{-1}(\alpha(t)) = \alpha^{-1}(t)$ or $t = \alpha^{-1}(t)$ for every $t \in T$ so that $\alpha^{-1} \in G_T$.

The proof for $G_{(T)}$ is similar; simply replace t by T in the obvious places. (See Problem 8.2.) □

Applications of the groups G_T will be given in Chapter VIII. At present it is the groups $G_{(T)}$ that we need.

Let P denote the set of all points in a plane, and let M denote the set of all permutations of P that preserve distance between points. Thus, if p and q are in P, and μ is in M, then the distance between $\mu(p)$ and $\mu(q)$ is equal to the distance between p and q. The permutations in M are called *motions* or *isometries* of the plane. It is not hard to show that M, with composition, is a subgroup of Sym(P) (Problem 8.4).

We met examples of motions in Example 5.7: the rotations of a plane about a fixed point. Another basic kind of motion is reflection through a line. Formally, *the reflection of the plane P through a line L in P* is the mapping that sends each point p in P to that point q such that L is the perpendicular bisector of the segment pq (Figure 8.1).

A third kind of motion is a *translation*, which is a mapping that sends all points the same distance in the same direction. For instance, a translation sending p_1 to q_1 (Figure 8.2) would send p_2 to q_2, and p_3 to q_3.

The following definition singles out the key to the use of groups in the study of geometrical symmetry.

Figure 8.1

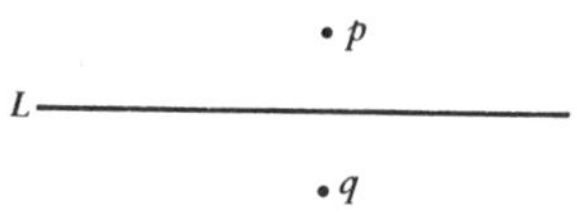

• p

L ————————

• q

Figure 8.2

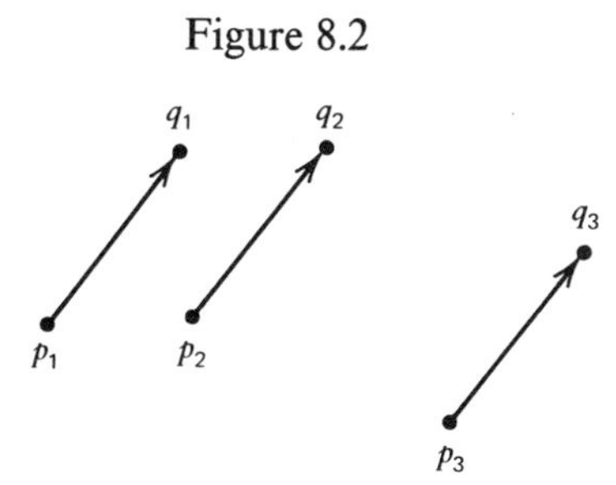

Definition. If T is a set of points in a plane, then $M_{(T)}$, the group of all motions leaving T invariant, is called the *group of symmetries* (or *symmetry group*†) of T.

Example 8.3. Consider a square, a rectangle, and a parallelogram (Figure 8.3). Notice that any motion of one of the figures will permute the vertices of the figure and the sides of the figure. Moreover, any motion will be completely determined by the way it permutes the vertices. It follows that in each case the group of symmetries will correspond to a subgroup of $\text{Sym}\{a,b,c,d\}$, and thus will have order at most $4! = 24$ (Theorem 6.2). In fact, the order must be less than 24, because some permutations of the vertices clearly cannot arise from motions of the plane (Problem 8.5). Here are the groups of symmetries for the three figures.

The elements are rotations around the center, and reflections through a horizontal line H, a vertical line V, and two diagonals D_1 and D_2. Notice that the more symmetric the figure, the larger its group of symmetries.

Group of symmetries of the square in Figure 8.3*a*

$$\begin{aligned}
\mu_1 &= \text{identity permutation}\\
\mu_2 &= \text{rotation } 90^\circ \text{ clockwise around } p\\
\mu_3 &= \text{rotation } 180^\circ \text{ clockwise around } p\\
\mu_4 &= \text{rotation } 270^\circ \text{ clockwise around } p\\
\mu_5 &= \text{reflection through } H\\
\mu_6 &= \text{reflection through } V\\
\mu_7 &= \text{reflection through } D_1\\
\mu_8 &= \text{reflection through } D_2
\end{aligned}$$

†Notice the difference between "symmetric" group, as used in Section 6, and "symmetry" group, as used here.

Figure 8.3

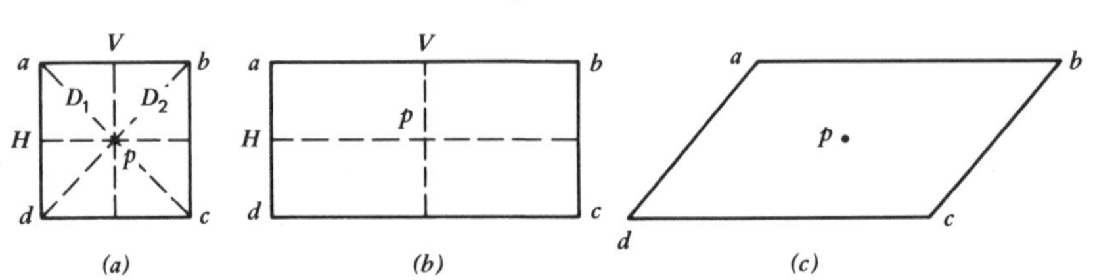

Group of symmetries of the rectangle in Figure 8.3*b*

$$\begin{aligned}\mu_1 &= \text{identity permutation}\\ \mu_3 &= \text{rotation } 180^\circ \text{ clockwise around } p\\ \mu_5 &= \text{reflection through } H\\ \mu_6 &= \text{reflection through } V\end{aligned}$$

Group of symmetries of the parallelogram in Figure 8.3*c*

$$\begin{aligned}\mu_1 &= \text{identity permutation}\\ \mu_3 &= \text{rotation } 180^\circ \text{ clockwise around } p\end{aligned}$$

Figure 8.4 illustrates how to compute entries for the Cayley tables of these groups. It shows that the result of $\mu_5 \circ \mu_2$ is the same as μ_8, reflection through D_2; and $\mu_2 \circ \mu_5$ is the same as μ_7, reflection through D_1. Notice that when we make such calculations we assume H, V, D_1, and D_2 to be fixed.

Table 8.1 is the Cayley table for the group of the square, which has each of the other two groups as a subgroup.

Figure 8.4

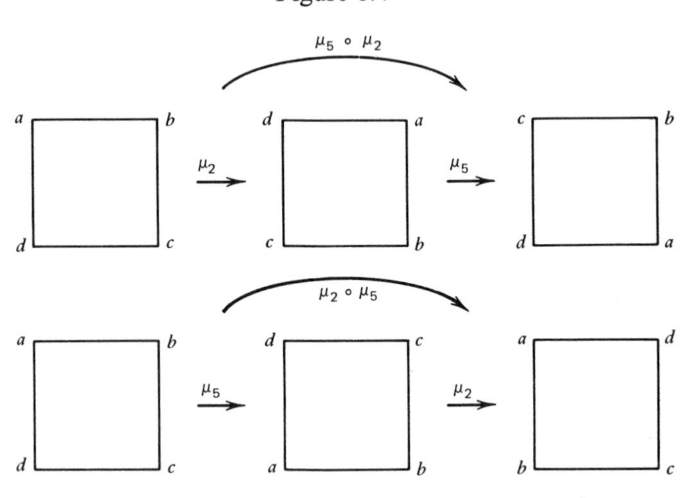

Table 8.1

$\circ$	μ_1	μ_2	μ_3	μ_4	μ_5	μ_6	μ_7	μ_8
μ_1	μ_1	μ_2	μ_3	μ_4	μ_5	μ_6	μ_7	μ_8
μ_2	μ_2	μ_3	μ_4	μ_1	μ_7	μ_8	μ_6	μ_5
μ_3	μ_3	μ_4	μ_1	μ_2	μ_6	μ_5	μ_8	μ_7
μ_4	μ_4	μ_1	μ_2	μ_3	μ_8	μ_7	μ_5	μ_6
μ_5	μ_5	μ_8	μ_6	μ_7	μ_1	μ_3	μ_4	μ_2
μ_6	μ_6	μ_7	μ_5	μ_8	μ_3	μ_1	μ_2	μ_4
μ_7	μ_7	μ_5	μ_8	μ_6	μ_2	μ_4	μ_1	μ_3
μ_8	μ_8	μ_6	μ_7	μ_5	μ_4	μ_2	μ_3	μ_1

PROBLEMS

8.1. Write the elements of G_T and $G_{(T)}$ for each of the following S, T, and G.
(a) $S=\{1,2,3\}$, $T=\{1\}$, and $G=S_3$
(b) $S=\{1,2,3\}$, $T=\{2,3\}$, and $G=S_3$
(c) $S=\{1,2,3,4\}$, $T=\{1\}$, and $G=S_4$
(d) $S=\{1,2,3,4\}$, $T=\{1,2,3\}$, and $G=S_4$

8.2. Prove that $G_{(T)}$ is a subgroup of G. (That is, complete the proof of Theorem 8.1.)

8.3. For $G=S_n$, state necessary and sufficient conditions on $T\subseteq S=\{1,2,\ldots,n\}$ for $G_{(T)}=G_T$.

8.4. Prove that M, the set of motions of a plane, is a group with respect to composition. [*Suggestion*: It may be helpful to introduce $d(p,q)$ to denote the distance between points p and q. Then $\alpha\in M$ iff $d(\alpha(p),\alpha(q))=d(p,q)$.]

8.5. The permutation $(a\quad b)(c)(d)$ of the vertices of the square *abcd* (Figure 8.3*a*) does not correspond to any motion of the plane. Why?

8.6. Draw figures like those in Figure 8.4 to verify the entries for $\mu_3\circ\mu_6$, $\mu_6\circ\mu_3$, $\mu_4\circ\mu_8$, and $\mu_8\circ\mu_4$ in Table 8.1.

8.7. Determine the group of symmetries of an equilateral triangle. (It will have order six.)

8.8. Determine the group of symmetries of an isosceles triangle.

8.9. Determine the group of symmetries of a regular pentagon. (It will have order 10.)

8.10. Determine the permutation of the vertices of the square *abcd* corresponding to each motion μ_i $(1\le i\le 8)$ in Table 8.1. [*Example*: μ_7 corresponds to (a)(c)(b d).]

8.11. Consider the mapping $T\mapsto M_{(T)}$ from the set of subsets of a plane to the set of symmetry groups. Is it one-to-one? Explain.

8.12. Determine the elements in each of the following subgroups of the group of symmetries of the square (Table 8.1).
(a) $\langle\mu_2\rangle$ (b) $\langle\mu_3\rangle$
(c) $\langle\mu_4\rangle$ (d) $\langle\mu_2,\mu_5\rangle$
(e) $\langle\mu_3,\mu_6\rangle$ (f) $\langle\mu_6,\mu_7\rangle$

8.13. Using the notation of Example 8.3, determine the group of symmetries of a rhombus (Figure 8.3*c* with $ab = ad$).

8.14. Considered as geometric objects, the 26 capital letters of the alphabet fall into five sets, with letters in each set having the same group of symmetries. Determine the five sets. (*Suggestion*: A, B, N, H, and F belong to different sets.)

8.15. Determine the group of symmetries of each of the following figures. (It suffices in each case to use motions similar to those in Example 8.3.)

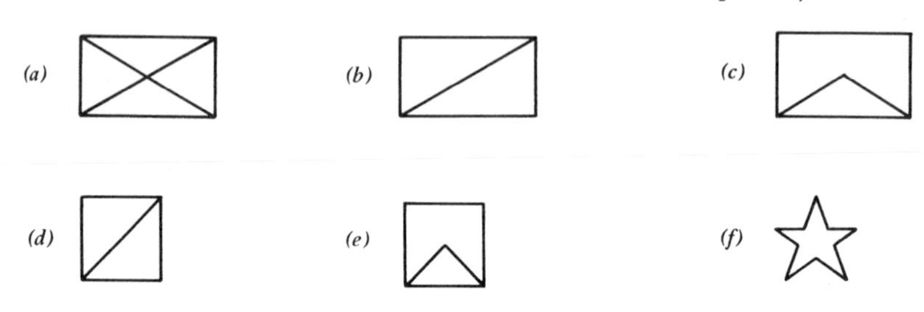

NOTES ON CHAPTER II

The origins of group theory can be found primarily in the theory of equations, number theory, and the study of geometrical transformations. The earliest connection was with the theory of equations in the late eighteenth and early nineteenth centuries, and will be discussed in Section 52. The connection with number theory was related to work that will be discussed in Section 47. The symmetry groups in Section 8 are examples of geometrical transformation groups; although we shall return to symmetry in Chapter IX, there are other kinds of transformation groups that we shall not be able to consider.

The history of group theory is considered in Chapter 49 of [1] and Chapter 7 of [3]. There are also a number of papers on the history of group theory in [2], the collected works of G. A. Miller (1865–1951), who was the first distinguished American group theorist.

1. Kline, M. *Mathematical Thought from Ancient to Modern Times*, Oxford University Press, London, 1972.
2. *The Collected Works of G. A. Miller*, 5 vols., University of Illinois, Urbana, 1935–1959.
3. Nový, L., *Origins of Modern Algebra*, Noordhoff, Leyden, The Netherlands, 1973.

CHAPTER III

Equivalence and Congruence

The first section in this chapter will be devoted to equivalence relations, which occur often, not only in algebra but throughout mathematics. The other two sections are devoted to some elementary facts about integers. These will be used to construct examples of groups, and of other algebraic systems yet to be introduced.

SECTION 9. EQUIVALENCE RELATIONS

Consider the following statements:

1. If $x,y \in \mathbb{R}$, then either

$$x = y \qquad \text{or} \qquad x \neq y.$$

2. If $x,y \in \mathbb{R}$, then either

$$x \leq y \qquad \text{or} \qquad x \nleq y.$$

3. If ABC and DEF are triangles, and $\cong$ denotes congruence, then either

$$ABC \cong DEF \qquad \text{or} \qquad ABC \ncong DEF.$$

In each statement, there is a set ($\mathbb{R}$, $\mathbb{R}$, all triangles) and a relation on that set ($=$, $\leq$, $\cong$), that is, a relationship that may or may not hold between ordered pairs of elements from the set. We are concerned now with relations of this type that satisfy three specific conditions.

Definition. A relation $\sim$ on a set S is an *equivalence relation* on S if it is *reflexive*

if $a \in S$, then $a \sim a$,

symmetric

if $a,b \in S$ and $a \sim b$, then $b \sim a$, and

transitive

if $a,b,c \in S$ and $a \sim b$ and $b \sim c$, then $a \sim c$.

Of the relations in 1, 2, and 3, the first and third are equivalence relations, but the second is not (because it is not symmetric). If $\sim$ is an equivalence relation and $a \sim b$, then we say that a and b are *equivalent*, or we use the specific term involved if there is one (such as equal or congruent).

Example 9.1. Let p denote a fixed point in a plane P, and for points x and y in P let $x \sim y$ mean that x and y are equidistant from p. This is an equivalence relation on the set of points in P. A point x will be equivalent to a given point q iff x lies on the circle through q with center p. (The point p is equivalent only to itself; think of $\{p\}$ as a circle with radius 0.)

Example 9.2. Let L denote the set of all lines in a plane with rectangular coordinate system. For $l_1, l_2 \in L$, let $l_1 \sim l_2$ mean that l_1 and l_2 have equal slopes (or that both slopes are undefined). This is an equivalence relation on L. The set of lines equivalent to a line l consists of l and all lines in L that are parallel to l.

Example 9.3. Let $\alpha : S \rightarrow T$ be a mapping. For $x,y \in S$, let $x \sim y$ mean that $\alpha(x) = \alpha(y)$. This is an equivalence relation on S. And, for example, with α the sine function (which has the real numbers as domain and the real numbers between -1 and 1 as image), the set of real numbers equivalent to π is the set $\{0, \pm\pi, \pm 2\pi, \ldots\}$.

Now return to Example 9.1. A different way to define the equivalence relation in that example is to say that $x \sim y$ iff x and y lie on a common circle with center p. These circles form a partition of the set of points in the plane, in the sense of the following definition.

Definition. A collection $\mathcal{P}$ of subsets of a set S forms a *partition* of S, provided that
(a) S is the union of the sets in $\mathcal{P}$, and
(b) if A and B are in $\mathcal{P}$ and $A \neq B$, then $A \cap B = \varnothing$.

Alternatively, the collection $\mathcal{P}$ forms a partition of S if each element of S is contained in one and only one of the sets in $\mathcal{P}$. To emphasize: each *element* of $\mathcal{P}$ is a *subset* of S.

The observation drawn from Example 9.1—that the equivalence relation yielded a partition—leads to something more general. In fact, the notions of equivalence relation on a set and partition of a set are in essence the same. Before proving this, we make the following definition.

Definition. Let $\sim$ be an equivalence relation on a set S, let $a \in S$, and let $[a] = \{x \in S \mid a \sim x\}$. This subset $[a]$ of S is called the *equivalence class* of a (relative to $\sim$).

In Example 9.1, the equivalence classes are the circles with centers at p. In Example 9.2, they are the sets of lines of equal slope. Notice that it is always true that $a \in [a]$, because $\sim$ is reflexive. And if $b \in [a]$ then $a \in [b]$, because $\sim$ is symmetric.

Theorem 9.1. *If $\sim$ is an equivalence relation on a set S, then the set of equivalence classes of $\sim$ forms a partition of S. Conversely, let $\mathcal{P}$ be a partition of S, and define a relation $\sim$ on S by $a \sim b$ iff there is a set in $\mathcal{P}$ that contains both a and b; then $\sim$ is an equivalence relation on S.*

PROOF. Let $\sim$ be an equivalence relation on S. If $a \in S$, then a belongs to at least one equivalence class, namely $[a]$, and thus S is indeed the union of the equivalence classes. It remains to be proved that if two equivalence classes are unequal, then they are disjoint; or, alternatively, if they are not disjoint, then they are equal. To this end, assume that $[a] \cap [b] \neq \varnothing$, and let c denote an element in the intersection. If x denotes any element in $[a]$, then we have both $a \sim c$ and $a \sim x$; thus $c \sim a$ and $a \sim x$, which imply $c \sim x$. But we also know that $b \sim c$; hence we can conclude that $b \sim x$, that is, $x \in [b]$. This shows that $[a] \subseteq [b]$. In the same way, it can be shown that $[a] \supseteq [b]$. Therefore $[a] = [b]$, which is what we were to prove.

Now assume $\sim$ to be defined as in the converse statement. If $a \in S$, then $a \sim a$ because there is some set in the partition containing a. If there is a set containing both a and b, then it contains both b and a, so that the symmetry of $\sim$ is trivial. Finally, assume that $a \sim b$ and $b \sim c$. Then there is a set in $\mathcal{P}$ containing both a and b, call it A; and also a set in $\mathcal{P}$ containing both b and c, call it B. Since $b \in A \cap B$, it must be true that $A = B$ because $\mathcal{P}$ is a partition. Both a and c belong to this set; therefore $a \sim c$. This proves that $\sim$ is transitive. □

Example 9.4. The set of even integers and the set of odd integers form a partition of the set of all integers. For the corresponding equivalence relation, $a \sim b$ iff a and b are both even or both odd, or, alternatively, iff $a - b$ is even.

In working with an equivalence relation on a set S, it is often useful to have a *complete set of equivalence class representatives*, that is, a subset of S such that each element of S is equivalent to precisely one element in that subset. In Example 9.4, $\{0,1\}$ is such a subset: each integer is equivalent to either 0 or 1, but no integer is equivalent to both. In Example 9.1, the set of points on any ray (half-line) with end point p would do, because each point in the plane is equivalent to precisely one point on any such ray. In Example 9.2, the set of all lines through the origin would do. For Example 9.3, see Problem 9.8.

PROBLEMS

9.1. For points with coordinates (x_1,y_1) and (x_2,y_2) in a plane with rectangular coordinate system, let $(x_1,y_1)\sim(x_2,y_2)$ mean that $y_1=y_2$.
(a) Prove that $\sim$ is an equivalence relation on the set of points in the plane. State clearly which properties of the relation $=$ on $\mathbb{R}$ are used in the proof.
(b) Describe the equivalence classes geometrically.
(c) Give a complete set of equivalence class representatives.

9.2. For points (x_1,y_1) and (x_2,y_2) in a plane with rectangular coordinate system, let $(x_1,y_1)\sim(x_2,y_2)$ mean that either $x_1=x_2$ or $y_1=y_2$. Is $\sim$ an equivalence relation on the set of points in the plane? Why?

9.3. For $x,y\in\mathbb{R}$, let $x\sim y$ mean that $x-y$ is an integer. Verify that $\sim$ is an equivalence relation. Describe the equivalence classes geometrically, with the elements of $\mathbb{R}$ identified with the points on a line in the usual way. Give a complete set of equivalence class representatives.

9.4. For points (x_1,y_1) and (x_2,y_2) in a plane with rectangular coordinate system, let $(x_1,y_1)\sim(x_2,y_2)$ mean that x_1-x_2 is an integer.
(a) Prove that $\sim$ is an equivalence relation.
(b) Give a geometric description of the equivalence class to which $(0,0)$ belongs.
(c) Give a complete set of equivalence class representatives.

9.5. Repeat Problem 9.4, except let $(x_1,y_1)\sim(x_2,y_2)$ mean that both x_1-x_2 and y_1-y_2 are integers.

9.6. For sets S and T, let $S\sim T$ mean that there is an invertible mapping of S onto T. Prove that $\sim$ is reflexive, symmetric, and transitive.

9.7. Let G be a permutation group on S. For $x,y\in S$, let $x\sim y$ mean that $\alpha(x)=y$ for some $\alpha\in G$.
(a) Prove that $\sim$ is an equivalence relation on S.
(b) What are the equivalence classes for $G=S_n$?
(c) What are the equivalence classes for $S=\{1,2,3,4\}$ and $G=\langle(2\ \ 3)\rangle$?

9.8. Determine a complete set of equivalence class representatives for the equivalence relation induced on $\mathbb{R}$ by the sine function (Example 9.3). What does this have to do with the inverse sine function?

9.9. For polynomials $f(x)$ and $g(x)$ with real coefficients, let $f(x)\sim g(x)$ mean that $f'(x)=g'(x)$ (where the primes denote derivatives). Prove that $\sim$ is an equivalence relation, and give a complete set of equivalence class representatives. (A polynomial with real coefficients is an expression of the form $a_0 + a_1x + \cdots + a_nx^n$, where $a_0, a_1, \ldots, a_n \in \mathbb{R}$.)

9.10. Let $\sim$ be a relation on a set S. Complete each of the following statements.
(a) $\sim$ is not reflexive iff....
(b) $\sim$ is not symmetric iff....
(c) $\sim$ is not transitive iff....

9.11. Find a flaw in the following "proof" that a relation on a set S is reflexive if it is both symmetric and transitive: Let $x \in S$. From $x \sim y$, by symmetry, we have $y \sim x$. By transitivity, $x \sim y$ and $y \sim x$ imply $x \sim x$. Therefore, $\sim$ is reflexive.

SECTION 10. CONGRUENCE. THE DIVISION ALGORITHM

We get an equivalence relation on the set of integers by agreeing that two integers are equivalent if either both are even or both are odd (Example 9.4). Another way to say this is to say that two integers are equivalent if their difference is even. The notion of congruence of integers generalizes this example. Congruence was first treated systematically at the beginning of the nineteenth century, by the eminent German mathematician Carl Friedrich Gauss (1777–1855); it has played a crucial role in the theory of numbers ever since. We shall see that congruence is also a fruitful source for examples in modern algebra. In fact, many concepts in modern algebra first arose in work relating to congruence.

Before defining congruence, we must make sure of some elementary facts about divisibility. An integer m is *divisible* by an integer n if there is an integer q (for quotient) such that $m = nq$. Thus 6 is divisible by 3 because $6 = 3 \cdot 2$. But 6 is not divisible by 4 or 5. If m is divisible by n, we also say that n *divides* m, and that m is a *multiple* of n, and we write $n|m$. So $3|6$ but $4\nmid 6$. If $n|m$, we also say that n if a *factor* of m. Notice that if $n|m$, then $n|(-m)$. Also, if $n|a$ and $n|b$, then $n|(a+b)$. [*Proof*: If $a = nq_1$ and $b = nq_2$, then $a + b = n(q_1 + q_2)$, and $q_1 + q_2$ is an integer if q_1 and q_2 are.] An integer p is a *prime* if $p > 1$ and p is divisible by no positive integer other than 1 and p itself.

Definition. Let n be a positive integer. Two integers a and b are said to be *congruent modulo n* if $a - b$ is divisible by n. This is written $a \equiv b \pmod{n}$.

Two integers are congruent modulo 2 iff either both are even or both are odd. This is the example from the introductory paragraph. Here are other examples: $17 \equiv 3 \pmod 7$ because 7 divides $17 - 3 = 14$; $4 \equiv 22 \pmod 9$ because 9 divides $4 - 22 = -18$; $19 \equiv 19 \pmod{11}$; $17 \not\equiv 3 \pmod 8$.

When working with congruences it is helpful to be able to move easily among the following equivalent statements:

$$a \equiv b \pmod{n}$$

$$n|(a-b)$$

$$a-b=un \qquad \text{for some } u \in \mathbb{Z}$$

$$a=b+un \qquad \text{for some } u \in \mathbb{Z}$$

(See Problem 10.10.)

Theorem 10.1. *Congruence modulo n is an equivalence relation on the set of integers, for each positive integer n.*

PROOF. *Reflexive*: If a is an integer, then $a \equiv a \pmod{n}$ because $n|(a-a) = 0$. *Symmetric*: If $a \equiv b \pmod{n}$, then $n|(a-b)$ so that $n|(b-a)$ and $b \equiv a \pmod{n}$. *Transitive*: If $a \equiv b \pmod{n}$ and $b \equiv c \pmod{n}$, then $n|(a-b)$ and $n|(b-c)$; but then $n|[(a-b)+(b-c)] = a-c$ so that $a \equiv c \pmod{n}$. □

The equivalence classes for this equivalence relation are called *congruence classes mod n*, or simply *congruence classes* if n is clear from the context. (These classes are sometimes called *residue classes*, but we shall not use this term.)

Example 10.1. (a) There are two congruence classes mod 2: the even integers and the odd integers.
(b) There are four congruence classes mod 4:

$$\{\ldots, -8, -4, 0, 4, 8, \ldots\}$$
$$\{\ldots, -7, -3, 1, 5, 9, \ldots\}$$
$$\{\ldots, -6, -2, 2, 6, 10, \ldots\}$$
$$\{\ldots, -5, -1, 3, 7, 11, \ldots\}.$$

Notice that in the last example there are four congruence classes and each integer is congruent to either 0, 1, 2, or 3 (mod 4). In the language of Section 9, $\{0,1,2,3\}$ is a complete set of equivalence class representatives. We are now going to show that this is typical, by showing that there are always n congruence classes modulo n and that each integer is congruent to either $0, 1, 2, \ldots,$ or $n-1 \pmod{n}$. But first we need more information about the integers—information, by the way, that is important far beyond our immediate need. We start from the following principle.

Least Integer Principle. *Every nonempty set of positive integers contains a least element.*

The Least Integer Principle is really an axiom, whose role will be clarified in Section 23. (Also see Appendix C.) At the moment we need it to prove the Division Algorithm.

If 11 is divided by 4, there is a quotient of 2 and a remainder of 3: $\frac{11}{4}=2+\frac{3}{4}$, or $11=4\cdot 2+3$. This illustrates the following result.

Division Algorithm. *If a and b are integers, with $b>0$, then there exist unique integers q and r such that*

$$a = bq + r, \qquad 0 \leq r < b.$$

Before giving the proof, we shall look at the idea behind it in the special case $11=4\cdot 2+3$ ($a=11$, $b=4$, $q=2$, and $r=3$). Consider the display in Example 10.1(b): $\mathbb{Z}$ has been partitioned into $b=4$ rows (congruence classes); $r=3$ is the smallest positive number in the row containing $a=11$; and $q=2$ is the number of positions (multiples of $b=4$) that we must move to the right to get from $r=3$ to $a=11$.

Here is another illustration: $-6=4\cdot(-2)+2$. Again, $b=4$; $r=2$ is the smallest positive number in the row containing $a=-6$; $q=-2$ is the number of positions that we must move (regarding left as negative) to get from $r=2$ to $a=-6$. In terms of such a display, with the integers partitioned into b rows (congruence classes), the set S in the proof that follows consists of the elements in the row to which a belongs.

Proof. We shall prove first that q and r exist, and then that they are unique.

Consider the set $S=\{a-bt \mid t \text{ is an integer}\}$, and let S' denote the set of nonnegative elements of S. Then $S'\neq\varnothing$, which can be seen as follows. If $a\geq 0$, then $t=0$ yields $a\in S'$. If $a<0$, then with $t=a$ we find $a-ba\in S$; and $a-ba=a(1-b)\geq 0$ because $a<0$ and $1-b\leq 0$ (recall $1\leq b$), and the product of two nonpositive integers is nonnegative.

Let r denote the least integer in S' (if $0\in S'$, then $r=0$; otherwise apply the Least Integer Principle). Let q denote the corresponding value of t, so that $a-bq=r$ and $a=bq+r$. Then $0\leq r$ by choice; therefore it suffices to show that $r<b$. Assume, on the contrary, that $r\geq b$. Then

$$a - b(q+1) = a - bq - b = r - b \geq 0,$$

and thus $a-b(q+1)\in S'$. But

$$a - b(q+1) = a - bq - b < a - bq = r$$

because $b>0$, and this contradicts the choice of r as the least element in S'. Thus we do have

$$a = bq + r, \qquad 0 \leq r < b.$$

To prove uniqueness, suppose that

$$a = bq_1 + r_1, \qquad 0 \leq r_1 < b$$

and

$$a = bq_2 + r_2, \qquad 0 \le r_2 < b.$$

We must show that $q_1 = q_2$ and $r_1 = r_2$. We have

$$bq_1 + r_1 = bq_2 + r_2$$
$$b(q_1 - q_2) = r_2 - r_1.$$

Thus $b|(r_2 - r_1)$. But $0 \le r_1 < b$ and $0 \le r_2 < b$, and so $-b < r_2 - r_1 < b$. The only multiple of b strictly between $-b$ and b is 0. Therefore $r_2 - r_1 = 0$, and $r_2 = r_1$. But then $b(q_1 - q_2) = 0$ with $b \ne 0$ so that $q_1 = q_2$. □

We can now return to congruences.

Theorem 10.2. *Let n be a positive integer. Then each integer is congruent modulo n to precisely one of the integers $0, 1, 2, \ldots, n-1$.*

PROOF. If a is an integer, then by the Division Algorithm there are unique integers q and r such that

$$a = nq + r, \qquad 0 \le r < n.$$

From this, $a - r = nq$ so that $n|(a - r)$ and $a \equiv r \pmod{n}$. Thus a is congruent to *at least* one of the integers $0, 1, 2, \ldots, n-1$. To show that r is unique, assume that $a \equiv s \pmod{n}$ with $0 \le s < n$. Then $a - s = nt$ (for some integer t), and

$$a = nt + s, \qquad 0 \le s < n,$$

so that $s = r$ by the uniqueness of r in the Division Algorithm. □

PROBLEMS

10.1. List all positive divisors of each of the following integers.

(a) 20 (b) 63 (c) 81 (d) -101

10.2. There are 25 primes less than 100. What are they?

10.3. Prove that if $a|b$ and $b|c$, then $a|c$.

10.4. Prove that if $a|b$ and $b|a$, then $a = \pm b$. (You may assume that the only divisors of 1 are ± 1.)

10.5. Example 10.1(b) shows the four congruence classes modulo 4. Make a similar array for the congruence classes modulo 3.

10.6. There are 10 integers x such that $-25 < x < 25$ and $x \equiv 3 \pmod{5}$. Find them all.

10.7. For which n is $25 \equiv 4 \pmod{n}$?

10.8. (a) For $a, b \in \mathbb{Z}$, let $a \sim b$ mean that the decimal representations of a and b have the same last (units) digit. This is an equivalence relation. How does it relate to congruence?

(b) Repeat part (a), except let k be a fixed positive integer and let $a \sim b$ mean that the decimal representations of a and b have the same digits in each of the last k positions. (*Example*: for $k = 3$, $4587 \sim 30587$.)

10.9. Prove that if $m|n$ and $a \equiv b \pmod{n}$, then $a \equiv b \pmod{m}$.

10.10. Prove the equivalence of the four statements just preceding Theorem 10.1. Also prove that $a \equiv b \pmod{n}$ iff a and b leave the same remainder upon division by n.

10.11. Prove that if n is a positive integer and $a, b \in \mathbb{Z}$, then there is an integer x such that $a + x \equiv b \pmod{n}$.

10.12. Find all x such that $0 \le x < 6$ and $2x \equiv 4 \pmod{6}$.

10.13. Disprove with a counterexample: If n is a positive integer and $a, b \in \mathbb{Z}$, then there is an integer x such that $ax \equiv b \pmod{n}$.

10.14. Assume that $a \equiv b \pmod{n}$ and $c \equiv d \pmod{n}$.

(a) Prove that $a + c \equiv b + d \pmod{n}$.

(b) Prove that $ac \equiv bd \pmod{n}$. (*Suggestion*: $a = b + un$, $c = d + vn$.)

10.15. Verify that each of the following statements is false. (Compare the Least Integer Principle.)

(a) Every nonempty set of integers contains a least element.

(b) Every nonempty set of positive rational numbers contains a least element.

10.16. For each pair a, b, find the unique integers q and r such that $a = bq + r$, $0 \le r < b$.

(a) $a = 19$, $b = 5$ (b) $a = -7$, $b = 5$

(c) $a = 13$, $b = 20$ (d) $a = 30$, $b = 1$

10.17. Use the Division Algorithm to prove that if a and b are integers, with $b \ne 0$, then there exist unique integers q and r such that

$$a = bq + r, \qquad 0 \le r < |b|.$$

(*Suggestion*: If $b > 0$, this is the Division Algorithm. Otherwise $|b| = -b$.)

10.18. Each subgroup of $\mathbb{Z}$ (operation $+$) is cyclic. Prove this by assuming that H is a subgroup of $\mathbb{Z}$ and giving a reason for each of the following statements.

(a) If $|H| = 1$, then H is cyclic.

(b) If $|H| > 1$, then H contains a least positive element; call it b.

(c) If $a \in H$ and $a = bq + r$, then $r \in H$.

(d) If $a \in H$ and $a = bq + r$, with $0 \le r < b$, then $r = 0$.

(e) $H = \langle b \rangle$.

SECTION 11. INTEGERS MODULO N

We have seen that if n is a positive integer then there are n congruence classes modulo n. With n fixed and k an integer, let $[k]$ denote the congruence class to which k belongs $(\text{mod}\, n)$. With $n = 5$, for example, $[2] = [7] = [-33] = \{\ldots, -8, -3, 2, 7, 12, \ldots\}$.† In general, $\{[0], [1], \ldots, [n-1]\}$ is a complete set of congruence classes modulo n, in the sense that each integer is in precisely one of these classes (Theorem 10.2). Let $\mathbb{Z}_n$ denote the set $\{[0], [1], \ldots, [n-1]\}$. We shall show that there is a natural operation on this set that makes it a group. This will enlarge our list of groups, and

†Notice that $[k]$ is ambiguous unless n has been specified. For example, $[3]$ means one thing in $\mathbb{Z}_5$, and something else in $\mathbb{Z}_6$. With reasonable attention to context there should be no confusion, however. In case of doubt, $[k]_n$ can be used in place of $[k]$.

will allow us to account for all cyclic groups—in a way that will be made precise in the next chapter.

Definition. For $[a] \in \mathbb{Z}_n$ and $[b] \in \mathbb{Z}_n$, define $[a] \oplus [b]$ by

$$[a] \oplus [b] = [a+b].$$

Example 11.1. Choose $n=5$. Then $[3] \oplus [4] = [3+4] = [7] = [2]$, and $[-29] \oplus [7] = [-22] = [3]$.

There is a question about the definition of $\oplus$: Is it really an operation on $\mathbb{Z}_n$? Or, as it is sometimes put, is $\oplus$ *well-defined*? Notice that $[a] \oplus [b]$ has been defined in terms of $a+b$. What if representatives other than a and b are chosen from $[a]$ and $[b]$? For example, with $n=5$ again, $[3]=[18]$ and $[4]=[-1]$; therefore, it should be that $[3] \oplus [4] = [18] \oplus [-1]$. Is that true? Yes, because $[3] \oplus [4] = [7] = [2]$ and $[18] \oplus [-1] = [17] = [2]$. The following lemma settles the question in general.

Lemma 11.1. *In $\mathbb{Z}_n$, if $[a_1]=[a_2]$ and $[b_1]=[b_2]$, then $[a_1+b_1]=[a_2+b_2]$.*

PROOF. If $[a_1]=[a_2]$ and $[b_1]=[b_2]$, then for some integers u and v

$$a_1 = a_2 + un \qquad \text{and} \qquad b_1 = b_2 + vn.$$

Addition yields

$$\begin{aligned} a_1 + b_1 &= (a_2 + un) + (b_2 + vn) \\ &= a_2 + b_2 + (u+v)n. \end{aligned}$$

Thus $(a_1+b_1)-(a_2+b_2)$ is n times an integer, $u+v$, and hence $a_1+b_1 \equiv a_2 + b_2 \pmod{n}$. Therefore $[a_1+b_1]=[a_2+b_2]$. □

Problem 11.17 is designed to help remove doubts as to whether Lemma 11.1 is really necessary.

Theorem 11.1. *$\mathbb{Z}_n$ is a cyclic group with respect to the operation $\oplus$.*

PROOF. Associativity:

$$\begin{aligned} [a] \oplus ([b] \oplus [c]) &= [a] \oplus [b+c] && \text{definition of } \oplus \\ &= [a+(b+c)] && \text{definition of } \oplus \\ &= [(a+b)+c] && \text{associativity of } + \\ &= [a+b] \oplus [c] && \text{definition of } \oplus \\ &= ([a] \oplus [b]) \oplus [c] && \text{definition of } \oplus. \end{aligned}$$

The identity is [0]:

$$[0]\oplus[a]=[0+a]=[a]$$
$$[a]\oplus[0]=[a+0]=[a].$$

Similarly, the inverse of $[a]$ is easily shown to be $[-a]$ (Problem 11.4).

Finally, if $0<k\leq n-1$, then $[k]=[1]\oplus\cdots\oplus[1]$ (k terms). Therefore $\mathbb{Z}_n=\langle[1]\rangle$, and $\mathbb{Z}_n$ is cyclic with generator [1]. □

The first parts of the preceding proof are typical of proofs of properties of $\oplus$, in that first the definition of $\oplus$ is used, then the corresponding property of $+$, and finally the definition of $\oplus$ again. Whenever $\mathbb{Z}_n$ is referred to as a group the operation is understood to be $\oplus$. This group is called the *group of integers modulo n* (or *mod n*).

Corollary. *There is a cyclic group of order n for each positive integer n.*

PROOF. $\mathbb{Z}_n$ contains n elements, $[0],[1],\ldots,[n-1]$. □

In order to appreciate this corollary, try to find another way to construct a group of order 20, for instance. Even for small orders, the associative law is especially hard to verify for an operation that is not "natural" in some sense.

Example 11.2. Table 11.1 is the Cayley table for $\mathbb{Z}_6$.

Using multiplication rather than addition, we obtain another operation on $\mathbb{Z}_n$ as follows:

$$[a]\odot[b]=[ab].$$

Example 11.3. Choose $n=6$. Then

$$[2]\odot[5]=[10]=[4]$$
$$[3]\odot[-4]=[-12]=[0].$$

As with $\oplus$, we must verify that $\odot$ is well-defined. Lemma 11.2 does that.

Table 11.1

$\oplus$	[0]	[1]	[2]	[3]	[4]	[5]
[0]	[0]	[1]	[2]	[3]	[4]	[5]
[1]	[1]	[2]	[3]	[4]	[5]	[0]
[2]	[2]	[3]	[4]	[5]	[0]	[1]
[3]	[3]	[4]	[5]	[0]	[1]	[2]
[4]	[4]	[5]	[0]	[1]	[2]	[3]
[5]	[5]	[0]	[1]	[2]	[3]	[4]

Lemma 11.2. *In $\mathbb{Z}_n$, if $[a_1]=[a_2]$ and $[b_1]=[b_2]$, then $[a_1b_1]=[a_2b_2]$.*

PROOF. If $[a_1]=[a_2]$ and $[b_1]=[b_2]$, then for some integers u and v

$$a_1 = a_2 + un \qquad \text{and} \qquad b_1 = b_2 + vn.$$

Therefore

$$\begin{aligned} a_1b_1 &= (a_2+un)(b_2+vn) \\ &= a_2b_2 + (a_2v + ub_2 + uvn)n. \end{aligned}$$

Thus $a_1b_1 - a_2b_2$ is an integer times n so that $[a_1b_1]=[a_2b_2]$. □

In contrast to $\mathbb{Z}_n$ with $\oplus$, $\mathbb{Z}_n$ with $\odot$ is not a group. (See Problem 11.11, for example.) The operation $\odot$ does have some important properties, however; the next lemma gives two of these, and Chapter V will give more.

Lemma 11.3. *The operation $\odot$ is associative on $\mathbb{Z}_n$ and has $[1]$ as an identity element.*

PROOF. Make the obvious changes in the proof of Theorem 11.1. □

Let $\mathbb{Z}_n^{\#}$ denote the set $\{[1],[2],\ldots,[n-1]\}$, that is, $\mathbb{Z}_n$ with $[0]$ deleted. Although $\mathbb{Z}_n$ with $\odot$ is never a group, the next example shows that $\mathbb{Z}_n^{\#}$ with $\odot$ can be a group.

Example 11.4. $\mathbb{Z}_5^{\#}$ is a group with respect to the operation $\odot$. Associativity is a consequence of Lemma 11.3. Table 11.2 shows closure, and also that [1] is an identity and that the inverses of [1], [2], [3], and [4], are [1], [3], [2], and [4], respectively.

Example 11.5. $\mathbb{Z}_6^{\#}$ is not a group with respect to $\odot$. For example, since $[2]\odot[3]=[6]=[0]$, there is not even closure. In Section 20 we shall uncover a general principle that will explain why $\mathbb{Z}_5^{\#}$ (the previous example) is a group but $\mathbb{Z}_6^{\#}$ is not a group.

Table 11.2

$\odot$	[1]	[2]	[3]	[4]
[1]	[1]	[2]	[3]	[4]
[2]	[2]	[4]	[1]	[3]
[3]	[3]	[1]	[4]	[2]
[4]	[4]	[3]	[2]	[1]

PROBLEMS

11.1. (a) Give five integers in [3] as an element of $\mathbb{Z}_5$.
(b) Give five integers in [3] as an element of $\mathbb{Z}_6$.

11.2. Simplify each of the following expressions in $\mathbb{Z}_5$; write the answer as [0], [1], [2], [3], or [4].

(a) $[3]\oplus[4]$ (b) $[2]\oplus[-7]$ (c) $[17]\oplus[76]$

(d) $[3]\odot[4]$ (e) $[2]\odot[-7]$ (f) $[17]\odot[76]$

(g) $[3]\odot([2]\oplus[4])$ (h) $([3]\odot[2])\oplus([3]\odot[4])$

11.3. (a) to (h). Simplify each of the expressions in Problem 11.2 after interpreting it in $\mathbb{Z}_6$ rather than $\mathbb{Z}_5$, and write the answer as [0], [1], [2], [3], [4], or [5].

11.4. Prove that $[-a]$ is an inverse for $[a]$ in $\mathbb{Z}_n$.

11.5. Construct the Cayley table for the group $\mathbb{Z}_3$.

11.6. Construct the Cayley table for the group $\mathbb{Z}_4$.

11.7. Prove that each group $\mathbb{Z}_n$ is Abelian.

11.8. For which k $(0 \leq k < 4)$ is $[k]$ a generator for the group $\mathbb{Z}_4$?

11.9. For which k $(0 \leq k < 5)$ is $[k]$ a generator for the group $\mathbb{Z}_5$?

11.10. In $\mathbb{Z}_6$, determine the elements in each of the subgroups $\langle [k] \rangle$ $(0 \leq k < 6)$.

11.11. There is no inverse for [0] relative to the operation $\odot$ on $\mathbb{Z}_n$. Why? (The element [1] is an identity element for $\odot$.)

11.12. Write the proof of Lemma 11.3 in detail.

11.13. Prove or disprove that $\mathbb{Z}_3^{\#}$ is a group with respect to $\odot$.

11.14. Prove or disprove that $\mathbb{Z}_4^{\#}$ is a group with respect to $\odot$.

11.15. Prove that the operation $\odot$ is commutative on $\mathbb{Z}_n$.

11.16. Prove that $[a]\odot([b]\oplus[c])=([a]\odot[b])\oplus([a]\odot[c])$ for all $[a],[b],[c]\in\mathbb{Z}_n$.

11.17. Define an equivalence relation on the set of integers by letting $a \sim b$ mean that either both a and b are negative or both a and b are nonnegative. There are two equivalence classes: $[-1]$, consisting of the negative integers; and $[0]$, consisting of the nonnegative integers. Attempt to define an operation $\boxed{+}$ on the set $\{[-1],[0]\}$ of equivalence classes by

$$[a]\boxed{+}[b] = [a+b],$$

in analogy with the definition of $\oplus$ on $\mathbb{Z}_n$. Show that $\boxed{+}$ is not well-defined.

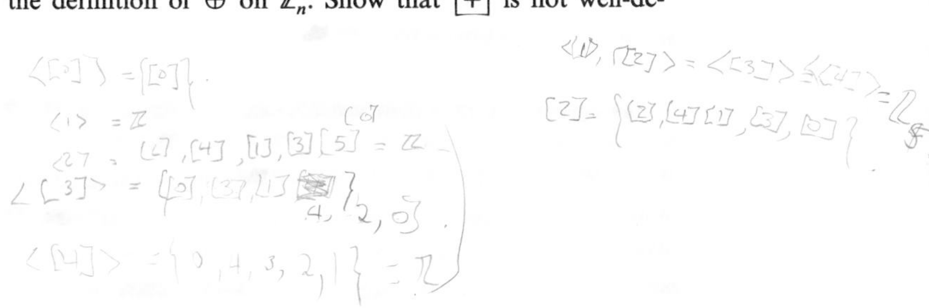

CHAPTER IV

Groups

In this chapter the emphasis will be less on giving examples of groups and more on proving general theorems about them. The focal points will be Lagrange's Theorem, which puts a strong restriction on which subsets of a group can be subgroups, and isomorphism, which makes precise the notion of what it means for groups to be "essentially" alike.

SECTION 12. ELEMENTARY PROPERTIES

Hereafter, whenever a group has no other specified operation we shall indicate the operation by juxtaposition, ab, and refer to it as a *product*. Also, if there is no other established notation then the identity element will be denoted by e, and the inverse of an element a by a^{-1}. When interpreting general statements in special cases this notation must be changed accordingly, of course. For instance, ab, e, and a^{-1} become, in additive notation, $a+b$, 0, and $-a$. The reasons for using ab, e, and a^{-1} consistently in general statements are economy and uniformity.

We begin with some remarks about associativity. There are two possible results from inserting parentheses in abc, and the associative law demands that these be equal: $a(bc)=(ab)c$. But what about $abcd$? For example, two of the possibilities here are $a(b(cd))$ and $(ab)(cd)$. One application of the associative law (for three elements) shows these possibilities to be equal: substitute a for x, b for y, and cd for z in $x(yz)=(xy)z$. The other possibilities for $abcd$ are given in Problem 12.6; they all give the same result. In fact, this is true for any number of elements, by what is known as the *generalized associative law*: If $a_1, a_2, \ldots, a_n$ $(n \geq 2)$ are elements of a group, then the product $a_1a_2 \cdots a_n$ is unambiguous; that is, the same group element will be obtained regardless of how parentheses are inserted in the product, as long as the elements $a_1, a_2, \ldots, a_n$ and their order of appearance are unchanged. (For a proof of this law see any of the last three references listed at the end of this chapter.)

Theorem 12.1. *Let G be a group.*

(*a*) *If $a,b,c \in G$ and $ab = ac$, then $b = c$* (left cancellation law).
(*b*) *If $a,b,c \in G$ and $ba = ca$, then $b = c$* (right cancellation law).
(*c*) *If $a,b \in G$, then each of the equations $ax = b$ and $xa = b$ has a unique solution in G. In the first, $x = a^{-1}b$; in the second, $x = ba^{-1}$.*
(*d*) *If $a \in G$, then $(a^{-1})^{-1} = a$.*
(*e*) *If $a,b \in G$, then $(ab)^{-1} = b^{-1}a^{-1}$.*

PROOF. (a) Assume that $ab = ac$. Upon multiplying both sides on the left by a^{-1}, we are led to $a^{-1}(ab) = a^{-1}(ac)$, $(a^{-1}a)b = (a^{-1}a)c$, $eb = ec$, and $b = c$.

(b) Similar to part (a) (Problem 12.1).

(c) To see that $x = a^{-1}b$ is a solution of $ax = b$, simply substitute: $a(a^{-1}b) = (a^{-1}a)b = eb = b$. To see that there is no other solution, assume that $ax = b$. Then multiplication on the left by a^{-1} leads to $a^{-1}(ax) = a^{-1}b$, $(a^{-1}a)x = a^{-1}b$, $ex = a^{-1}b$, $x = a^{-1}b$; thus $x = a^{-1}b$ is indeed the only solution.

The proof for the equation $xa = b$ is similar (Problem 12.2).

(d) The inverse of a^{-1} is the unique element x such that $a^{-1}x = e$. But $a^{-1}a = e$; therefore, the inverse of a^{-1} must be a.

(e) The inverse of ab is the unique element x such that $(ab)x = e$. But $(ab)(b^{-1}a^{-1}) = a(bb^{-1})a^{-1} = aea^{-1} = aa^{-1} = e$; thus the inverse of ab must be $b^{-1}a^{-1}$. □

Here are some observations about the theorem. If a and x are elements of a finite group, then in the Cayley table for the group ax will be in the row labeled by a. If b is also an element of the group, then the existence of a unique solution of $ax = b$ [Theorem 12.1(c)] implies that b appears exactly once in the row labeled by a. Thus

each element of a finite group appears exactly once in each row of the Cayley table for the group.

(This ignores the row labels at the outside of the table.) Similarly, because there is a unique solution of $xa = b$,

each element of a finite group appears exactly once in each column of the Cayley table for the group.

Part (e) of the theorem indicates that the inverse of a product is the product of the inverses, *in reverse order.*

Integral powers of group elements are defined as follows: $a^0 = e$, $a^1 = a$, $a^2 = aa$, $a^3 = aaa$, and so on, so that a^n is equal to the product of n a's for each positive integer n. Also, $a^{-n} = (a^{-1})^n$ for each positive integer n. The familiar laws of exponents can be proved by mathematical induction:

$$a^m a^n = a^{m+n}$$
$$(a^m)^n = a^{mn}$$

for all integers m and n. (See Appendix C.) In additive notation, for n a positive integer, a^n becomes $na = a + a + \cdots + a$ (n times), and a^{-n} becomes $(-n)a = n(-a)$. In this case the laws above become

$$(ma) + (na) = (m+n)a$$
$$n(ma) = (mn)a$$

for all integers m and n.

Now assume that a is an element of a group G, and consider the set of all integral powers of $a: \{\ldots, a^{-2}, a^{-1}, e, a, a^2, \ldots\}$. This is easily seen to be a subgroup of G by Theorem 7.1: it is nonempty (it contains a); it is closed ($a^m a^n = a^{m+n}$); and it contains the inverse of each of its elements (the inverse of a^m is a^{-m}, whatever the integer m). In fact, this subgroup is $\langle a \rangle$, the subgroup generated by a. On the surface, such subgroups may appear to be infinite. But in many cases they will be finite. For instance, if the group G is finite, then $\langle a \rangle$ must be finite, since $\langle a \rangle$ cannot possibly contain more elements than G. The point is, of course, that different powers of an element may be equal.

Example 12.1. In S_3 (Example 6.1), $(1\ 2\ 3)^3 = (1)$, $(1\ 2\ 3)^{-6} = (1)$, and, in general, $(1\ 2\ 3)^t = (1)$ for every multiple t of 3. Notice that 3 is the smallest postive integer n such that $(1\ 2\ 3)^n = (1)$. Also, $\langle (1\ 2\ 3) \rangle = \{(1), (1\ 2\ 3), (1\ 3\ 2)\}$. The following theorem analyzes this and all similar cases.

Theorem 12.2. *Assume that G is a group, that $a \in G$, and that there exist unequal integers r and s such that $a^r = a^s$.*

(a) *There is a smallest positive integer n such that* $a^n = e$.
(b) *If t is an integer, then* $a^t = e$ *iff n is a divisor of t.*
(c) *The elements* $e = a^0, a, a^2, \ldots, a^{n-1}$ *are distinct, and* $\langle a \rangle = \{e, a, a^2, \ldots, a^{n-1}\}$.

PROOF. (a) To prove part (a), it suffices to show that $a^t = e$ for *some* positive integer t; for the Least Integer Principle will then tell us that there is a smallest such integer, which we can call n. Assume $r > s$. (If $s > r$, just interchange r and s in the next sentence.) After multiplying both sides of $a^r = a^s$ by a^{-s}, we obtain $a^{r-s} = e$ with $r - s > 0$. And that is sufficient.

(b) If n is a divisor of t, say $t = nv$, then $a^t = a^{nv} = (a^n)^v = e^v = e$. To prove the other half of part (b), suppose that $a^t = e$. By the Division Algorithm there are integers q and r such that $t = nq + r, 0 \leq r < n$. Thus $a^t = a^{nq+r} = (a^n)^q a^r = e^q a^r = a^r$. But $a^t = e$, so $a^r = e$. This implies that $r = 0$, since $0 \leq r < n$ and n is the smallest positive integer such that $a^n = e$. Therefore, $t = nq$ so that n is a divisor of t.

(c) To prove that $a^0, a^1, a^2, \ldots, a^{n-1}$ are distinct, suppose that $a^u = a^v$ with $0 \leq u < n$ and $0 \leq v < n$. We shall prove that u must equal v. Interchanging u and v if necessary, we can assume that $u \geq v$. This leads to $a^{u-v} = e$ with $u - v \geq 0$. Therefore, by part (b), n must be a divisor of $u - v$. But $u - v < n$, since $0 \leq u < n$ and $0 \leq v < n$. Thus n is a divisor of $u - v$ and $0 \leq u - v < n$, which can happen only if $u - v = 0$, or $u = v$.

Certainly any power of a is in $\langle a \rangle$; hence the proof will be complete if we show that each power of a is in the set $\{e, a, a^2, \ldots, a^{n-1}\}$. Consider a power a^m. By the Division Algorithm there are integers q and r such that $m = nq + r$. This leads to $a^m = a^{nq+r} = (a^n)^q a^r = e^q a^r = a^r$, with $0 \leq r < n$, which is just what we need. □

If a is an element of a group, then the smallest positive integer n such that $a^n = e$, if it exists, is called the *order* of a. If there is no such integer, then a is said to have *infinite order*. The order of an element a will be denoted by $o(a)$. In S_3, for instance, $o((1\ 2\ 3)) = 3$ (Example 12.1). In the group of nonzero rationals (operation multiplication), 2 has infinite order: $2^n \neq 1$ for all positive integers n.

In additive notation, the condition $a^n = e$ becomes $na = 0$. This applies to $\mathbb{Z}_n$, also. For example, in $\mathbb{Z}_6$, $o([2]) = 3$, because $3[2] = [2] \oplus [2] \oplus [2] = [6] = [0]$ (and $[2] \neq [0]$ and $2[2] = [4] \neq [0]$).

The word *order* has been used in two senses: the *order* of a group (Section 5) and the *order* of an element. The next corollary shows how the two are related.

Corollary. *If a is an element of a group, then* $o(a)=|\langle a\rangle|$.

PROOF. If $o(a)=n$ is finite, then $\langle a\rangle=\{e,a,a^2,\ldots,a^{n-1}\}$ by Theorem 12.2(c), and thus $|\langle a\rangle|=n$. Otherwise, a is of infinite order, all integral powers of a are distinct, and $|\langle a\rangle|$ is infinite. □

PROBLEMS

12.1. Prove Theorem 12.1(b).

12.2. Prove that $xa=b$ has a unique solution in a group. [This is the omitted part of the proof of Theorem 12.1(c).]

12.3. Prove that $axb=c$ has a unique solution in a group (given a,b,c).

12.4. Determine the order of each element of S_3.

12.5. Determine the order of each element of $\mathbb{Z}_6$.

12.6. Parentheses can be inserted in the product $abcd$ in the following ways: $a(b(cd))$, $a((bc)d)$, $(ab)(cd)$, $((ab)c)d$, $(a(bc))d$. Prove that in a group these five products are equal. Which of the group axioms does the proof require?

12.7. Prove that if a and b are elements of an Abelian group G, with $o(a)=m$ and $o(b)=n$, then $(ab)^{mn}=e$. Indicate where you use that G is Abelian.

12.8. Show with an example that if G is not Abelian, then the statement in Problem 12.7 may be false. [Consider $a=(1\quad 2)$ and $b=(1\quad 3)$ in S_3.]

12.9. (a) Use Problem 12.7 to prove that in an Abelian group the elements of finite order form a subgroup.

(b) What are the elements of finite order in the group of positive rationals (operation multiplication)?

12.10. Construct a Cayley table for a group G given that $G=\langle a\rangle$, $a\neq e$, and $a^5=e$.

12.11. Rewrite Theorem 12.1 (not its proof) for a group written additively, that is, with operation $+$, identity 0, and $-a$ for the inverse of a.

12.12. (a) If a, b, and c are elements of a group, then any one of the following three equations implies the other two:

$$ab=c,\qquad a=cb^{-1},\qquad b=a^{-1}c.$$

Why?

(b) Show with an example that $ab=c$ does not always imply $a=b^{-1}c$. (Look in S_3.)

12.13. Prove that a nonidentity element of a group has order 2 iff it is its own inverse.

12.14. Prove that every group of even order has an element of order 2. (Problem 12.13 may help.)

12.15. Prove that a group G is Abelian iff $(ab)^{-1}=a^{-1}b^{-1}$ for all $a,b\in G$.

12.16. If G is a group, $a,b\in G$, and $ab=b$, then a must be the identity of the group.

12.17. There is only one way to complete the following Cayley table so as to get a group. Find it. Why is it unique? (Problem 12.16 may help.)

*	a	b	c
a		b	
b			
c			

12.18. Assume that $\{x,y,z,w\}$ is to be a group, with identity x (operation juxtaposition). With any one of the following additional assumptions (a), (b), (c), or (d), there is only one Cayley table yielding a group. Determine that Cayley table in each case.
(a) $y^2=z$ (b) $y^2=w$ (c) $y^2=x$ and $z^2=x$ (d) $y^2=x$ and $z^2=y$

12.19. If a is a fixed element of a group G, and $\theta: G \to G$ is defined by $\theta(x)=ax$ for each $x \in G$, then θ is one-to-one and onto. Why?

12.20. Prove that a group is Abelian if each of its nonidentity elements has order 2.

12.21. Prove $o(a^{-1})=o(a)$.

12.22. Prove or give a counterexample: If a group has a subgroup of order n, then it has an element of order n.

12.23. Prove that if a group G has no subgroup other than G and $\{e\}$, then G is cyclic.

12.24. Prove that if G is a finite group, then Theorem 7.1 is true with condition (c) omitted. Also give an example to show that (c) cannot be omitted if Theorem 7.1 is to be true for all groups.

12.25. Prove that if A and B are subgroups of a group G, and $A \cup B$ is also a subgroup, then $A \subseteq B$ or $A \supseteq B$.

SECTION 13. COSETS. DIRECT PRODUCTS

We know that congruence modulo n is an equivalence relation on the group of integers. By viewing this in an appropriate way we are led to an idea that is important in the study of *all* groups. To do this, we first recall that if $n \in \mathbb{Z}$ then $\langle n \rangle$ is the subgroup consisting of all multiples of n. Because $a \equiv b \pmod{n}$ iff $a-b$ is a multiple of n, we see that $a \equiv b \pmod{n}$ iff $a-b \in \langle n \rangle$. If $\mathbb{Z}$ is replaced by an arbitrary group G, $\langle n \rangle$ by an arbitrary subgroup H of G, and $a-b$ by the corresponding expression ab^{-1} in our general multiplicative notation, then the following theorem results.

Theorem 13.1. *Let H denote a subgroup of a group G, and define a relation $\sim$ on G as follows:*

$$a \sim b \quad \textit{iff} \quad ab^{-1} \in H.$$

Then $\sim$ is an equivalence relation on G.

PROOF. *Reflexive*: if $a \in G$, then $a \sim a$ because $aa^{-1} = e \in H$. *Symmetric*: if $a \sim b$, then $ab^{-1} \in H$, and so $ba^{-1} = (ab^{-1})^{-1} \in H$ because H contains the inverse of each of its elements; thus $b \sim a$. *Transitive*: if $a \sim b$ and $b \sim c$, then $ab^{-1} \in H$ and $bc^{-1} \in H$, and so $ac^{-1} = (ab^{-1})(bc^{-1}) \in H$ because H is closed under products; thus $a \sim c$. □

The equivalence classes for this equivalence relation are called the *right cosets* of H in G. (There are also *left cosets*. See Problem 13.6.) Looking back at our motivating example, $\mathbb{Z}_n$, we see that the right cosets of $\langle n \rangle$ in $\mathbb{Z}$ are simply the congruence classes mod n. Lemma 13.1 will show that the right cosets of H in G have the form described by the following notation: for H a subgroup of a group G and $a \in G$, let

$$Ha = \{ha \mid h \in H\}.$$

If the group operation is $+$, then $H + a$ is written in place of Ha; similarly for other operations.

Example 13.1. Let $G = \mathbb{Z}$ and $H = \langle 7 \rangle$. Then

$$\begin{aligned} H + 3 = \langle 7 \rangle + 3 &= \{\ldots, -14, -7, 0, 7, 14, \ldots\} + 3 \\ &= \{\ldots, -11, -4, 3, 10, 17, \ldots\}. \end{aligned}$$

This is the congruence class [3] in $\mathbb{Z}_7$.

Example 13.2. Let $G = S_3$ and $H = \{(1), (1\ \ 2)\}$. Then

$$\begin{aligned} H(1) &= \{(1)(1), (1\ \ 2)(1)\} = \{(1), (1\ \ 2)\} \\ H(1\ \ 2\ \ 3) &= \{(1)(1\ \ 2\ \ 3), (1\ \ 2)(1\ \ 2\ \ 3)\} = \{(1\ \ 2\ \ 3), (2\ \ 3)\} \\ H(1\ \ 3\ \ 2) &= \{(1)(1\ \ 3\ \ 2), (1\ \ 2)(1\ \ 3\ \ 2)\} = \{(1\ \ 3\ \ 2), (1\ \ 3)\}. \end{aligned}$$

Notice that these three sets form a partition of G. In fact, by the following lemma, they are the right cosets of H in G.

Lemma 13.1. *If H is a subgroup of a group G, and $a, b \in G$, then the following four conditions are equivalent.*
(a) $ab^{-1} \in H$.
(b) $a = hb$ for some $h \in H$.
(c) $a \in Hb$.
(d) $Ha = Hb$.
As a consequence, the right coset of H to which a belongs is Ha.

PROOF. Problem 13.5. □

One right coset of H in G will be $H = He$. To compute all of the right cosets of a subgroup H in a finite group G, first write H, and then choose any element $a \in G$ such that $a \notin H$, and compute Ha. Next, choose any element $b \in G$ such that $b \notin H \cup Ha$, and compute Hb. Continue in this way until the elements of G have been exhausted.

Example 13.3. Let $G = \mathbb{Z}_{12}$ and $H = \langle [4] \rangle$. The right cosets of H in G are

$$H = \{[0], [4], [8]\}$$
$$H \oplus [1] = \{[1], [5], [9]\}$$
$$H \oplus [2] = \{[2], [6], [10]\}$$
$$H \oplus [3] = \{[3], [7], [11]\}.$$

One application of cosets will appear in the next section, and others in Chapters VII and XVI. We now leave cosets to look at a useful way to construct more examples.

If A and B are groups, then $A \times B$ is the Cartesian product of A and B:

$$A \times B = \{(a,b) | a \in A \quad \text{and} \quad b \in B\}$$

(Appendix A). With the following operation $A \times B$ becomes a group:

$$(a_1, b_1)(a_2, b_2) = (a_1 a_2, b_1 b_2)$$

for all $a_1, a_2 \in A$ and $b_1, b_2 \in B$. The identity element is (e_A, e_B), where e_A and e_B denote the identity elements of A and B, respectively. The inverse of (a,b) is (a^{-1}, b^{-1}). Verification of the associative law is left as an exercise (Problem 13.18). The group $A \times B$ (with this operation) is called the *direct product* of A and B. Notice that if A and B are finite, then so is $A \times B$, with $|A \times B| = |A| \cdot |B|$.

In defining the direct product we have followed the convention of writing groups multiplicatively. If the operation on either A or B is something other than juxtaposition, then that is taken into account in working with $A \times B$.

Example 13.4. With $B = \{(1), (1 \quad 2)\}$ (subgroup of S_3)

$$\begin{aligned} \mathbb{Z}_3 \times B = \{&([0],(1)), ([0],(1 \quad 2)), ([1],(1)), \\ &([1],(1 \quad 2)), ([2],(1)), ([2],(1 \quad 2))\} \end{aligned}$$

And, for example, $([1],(1 \quad 2))([2],(1)) = ([1] \oplus [2],(1 \quad 2)(1)) = ([0],(1 \quad 2))$.

If A and B are groups, then both $A \times \{e\} = \{(a,e) | a \in A\}$ and $\{e\} \times B = \{(e,b) | b \in B\}$ are subgroups of $A \times B$ (Problem 13.19). With either of these viewed as a subgroup of $A \times B$, the other furnishes a *complete set of right coset representatives*. Thus, for example, each right coset of $A\times\{e\}$ in $A\times B$ contains precisely one element from $\{e\}\times B$ (Problem 13.21).

PROBLEMS

13.1. Determine the right cosets of $\langle[4]\rangle$ in $\mathbb{Z}_8$.

13.2. Determine the right cosets of $\langle[3]\rangle$ in $\mathbb{Z}_{12}$.

13.3. For Example 8.3 (the group of symmetries of a square), determine the right cosets of $\langle\mu_5\rangle$.

13.4. Prove that if H is a subgroup of a group G and $a\in G$, then $Ha=H$ iff $a\in H$. (In particular, $He=H$.)

13.5. Prove Lemma 13.1.

13.6. State and prove Theorem 13.1 with the condition $ab^{-1}\in H$ replaced by $b^{-1}a\in H$. The equivalence classes in this case are called the *left cosets* of H in G.

13.7. State and prove the analogue of Lemma 13.1 for left cosets in place of right cosets. (See Problem 13.6.)

13.8. Verify that if H is a subgroup of an Abelian group G, and $a\in G$, then the right coset of H to which a belongs is the same as the left coset of H to which a belongs.

13.9. Compute the left cosets of H in G for H and G as in Example 13.2. (See Problems 13.6 and 13.7.) Verify that in this case the collection of left cosets is different from the collection of right cosets.

13.10. Let $G=S_3$ and $H=\langle(1\ \ 3)\rangle$.
- (a) Determine the right cosets of H in G.
- (b) Determine the left cosets of H in G. (See Problems 13.6 and 13.7.)
- (c) Verify that the collection of right cosets is different from the collection of left cosets.

13.11.
- (a) Compute the right and left cosets of $\langle(1\ \ 2\ \ 3)\rangle$ in S_3.
- (b) Verify that for each element π of S_3 the right coset of $\langle(1\ \ 2\ \ 3)\rangle$ to which π belongs is the same as the left coset of $\langle(1\ \ 2\ \ 3)\rangle$ to which π belongs.

13.12. List the elements of $\mathbb{Z}_2\times\mathbb{Z}_2$.

13.13. What is the order of $\mathbb{Z}_4\times\mathbb{Z}_7$?

13.14. What is the order of $S_4\times S_4$?

13.15. Show that $\mathbb{Z}_2\times\mathbb{Z}_3$ is cyclic.

13.16. Prove or disprove that if A and B are cyclic, then $A\times B$ is cyclic. (Look at Problem 13.12.)

13.17. Construct a Cayley table for the group in Example 13.4.

13.18. Prove the associative law for the direct product $A\times B$ of groups A and B.

13.19. Prove that $A\times\{e\}$ is a subgroup of $A\times B$.

13.20. Prove that $A\times B$ is Abelian iff both A and B are Abelian.

13.21. Prove that each right coset of $A\times\{e\}$ in $A\times B$ contains precisely one element from $\{e\}\times B$.

13.22. Prove that if A is a subgroup of G and B is a subgroup of H, then $A\times B$ is a subgroup of $G\times H$.

13.23. (a) List the elements of $S_3 \times \mathbb{Z}_2$.
(b) List the elements of the cyclic subgroup $\langle((1\ \ 2),[1])\rangle$ of $S_3 \times \mathbb{Z}_2$.
(c) List the elements of the cyclic subgroup $\langle((1\ \ 2\ \ 3),[1])\rangle$ of $S_3 \times \mathbb{Z}_2$.

13.24. (a) List the elements in the subgroup $\langle([2],[2])\rangle$ of $\mathbb{Z}_4 \times \mathbb{Z}_8$. (The first [2] is in $\mathbb{Z}_4$; the second is in $\mathbb{Z}_8$.)
(b) List the elements in the subgroup $\langle[2]\rangle \times \langle[2]\rangle$ of $\mathbb{Z}_4 \times \mathbb{Z}_8$. (Again, the first [2] is in $\mathbb{Z}_4$, and the second [2] is in $\mathbb{Z}_8$.)

13.25. Prove that if A is a group then $\{(a,a) | a \in A\}$ is a subgroup of $A \times A$. This is called the *diagonal subgroup* of $A \times A$. What is it, geometrically, for $A = \mathbb{R}$?

SECTION 14. LAGRANGE'S THEOREM

Of all the subsets of a finite group, only some will be subgroups. Lagrange's Theorem narrows the field.

Lagrange's Theorem. *If H is a subgroup of a finite group G, then the order of H is a divisor of the order of G.*

Thus, since S_3 has order $3! = 6$, any subgroup of S_3 must have order 1, 2, 3, or 6; S_3 cannot have subgroups of order 4 or 5. A group of order 7 can have only the two obvious subgroups: $\{e\}$, of order 1; and the group itself, of order 7.

The proof of Lagrange's Theorem comes from looking at the number of elements in the cosets of a subgroup. In doing this, it is convenient to extend the notation $|H|$ to include subsets, so that for S a finite set $|S|$ will denote the number of elements in S. Then $|S| = |T|$ iff there exists a one-to-one mapping of S onto T. (This is, in fact, the definition of $|S| = |T|$.)

Lemma 14.1. *If H is a finite subgroup of a group G, and $a \in G$, then $|H| = |Ha|$.*

PROOF. By the remark preceding the lemma, it suffices to find a one-to-one mapping of H onto Ha. Define $\alpha : H \to Ha$ by $\alpha(h) = ha$ for each $h \in H$. This is a mapping because ha is uniquely determined by h and a. It is onto because Ha consists precisely of the elements of the form ha for $h \in H$. To show that α is one-to-one, assume that $h_1, h_2 \in H$ and $\alpha(h_1) = \alpha(h_2)$. Then $h_1 a = h_2 a$, and therefore, by right cancellation, $h_1 = h_2$. Thus α is one-to-one. $\square$

Proof of Lagrange's Theorem. The right cosets of H, being equivalence classes, form a partition of G (Theorem 9.1). Thus two right cosets of H are either equal or disjoint. Moreover, since G is finite, there can be only finitely many of these cosets. Choose one element from each coset and let the elements chosen be $a_1, a_2, \ldots, a_k$. Then

$$G = Ha_1 \cup Ha_2 \cup \cdots \cup Ha_k,$$

with each coset Ha_i containing $|H|$ elements (Lemma 14.1), and no element in more than one coset. It follows that $|G| = |H| \cdot k$; therefore, $|H|$ is a divisor of $|G|$. □

The integer k appearing in the proof of Lagrange's Theorem is called the *index* of H in G, and will be denoted $[G:H]$. Thus $[G:H]$ is the number of right cosets of H in G, and

$$|G| = |H| \cdot [G:H].$$

Notice that this equation shows that $[G:H]$, as well as $|H|$, is a divisor of $|G|$.

Corollary. *If G is a finite group and $a \in G$, then the order of a is a divisor of the order of G.*

Proof. By the corollary of Theorem 12.2, $o(a) = |\langle a \rangle|$. But $\langle a \rangle$ is a subgroup of G, and thus $|\langle a \rangle|$ is a divisor of $|G|$ by Lagrange's Theorem. Therefore $o(a)$ is a divisor of $|G|$. □

Corollary. *A group G of prime order contains no subgroup other than $\{e\}$ and G.*

Proof. This is a direct consequence of Lagrange's Theorem, since a prime has no divisor other than 1 and itself. □

Corollary. *Each group of prime order is cyclic.*

Proof. Problem 14.1. □

Example 14.1. In contrast to groups of prime order (the preceding corollary), groups of prime-squared order need not be cyclic. For example, a direct product $\mathbb{Z}_p \times \mathbb{Z}_p$ has order p^2 but is not cyclic because it has no element of order p^2; each of its nonidentity elements has order p (Problem 14.2).

Example 14.2. Lagrange's Theorem greatly simplifies the problem of determining all subgroups of a finite group. Consider the group of symmetries of a square (Example 8.3). Aside from $\{\mu_1\}$ and the whole group, any subgroup must have order 2 or 4. Subgroups of order 2 are easy to determine—each contains the identity together with an element of order 2 (these correspond to appearances of μ_1 on the diagonal of the Cayley table). In this case $\langle\mu_3\rangle$, $\langle\mu_5\rangle$, $\langle\mu_6\rangle$, $\langle\mu_7\rangle$, and $\langle\mu_8\rangle$ all have order 2. This leaves only order 4, and inspection gives three subgroups of this order: $\langle\mu_2\rangle$, $\langle\mu_5,\mu_6\rangle$, and $\langle\mu_7,\mu_8\rangle$.

Notice that Lagrange's Theorem does not say that if n is a divisor of the order of G then G will have a subgroup of order n. That would be false. For example, there is a group of order 12 having no subgroup of order 6 (Problem 14.16). On the other hand, the Norwegian mathematician Ludwig Sylow (1832–1918) proved in 1872 that if p^k is any power of a prime and p^k is a divisor of $|G|$, then G must have a subgroup of order p^k. Thus, for example, a group of order 12 must have subgroups of orders 2, 3, and 4. A proof of Sylow's Theorem, for the case of the highest power of each prime dividing the order of a finite group, will be given in Section 33.

Lagrange's Theorem is named for the French mathematician Joseph Louis Lagrange (1736–1813), generally regarded as one of the two foremost mathematicians of the eighteenth century, the other being the Swiss-born mathematician Leonhard Euler (1707–1783). Lagrange did not prove this theorem in the form applying to all finite groups; indeed, the general concept of *group* was not studied until after Lagrange. But he did use the theorem in a significant special case, and therefore it is fitting that it be named for him.

PROBLEMS

14.1. Prove that each group of prime order is cyclic.

14.2. Show that each nonidentity element of $\mathbb{Z}_p\times\mathbb{Z}_p$ has order p.

14.3. Determine all subgroups of $\mathbb{Z}_6$. (Compare Example 14.2.)

14.4. Determine all subgroups of S_3. (Compare Example 14.2.)

14.5. Prove that if H is a subgroup of a finite group G, p is the smallest prime divisor of $|G|$, and $|H|>(1/p)|G|$, then $H=G$.

14.6. Verify that $\mathbb{Z}_{12}$ has a subgroup of order k for each divisor k of 12.

14.7. Determine all subgroups of $\mathbb{Z}_2\times\mathbb{Z}_2$.

14.8. Determine all subgroups of $\mathbb{Z}_3\times\mathbb{Z}_3$.

14.9. Determine the number of subgroups of $\mathbb{Z}_p\times\mathbb{Z}_p$, for p a prime.

14.10. The *exponent* of a group G is the smallest positive integer n such that $a^n=e$ for each $a\in G$, if such an integer n exists. Prove that every finite group has an exponent, and that this exponent divides the order of the group.

14.11. Determine the exponent of each of the following groups. (See Problem 14.10.)

(a) S_3 (b) $\mathbb{Z}_n$ (c) $\mathbb{Z}_2\times\mathbb{Z}_2$ (d) $\mathbb{Z}_2\times\mathbb{Z}_3$ (e) $\mathbb{Z}_m\times\mathbb{Z}_n$

14.12. Prove that if G is a group of order p^2 (p a prime), and G is not cyclic, then $a^p = e$ for each $a \in G$.

14.13. If H is a subgroup of G, $[G:H]=2$, $a,b \in G$, $a \notin H$, and $b \notin H$, then $ab \in H$.

14.14. Verify that S_4 has at least one subgroup of order k for each divisor k of 24.

14.15. Prove that if A and B are finite subgroups of a group G, and $|A|$ and $|B|$ have no common divisor greater than 1, then $A \cap B = \{e\}$.

14.16. The subgroup $\langle (1\ \ 2\ \ 3), (1\ \ 2)(3\ \ 4) \rangle$ of S_4 has order 12. Determine its elements. Verify that it has no subgroup of order 6.

14.17. State and prove Lemma 14.1 with right cosets replaced by left cosets. (Left cosets were introduced in Problem 13.6.)

14.18. If H is a subgroup of G and $[G:H]=2$, then the right cosets of H in G are the same as the left cosets of H in G. Why?

14.19. Using Problem 14.17, write a proof of Lagrange's Theorem using left cosets rather than right cosets.

14.20. Prove that if H is a subgroup of a finite group G, then the number of right cosets of H in G equals the number of left cosets of H in G.

SECTION 15. ISOMORPHISM

We speak of *the* set of integers, but if we were to allow ourselves to be distracted by things that are mathematically irrelevant, we might think that there were many such sets. The integers can appear in Arabic notation $\{\ldots,1,2,3,\ldots\}$, in Roman notation $\{\ldots,\mathrm{I},\mathrm{II},\mathrm{III},\ldots\}$, in German $\{\ldots,$ein, zwei, drei$,\ldots\}$, and so on; but mathematically we want to think of all these sets as being the same. The idea that filters out such irrelevant differences as names and notation is *isomorphism*. Isomorphism allows us to treat certain groups as being alike just as geometrical congruence allows us to treat certain triangles as being alike. The idea also applies in many cases that are less obvious than that of the integers presented in different language or notation. As a hint of this, consider the subgroup $\langle \pi_2 \rangle$ of S_3 (Example 7.2), and the group $\mathbb{Z}_3$, whose tables are shown here. The elements of Table 15.1 are *permutations* and the operation is *composition*; the elements of Table 15.2 are *congruence classes* and the operation is *addition modulo* 3. So the underlying sets and operations arise in totally different ways. Still, these groups are obviously somehow alike: given the

Table 15.1

$\circ$	π_1	π_2	π_3
π_1	π_1	π_2	π_3
π_2	π_2	π_3	π_1
π_3	π_3	π_1	π_2

Table 15.2

$\oplus$	[0]	[1]	[2]
[0]	[0]	[1]	[2]
[1]	[1]	[2]	[0]
[2]	[2]	[0]	[1]

correspondence $\pi_1 \leftrightarrow [0]$, $\pi_2 \leftrightarrow [1]$, and $\pi_3 \leftrightarrow [2]$, we could fill in all of one table just by knowing the other. Here is the idea behind this example.

Definition. Let G be a group with operation $*$, and let H be a group with operation $\#$. An *isomorphism of G onto H* is a mapping $\theta: G \to H$ that is one-to-one and onto, and that satisfies

$$\theta(a*b) = \theta(a)\#\theta(b)$$

for all $a, b \in G$. If there is an isomorphism of G onto H, then G and H are said to be *isomorphic*, and we write $G \approx H$.

The condition $\theta(a*b) = \theta(a)\#\theta(b)$ is sometimes described by saying that *θ preserves the operations*. It makes no difference whether we operate in G and then apply θ, or apply θ first and then operate in H—we get the same result either way.

Example 15.1. With the obvious mapping $(\ldots, 1 \mapsto \text{I}, 2 \mapsto \text{II}, 3 \mapsto \text{III}, \ldots)$ from the integers in Arabic notation to the integers in Roman notation, we get the same answer whether we add in Arabic $(1+2=3)$ and then translate into Roman $(3 \mapsto \text{III})$, or translate first $(1 \mapsto \text{I},\ 2 \mapsto \text{II})$ and then add $(\text{I}+\text{II}=\text{III})$. (And this is true for all $m+n$, not just $1+2$.)

Example 15.2. To illustrate the definition for the case of $\langle \pi_2 \rangle$ and S_3, already considered, use $\theta(\pi_1) = [0]$, $\theta(\pi_2) = [1]$, and $\theta(\pi_3) = [2]$. Then, for example,

$$\theta(\pi_2 \circ \pi_3) = \theta(\pi_1) = [0]$$

and

$$\theta(\pi_2) \oplus \theta(\pi_3) = [1] \oplus [2] = [0]$$

so that

$$\theta(\pi_2 \circ \pi_3) = \theta(\pi_2) \oplus \theta(\pi_3).$$

There are nine such equations to be checked in this case (one for each entry of the Cayley table), and in general there will be n^2 total equations for G and H finite of order n.

Example 15.3. Define a mapping θ from the set of all integers to the set of even integers by $\theta(n)=2n$ for each n. This mapping is one-to-one and onto, and, moreover, it preserves addition:

$$\theta(m+n) = 2(m+n) = 2m + 2n = \theta(m) + \theta(n)$$

for all integers m and n. Thus θ is an isomorphism, and the group of all integers (operation $+$) is isomorphic to the group of even integers (operation $+$).

The preceding example may seem puzzling—isomorphic groups are supposed to be essentially alike, but surely there is a difference between the integers and the even integers. This example shows that, *as groups*, each with addition as operation, there is in fact no essential difference between the integers and the even integers. Remember, however, that we are ignoring multiplication at present; when we take both addition and multiplication into account in the next chapter, we shall be able to detect a difference between the two sets. In Section 23 we shall see precisely what distinguishes the integers mathematically.

Example 15.4. Let $\mathbb{R}^p$ denote the set of positive real numbers, and define $\theta: \mathbb{R}^p \to \mathbb{R}$ by $\theta(x)=\log_{10}x$ for each $x \in \mathbb{R}^p$. Here $\mathbb{R}^p$ is a group with respect to multiplication; $\mathbb{R}$ is a group with respect to addition; and θ is an isomorphism. Problem 15.9 gives a way to verify that θ is one-to-one and onto; this can also be seen from the graph of $y=\log_{10}x$. The mapping θ preserves the operations because

$$\theta(xy) = \log_{10}(xy) = \log_{10}x + \log_{10}y = \theta(x) + \theta(y)$$

for all $x, y \in \mathbb{R}^p$. This is what makes logarithms useful in simplifying calculations: when used in conjunction with a table of logarithms, it allows us to replace a problem in multiplication by a problem in addition.

The following theorem shows that any group isomorphic to an Abelian group must also be Abelian. This can be taken as an illustration that isomorphic groups share significant properties. On the other hand, it can also be taken as showing that the property of being Abelian (or non-Abelian) is a significant property of groups. For the significant properties of groups, as groups, are those properties that are shared by isomorphic groups—that is what isomorphism is all about.

Theorem 15.1. *If G and H are isomorphic groups and G is Abelian, then H is Abelian.*

PROOF. Let the operations on G and H be $*$ and $\#$, respectively, and let $\theta: G \to H$ be an isomorphism. If $x, y \in H$, then there are elements $a, b \in G$ such that $\theta(a) = x$ and $\theta(b) = y$. Since θ preserves the operations and G is Abelian,

$$x \# y = \theta(a) \# \theta(b) = \theta(a * b) = \theta(b * a) = \theta(b) \# \theta(a) = y \# x,$$

which proves that H is Abelian. □

Other examples of properties shared by isomorphic groups will be given in the next section. The following theorem gives some technical facts about isomorphisms.

Theorem 15.2. *Let G be a group with operation $*$, let H be a group with operation $\#$, and let $\theta: G \to H$ be a mapping such that $\theta(a * b) = \theta(a) \# \theta(b)$ for all $a, b \in G$. Then*

(a) $\theta(e_G) = e_H$,
(b) $\theta(a^{-1}) = \theta(a)^{-1}$ for each $a \in G$,
(c) $\theta(G)$, the image of θ, is a subgroup of H, and
(d) if θ is one-to-one, then $G \approx \theta(G)$.

PROOF. (a) Because θ preserves the operations and $e_G * e_G = e_G$, we have $\theta(e_G) \# \theta(e_G) = \theta(e_G * e_G) = \theta(e_G)$. But $\theta(e_G) \in H$, so that $\theta(e_G) = \theta(e_G) \# e_H$. This gives $\theta(e_G) \# \theta(e_G) = \theta(e_G) \# e_H$, from which $\theta(e_G) = e_H$ by left cancellation.

(b) Using, in order, the properties of θ and a^{-1}, and part (a), we can write $\theta(a) \# \theta(a^{-1}) = \theta(a * a^{-1}) = \theta(e_G) = e_H$. Therefore, $\theta(a^{-1})$ must equal $\theta(a)^{-1}$ because $\theta(a)^{-1}$ is the unique solution of $\theta(a) \# x = e_H$.

(c) By parts (a) and (b), $\theta(G)$ contains e_H, and also along with any element the inverse of that element. Thus it now suffices to show that $\theta(G)$ is closed with respect to $\#$. Assume that $x, y \in \theta(G)$. Then $x = \theta(a)$ and $y = \theta(b)$ for some $a, b \in G$; thus $x \# y = \theta(a) \# \theta(b) = \theta(a * b) \in \theta(G)$, which establishes closure.

(d) By assumption, θ preserves the operations and is one-to-one. Also, thought of as a mapping from G to $\theta(G)$, θ is onto. Therefore $\theta: G \to \theta(G)$ is an isomorphism. □

PROBLEMS

15.1. Prove that $\mathbb{Z}$ is isomorphic to the multiplicative group of all rational numbers of the form $2^m (m \in \mathbb{Z})$.

15.2. Prove that $\mathbb{Z} \times \mathbb{Z}$ is isomorphic to the multiplicative group of all rational numbers of the form $2^m 3^n (m, n \in \mathbb{Z})$.

15.3. Fill in the blanks in the following table to obtain a group isomorphic to $\mathbb{Z}_4$. What is the isomorphism?

$*$	a	b	c	d
a				
b				
c				
d				

15.4. Repeat Problem 15.3, with $\mathbb{Z}_4$ replaced by $\mathbb{Z}_2 \times \mathbb{Z}_2$.

15.5. One of the conditions in the definition of isomorphism was not used in the proof of Theorem 15.1. Which one?

15.6. Describe an isomorphism between the two groups in Example 5.5.

15.7. Prove that if G, H, and K are groups, and $\theta: G \to H$ and $\phi: H \to K$ are isomorphisms, then $\phi \circ \theta: G \to K$ is an isomorphism. (Use juxtaposition for all group operations.)

15.8. Prove that if G and H are groups, then $G \times H \approx H \times G$. (Let the operations on G and H be $*$ and $\#$, respectively.)

15.9. Verify that $\phi(x) = 10^x$ defines an isomorphism of the group of real numbers (operation addition) onto the group of positive real numbers (operation multiplication). What is the inverse of the mapping ϕ? Is the inverse an isomorphism?

15.10. Verify that $\mathbb{Z}_4$ (operation $\oplus$) is isomorphic to $\mathbb{Z}_5^{\#}$ (operation $\odot$). (See Example 11.4.)

15.11. Use the mapping $\theta([a]_6) = ([a]_2, [a]_3)$ to show that $\mathbb{Z}_6 \approx \mathbb{Z}_2 \times \mathbb{Z}_3$.

SECTION 16. MORE ON ISOMORPHISM

If two finite groups are isomorphic then they must have the same order, because an isomorphism is, among other things, one-to-one and onto. Turning this around, we get the simplest of all tests for showing that two groups are *not* isomorphic: groups G and H are not isomorphic if $|G| \neq |H|$. It is useful to have a list of other properties that are shared by isomorphic groups—such a list will frequently make it much easier to determine quickly if two groups are not isomorphic.

If one of two isomorphic groups has any of the following properties, then the other group must also have that property:

Order n
Abelian
Cyclic
Has a subgroup of order n
Has an element of order n
Each element is its own inverse
Each element has finite order

(For proofs, see Theorem 15.1 and Problem 16.10.) It is equally important to be able to determine if groups *are* isomorphic, of course. This problem is considered in the following more general discussion.

In Example 15.2 we saw that two particular groups of order 3 are isomorphic. It will follow from Theorem 16.2 that *any* two groups of order 3 are isomorphic. This means that in essence there is only one group of order 3. More precisely, it means that there is only one *isomorphism class* of groups of order 3, where by an isomorphism class we mean an equivalence class for the equivalence relation imposed on groups by isomorphism, as described by the following theorem. We revert to the convention of using juxtaposition for unspecified group operations.

Theorem 16.1. *Isomorphism is an equivalence relation on the class of all groups.*

PROOF. If G is a group, then it is easy to verify that the identity mapping $\iota: G \to G$ is an isomorphism. Thus $G \approx G$, and $\approx$ is reflexive.

Assume that $G \approx H$ and that $\theta: G \to H$ is an isomorphism. Then θ is one-to-one and onto, and so there is an inverse mapping $\theta^{-1}: H \to G$ (Theorem 2.2). We shall show that θ^{-1} is an isomorphism; it is necessarily one-to-one and onto. Suppose that $a, b \in H$. We must show that $\theta^{-1}(ab) = \theta^{-1}(a)\theta^{-1}(b)$. Let $\theta^{-1}(a) = x$ and $\theta^{-1}(b) = y$. Then $a = \theta(x)$ and $b = \theta(y)$ so that $ab = \theta(x)\theta(y) = \theta(xy)$. This implies that $\theta^{-1}(ab) = xy = \theta^{-1}(a)\theta^{-1}(b)$, as required. Therefore $H \approx G$, and $\approx$ is symmetric.

Finally, it is easy to show that if G, H, and K are groups, and $\theta: G \to H$ and $\phi: H \to K$ are isomorphisms, then $\phi \circ \theta: G \to K$ is also an isomorphism (Problem 15.7). Thus $\approx$ is transitive. □

Theorem 16.2. *If p is a prime and G is a group of order p, then G is isomorphic to $\mathbb{Z}_p$.*

PROOF. Let the operation on G be $*$, and let a be an element of G, not the identity. Then $\langle a \rangle$ is a subgroup of G and $\langle a \rangle \neq \{e\}$, so that $\langle a \rangle = G$ by the second corollary of Lagrange's Theorem (Section 14). Thus $G = \{e, a, a^2, \ldots, a^{p-1}\}$. Define $\theta: G \to \mathbb{Z}_p$ by $\theta(a^k) = [k]$ $(0 \leq k < p-1)$. This mapping is one-to-one and onto, and, if $a^m, a^n \in G$, then $\theta(a^m * a^n) = \theta(a^{m+n}) = [m+n] = [m] \oplus [n] = \theta(a^m) \oplus \theta(a^n)$. Therefore θ is an isomorphism, and $G \approx \mathbb{Z}_p$. □

With Theorem 16.2 we have completely classified all groups of prime order. The principal problem of finite group theory is to do the same for groups of all finite orders. An immense amount of work has been done on this problem. Although much of this work is well beyond the range of this book, it will still be interesting to look at what is known in some special cases. Proofs will be omitted.

Table 16.1 shows the number of isomorphism classes of groups of order n for each n from 1 to 32. The label "number of groups" is what is conventionally used in place of the more accurate but longer phrase "number of isomorphism classes of groups." Whenever we ask for "all" groups having a property (such as being Abelian and of order n, for example), we are really asking for one group from each isomorphism class of groups with that property.

Notice from Table 16.1 that there is just one group of each prime order —Theorem 16.2 guarantees that. But notice also that there is just one group of order 15, and 15 is not a prime. The key is this:

There is just one group of order n
iff
n is a product of distinct primes $p_1, p_2, \ldots, p_k$
such that $p_j \nmid (p_i - 1)$ *for* $1 \le i \le k$, $1 \le j \le k$.

Thus, for instance, there is also only one group of order 33, since $33 = 3 \cdot 11$ and $3 \nmid 10$ and $11 \nmid 2$. For such n, any group of order n will be isomorphic to $\mathbb{Z}_n$.

Another easy-to-describe case is $n = p^2$, a square of a prime. There are two isomorphism classes of groups of order p^2: $\mathbb{Z}_{p^2}$ is in one class, and $\mathbb{Z}_p \times \mathbb{Z}_p$ is in the other. (Compare the entries for $n = 4$, 9, and 25 in Table 16.1.) Notice that both of these groups are Abelian. If $n = p^3$, a cube of a prime, then there are five isomorphism classes: three of these classes consist of Abelian groups and are represented by $\mathbb{Z}_{p^3}$, $\mathbb{Z}_{p^2} \times \mathbb{Z}_p$, and

Table 16.1

Order	1	2	3	4	5	6	7	8	9	10	11	12	13	14	15	16
Number of groups	1	1	1	2	1	2	1	5	2	2	1	5	1	2	1	14

Order	17	18	19	20	21	22	23	24	25	26	27	28	29	30	31	32
Number of groups	1	5	1	5	2	2	1	15	2	2	5	4	1	4	1	51

$\mathbb{Z}_p \times \mathbb{Z}_p \times \mathbb{Z}_p$; the other two classes consist of non-Abelian groups. (For direct products of more than two groups, see Problem 16.11.)

If only Abelian groups are considered, then the problem of determining all isomorphism classes is completely settled by the following theorem, which has been known since at least the 1870s.

Fundamental Theorem of Finite Abelian Groups. *If G is a finite Abelian group, then G is the direct product of cyclic groups of prime power order. Moreover, if*

$$G \approx A_1 \times A_2 \times \cdots \times A_s$$

and

$$G \approx B_1 \times B_2 \times \cdots \times B_t,$$

where each A_i and each B_j is cyclic of prime power order, then $s = t$ and, after suitable relabeling of subscripts, $|A_i| = |B_i|$ for $1 \leq i \leq s$.

Because each cyclic group of prime power order p^k is isomorphic to $\mathbb{Z}_{p^k}$, we can use this theorem to exhibit one group from each isomorphism class of finite Abelian groups.

Example 16.1. Let $n = 125 = 5^3$. To apply the theorem, first determine all possible ways of factoring 125 as a product of (not necessarily distinct) prime powers: 5^3, $5^2 \cdot 5$, $5 \cdot 5 \cdot 5$. Each factorization gives a different isomorphism class so that there are three isomorphism classes of Abelian groups of order 125. Here is one representative from each class:

$$\mathbb{Z}_{5^3} \qquad \mathbb{Z}_{5^2} \times \mathbb{Z}_5 \qquad \mathbb{Z}_5 \times \mathbb{Z}_5 \times \mathbb{Z}_5.$$

Notice that this same idea accounts for the general statement that was made about Abelian groups of order p^3.

Example 16.2. There are six isomorphism classes of Abelian groups of order $200 = 2^3 \cdot 5^2$. Here is one representative from each class:

$$\mathbb{Z}_{2^3} \times \mathbb{Z}_{5^2} \qquad \mathbb{Z}_{2^3} \times \mathbb{Z}_5 \times \mathbb{Z}_5 \qquad \mathbb{Z}_{2^2} \times \mathbb{Z}_2 \times \mathbb{Z}_{5^2}$$

$$\mathbb{Z}_{2^2} \times \mathbb{Z}_2 \times \mathbb{Z}_5 \times \mathbb{Z}_5 \qquad \mathbb{Z}_2 \times \mathbb{Z}_2 \times \mathbb{Z}_2 \times \mathbb{Z}_{5^2} \qquad \mathbb{Z}_2 \times \mathbb{Z}_2 \times \mathbb{Z}_2 \times \mathbb{Z}_5 \times \mathbb{Z}_5$$

Further remarks about the problem of classifying finite groups can be found in Section 30. As the case of Abelian groups suggests, the number of groups of order n is influenced largely by the character of the prime factorization of n, and not by the size of n alone. Table 16.1 gives a further hint of this, and here is one more example to make it even clearer: there are 267 groups of order 64.

PROBLEMS

16.1. In each of the following pairs, the two groups are not isomorphic. Give a reason in each case.

(a) $\mathbb{Z}_5$, $\mathbb{Z}_6$
(b) $\mathbb{Z}_2 \times \mathbb{Z}_2$, $\mathbb{Z}_4$
(c) $\mathbb{Z}_6$, S_3
(d) $\mathbb{Z}_4 \times \mathbb{Z}_2$, symmetry group of a square
(e) $\mathbb{Z}$, $\mathbb{Q}$ (both with operation $+$)
(f) $\mathbb{Q}$ (operation $+$), positive rational numbers (operation $\cdot$)
(g) $S_3 \times \mathbb{Z}_2$, the group in Problem 14.16
(h) $\mathbb{Z}_8 \times \mathbb{Z}_4$, $\mathbb{Z}_{16} \times \mathbb{Z}_2$

16.2. Is there a noncyclic group of order 59? Why?

16.3. Give an example of a noncyclic group of order 49.

16.4. Is there a noncyclic group of order 39? Why?

16.5. Give one example from each isomorphism class of Abelian groups of order 36.

16.6. If p is a prime, then there are five isomorphism classes of Abelian groups of order p^4. Give one group from each class. (Compare Example 16.1.)

16.7. If p is a prime, then there are seven isomorphism classes of Abelian groups of order p^5. Give one group from each class. (Compare Example 16.1 and Problem 16.6.)

16.8. Prove that if G is a cyclic group of order n, then $G \approx \mathbb{Z}_n$. (*Suggestion*: Write G as $G = \langle a \rangle = \{e, a, \ldots, a^{n-1}\}$.)

16.9. Prove that if G is an infinite cyclic group, then $G \approx \mathbb{Z}$. (Compare Problem 16.8.)

16.10. Assume that G and H are groups and that $G \approx H$. Prove each of the following statements.

(a) If G is cyclic, then H is cyclic.
(b) If G has a subgroup of order n, then H has a subgroup of order n.
(c) If G has an element of order n, then H has an element of order n.
(d) If $x^{-1} = x$ for each $x \in G$, then $x^{-1} = x$ for each $x \in H$.
(e) If each element of G has finite order, then each element of H has finite order.

16.11. If $\{A_1, A_2, \ldots, A_n\}$ is any collection of groups, each with juxtaposition as operation, then the *direct product* $A_1 \times A_2 \times \cdots \times A_n$ is $\{(a_1, a_2, \ldots, a_n) | a_i \in A_i\}$ with operation

$$(a_1, a_2, \ldots, a_n)(b_1, b_2, \ldots, b_n) = (a_1 b_1, a_2 b_2, \ldots, a_n b_n).$$

(a) Verify that this direct product is a group.
(b) If each group A_i is finite, what is the order of the direct product?

16.12. An isomorphism of a group onto itself is called an *automorphism*. Prove that the set of all automorphisms of a group is itself a group with respect to composition. [The group of all automorphisms of a group G is called the *automorphism group* of G, and will be denoted $\mathrm{Aut}(G)$. Notice that $\mathrm{Aut}(G)$ is a subgroup of $\mathrm{Sym}(G)$. Examples are given in the problems that follow, and also in Problem 31.8.]

16.13. Verify that if G is an Abelian group, then $\theta: G \to G$ defined by $\theta(a) = a^{-1}$ for each $a \in G$ is an automorphism of G. (See Problem 16.12.)

16.14. Prove that if $[a]$ is a generator for $\mathbb{Z}_n$, and $\theta: \mathbb{Z}_n \to \mathbb{Z}_n$ is defined by $\theta([k]) = [ka]$, then $\theta \in \text{Aut}(\mathbb{Z}_n)$. (See Problem 16.12.)

16.15. (a) Prove that $|\text{Aut}(\mathbb{Z}_2)| = 1$.

(b) Prove that $\text{Aut}(\mathbb{Z}_3) \approx \mathbb{Z}_2$. (See Problem 16.14.)

(c) Prove that $\text{Aut}(\mathbb{Z}_4) \approx \mathbb{Z}_2$.

(d) Prove that if p is a prime, then $\text{Aut}(\mathbb{Z}_p) \approx \mathbb{Z}_{p-1}$.

(e) Prove that $\text{Aut}(\mathbb{Z}) \approx \mathbb{Z}_2$.

SECTION 17. CAYLEY'S THEOREM

We have seen that the nature of groups can vary widely—from groups of numbers to groups of permutations to groups defined by tables. Cayley's Theorem asserts that in spite of this broad range of possibilities each group is isomorphic to some group of permutations. This is an example of what is known as a *representation theorem*—it tells us that any group can be represented as (is isomorphic to) something reasonably concrete. In place of studying the given group, we can just as well study the concrete object (permutation group) representing it; and this can be an advantage. On the other hand, it can also be a disadvantage, for part of the power of abstraction comes from the fact that abstraction filters out irrelevancies, and in concentrating on any concrete object we run the risk of being distracted by irrelevancies. Still, Cayley's Theorem has proved to be useful, and its proof will tie together several of the important ideas that we have studied.

In proving Cayley's Theorem, we shall associate with each element of a group G a permutation of the set G. The way in which this is done is suggested by looking at the Cayley table for a finite group. As we observed after Theorem 12.1, each element of a finite group appears exactly once in each row of the Cayley table for the group. Thus the elements in each row of the table are merely a permutation of the elements in the first row. What we shall do is simply associate with each element in the left column of the table the permutation given by the elements in its row. If the elements in the first row are $a_1, a_2, \ldots, a_n$ (in that order), then the elements in the row labeled by a will be $aa_1, aa_2, \ldots, aa_n$ (in that order).

Example 17.1. Consider the Cayley table for $\mathbb{Z}_6$, given in Example 11.2. The permutation associated with [3] by the idea just described is

$$\begin{pmatrix} [0] & [1] & [2] & [3] & [4] & [5] \\ [3] & [4] & [5] & [0] & [1] & [2] \end{pmatrix}.$$

Cayley's Theorem extends this idea to groups that are not necessarily finite, and also establishes that this association of group elements with permutations is an isomorphism.

Cayley's Theorem. *Every group is isomorphic to a permutation group on its set of elements.*

PROOF. The isomorphism will be a mapping $\theta: G \to \text{Sym}(G)$. We begin by describing the permutation that θ will assign to an element $a \in G$.

For $a \in G$, define $\lambda_a: G \to G$ by $\lambda_a(x) = ax$ for each $x \in G$. Each such mapping λ_a is one-to-one and onto because each equation $ax = b$ $(b \in G)$ has a unique solution in G [Theorem 12.1(c)]. Thus $\lambda_a \in \text{Sym}(G)$ for each $a \in G$.

Now define $\theta: G \to \text{Sym}(G)$ by $\theta(a) = \lambda_a$ for each $a \in G$. To prove that θ is one-to-one, suppose that $\theta(a) = \theta(b)$; we shall deduce that $a = b$. From $\theta(a) = \theta(b)$ we have $\lambda_a = \lambda_b$, and thus in particular $\lambda_a(e) = \lambda_b(e)$, since the mappings λ_a and λ_b can be equal only if they give the same image for each element in their common domain. But $\lambda_a(e) = ae = a$ and $\lambda_b(e) = be = b$, so that $a = b$. Thus θ is one-to-one.

Finally, if a, $b \in G$, then $\theta(ab) = \lambda_{ab}$ and $\theta(a) \circ \theta(b) = \lambda_a \circ \lambda_b$, implying that $\theta(ab) = \theta(a) \circ \theta(b)$ if $\lambda_{ab} = \lambda_a \circ \lambda_b$. To verify the latter, let $x \in G$ and write $\lambda_{ab}(x) = (ab)x = a(bx) = \lambda_a(bx) = \lambda_a(\lambda_b x) = (\lambda_a \circ \lambda_b)(x)$. Thus we have proved that $\theta(ab) = \theta(a) \circ \theta(b)$ for all a, $b \in G$.

It follows from Theorem 15.2(d) that G is isomorphic to $\theta(G)$, a subgroup of $\text{Sym}(G)$, which completes the proof. □

Corollary. *Every group of finite order n is isomorphic to a subgroup of S_n.*

PROOF. Label the elements of the group $a_1, a_2, \ldots, a_n$. The construction in the proof of Cayley's Theorem will assign to an element a in the group the permutation

$$\begin{pmatrix} a_1 & a_2 & \cdots & a_n \\ aa_1 & aa_2 & \cdots & aa_n \end{pmatrix}.$$

As remarked preceding Example 17.1, the elements $aa_1, aa_2, \ldots, aa_n$ are just $a_1, a_2, \ldots, a_n$ in some order. If we replace each aa_i by that a_j such that $aa_i = a_j$, and then replace each a_k in the permutation by the number k, we obtain an element of S_n. (See Example 17.2, for instance.) By assigning each a to the element of S_n obtained in this way, we get the desired isomorphism. □

Example 17.2. If we begin with S_3, whose Cayley table is given in Example 6.1, the construction in the proof of Cayley's Theorem yields

$$\theta(\pi_4) = \begin{pmatrix} \pi_1 & \pi_2 & \pi_3 & \pi_4 & \pi_5 & \pi_6 \\ \pi_4 & \pi_6 & \pi_5 & \pi_1 & \pi_3 & \pi_2 \end{pmatrix}.$$

The idea in the corollary is simply to delete the π's and keep the subscripts, so that

$$\begin{pmatrix} 1 & 2 & 3 & 4 & 5 & 6 \\ 4 & 6 & 5 & 1 & 3 & 2 \end{pmatrix}$$

is assigned to π_4. Notice that this is an element of S_6 because $|S_3|=6$.

Corollary. *For each positive integer n, there are only finitely many isomorphism classes of groups of order n.*

PROOF. By the previous corollary every group of order n is isomorphic to a subgroup of S_n. But S_n is finite and thus has only finitely many subgroups. Hence there can be only finitely many isomorphism classes of groups of order n. □

PROBLEMS

17.1. Write the permutation associated with each element of $\mathbb{Z}_5$ by the isomorphism θ in the proof of Cayley's Theorem.

17.2. Write the permutation associated with each element of the symmetry group of a square (Example 8.3) by the isomorphism θ in the proof of Cayley's Theorem.

17.3. Give an alternate proof of Cayley's Theorem by replacing λ_a (lambda for "left" multiplication) by ρ_a (rho for "right" multiplication) defined as follows: $\rho_a(x)=xa^{-1}$ for each $x \in G$. Why is xa^{-1} used here, rather than xa?

17.4. Assume that G is a group, and that $\lambda_a : G \to G$ and $\rho_a : G \to G$ are defined by $\lambda_a(x)=ax$ and $\rho_a(x)=xa^{-1}$ for each $x \in G$. For each $a \in G$, define $\gamma_a : G \to G$ by $\gamma_a = \rho_a \circ \lambda_a$. Prove that each γ_a is an isomorphism of G onto G. [In the language of Problem 16.12, each γ_a is an automorphism of G. Notice that, in particular, each $\gamma_a \in \text{Sym}(G)$.]

17.5. Let γ_a be defined as in Problem 17.4, and define $\phi : G \to \text{Sym}(G)$ by $\phi(a)=\gamma_a$ for each $a \in G$. Prove that $\phi(a)=\iota_G$ iff $ax=xa$ for all $a \in G$.

NOTES ON CHAPTER IV

Here are four standard references on group theory. The first is a classic and is listed because of its historical importance. The second emphasizes infinite groups.

1. Burnside, W., *Theory of Groups of Finite Order*, 2nd ed., Cambridge University Press, Cambridge, England, 1911; Dover, New York, 1955.
2. Kurosh, A., *Theory of Groups*, Vols. I and II, trans. from the Russian by K. A. Hirsch, Chelsea, New York, 1955.
3. Rotman, J. J., *The Theory of Groups*, 2nd ed., Allyn and Bacon, Boston, 1973.
4. Scott, W. R., *Group Theory*, Prentice-Hall, Englewood Cliffs, N. J., 1964.

CHAPTER V

Introduction to Rings

In considering the integers as a group, we have used addition but not multiplication. In doing that, we have ignored more than just multiplication. We have also ignored properties that combine the two operations, such as the law $a(b+c)=ab+ac$. The same thing has happened with other groups—the rational numbers and the integers mod n, for example. The concept of a *ring* covers all of these systems, as well as many others with two operations—rings, just like groups, arise in widely varying contexts. Some of the most basic ideas about rings will be used in the next chapter to analyze the integers and other familiar number systems. Rings will also play some role in each of Chapters XI through XVII.

SECTION 18. DEFINITION AND EXAMPLES

A ring consists of a set with two operations, which are nearly always written as a sum $(a+b)$ and a product (ab). For rings whose elements are numbers, this notation always has its usual meaning unless explicitly stated to the contrary. In some cases it is necessary to specify what is meant by the two operations. In reading the axioms for a ring, which follow, it may be helpful to keep the integers in mind—they do form a ring. Finally, do not confuse R (which may be any nonempty set) with $\mathbb{R}$ (which we reserve for the real numbers).

Definition. A *ring* is a set R together with two operations on R, called *addition* $(a+b)$ and *multiplication* (ab), such that each of the following axioms is satisfied:

R with addition is an Abelian group

$a+(b+c)=(a+b)+c$ for all $a,b,c \in R$,

there is an element $0 \in R$ such that

$$a+0=0+a=a \text{ for each } a \in R,$$

for each $a \in R$ there is an element $-a \in R$ such that

$$a+(-a)=(-a)+a=0,$$

$a+b=b+a$ for all $a,b \in R$,

multiplication is associative

$a(bc)=(ab)c$ for all $a,b,c \in R$, and

distributive laws

$a(b+c)=ab+ac$ and $(a+b)c=ac+bc$ for all $a,b,c \in R$.

The group formed by R with addition is referred to as the *additive group* of R. Its identity element, 0, is called the *zero* of the ring; the context will usually make clear whether this or the integer zero is meant. In expressions such as $ab+ac$, multiplications are to be performed first, just as in elementary arithmetic; that is, $ab+ac$ means $(ab)+(ac)$.

Example 18.1. The integers form a ring with respect to the usual addition and multiplication. The same is true for the rational numbers, the real numbers, and also the even integers.

Example 18.2. For any positive integer n, $\mathbb{Z}_n$, the integer mod n, forms a ring with respect to the operations $\oplus$ and $\odot$ introduced in Section 11. Theorem 11.1 shows $\mathbb{Z}_n$ with $\oplus$ is a group, and Problem 11.7 shows that this group is Abelian. Associativity of $\odot$ was recorded in Lemma 11.3. Here is a verification of the first distributive law:

$$\begin{aligned}
[a]\odot([b]\oplus[c]) &= [a]\odot[b+c] && \text{definition of } \oplus \\
&= [a(b+c)] && \text{definition of } \odot \\
&= [ab+ac] && \text{distributivity for } +,\cdot \\
&= [ab]\oplus[ac] && \text{definition of } \oplus \\
&= ([a]\odot[b])\oplus([a]\odot[c]) && \text{definition of } \odot.
\end{aligned}$$

The proof of the other distributive law is similar (Problem 18.1).

Example 18.3. Let $M(2,\mathbb{Z})$ denote the set of all 2×2 matrices with integers as entries. The sum of two matrices in $M(2,\mathbb{Z})$ is defined by

$$\begin{bmatrix} a & b \\ c & d \end{bmatrix} + \begin{bmatrix} w & x \\ y & z \end{bmatrix} = \begin{bmatrix} a+w & b+x \\ c+y & d+z \end{bmatrix}.$$

Products are defined as in Example 3.5. With these operations $M(2,\mathbb{Z})$ is a ring (Problem 18.2). Appendix D has more about groups and rings of matrices, which are very important in linear algebra.

Example 18.4. Let $\mathbb{Z}[\sqrt{2}\,]$ denote the set of all numbers $a+b\sqrt{2}$ with $a,b\in\mathbb{Z}$. The sum of two such numbers is also in $\mathbb{Z}[\sqrt{2}\,]$:

$$(a+b\sqrt{2}\,)+(c+d\sqrt{2}\,)=(a+c)+(b+d)\sqrt{2}\,.$$

The same is true for products:

$$\begin{aligned}(a+b\sqrt{2}\,)(c+d\sqrt{2}\,) &= ac+ad\sqrt{2}+bc\sqrt{2}+bd\sqrt{2}\,\sqrt{2}\\ &= (ac+2bd)+(ad+bc)\sqrt{2}\,.\end{aligned}$$

With these operations $\mathbb{Z}[\sqrt{2}\,]$ is a ring (Problem 18.3).

Example 18.5. Let $F=M(\mathbb{R})$ denote the set of all functions (mappings) $f:\mathbb{R}\to\mathbb{R}$. We can define $f+g$ and fg for $f,g\in F$ in a way that will give a ring. If $f+g$ is to be in F, then it must be a function from $\mathbb{R}$ to $\mathbb{R}$. Thus we must specify $(f+g)(x)$ for each $x\in\mathbb{R}$. Similarly for fg. The definitions are

$$(f+g)(x)=f(x)+g(x) \qquad \text{for each } x\in\mathbb{R}$$

and (18.1)

$$(fg)(x)=f(x)g(x) \qquad \text{for each } x\in\mathbb{R}.$$

To verify that this operation + is associative, $f+(g+h)=(f+g)+h$, we observe that since each side is a function with domain $\mathbb{R}$, what must be shown is that for all $f,g,h\in F$

$$[f+(g+h)](x)=[(f+g)+h](x) \qquad \text{for each } x\in\mathbb{R}.$$

To do this, write

$$\begin{aligned}[f+(g+h)](x) &= f(x)+(g+h)(x) && \text{definition of + on } F\\ &= f(x)+[g(x)+h(x)] && \text{definition of + on } F\\ &= [f(x)+g(x)]+h(x) && \text{associativity of + on } \mathbb{R}\\ &= (f+g)(x)+h(x) && \text{definition of + on } F\\ &= [(f+g)+h](x) && \text{definition of + on } F.\end{aligned}$$

The 0 (identity element for +) for this ring is the function defined by $0(x)=0$ for each $x\in\mathbb{R}$, where the 0 on the right is the zero of $\mathbb{R}$: if $f\in F$, then

$$(f+0)(x)=f(x)+0(x)=f(x)+0=f(x)$$

for each $x\in\mathbb{R}$, and so $f+0=f$. The negative of a function f is the function $-f$ defined by $(-f)(x)=-f(x)$ for each $x\in\mathbb{R}$. Verification of the remaining axioms is left to Problem 18.4.

Notice that the product fg in this example is *not* $f \circ g$. See Problem 18.6 for what happens when $f \circ g$ is used.

Example 18.6. Let R and S be rings, and $R \times S$ the Cartesian product of R and S, that is, the set of all ordered pairs (r,s) with $r \in R$ and $s \in S$. Then $R \times S$ becomes a ring with the following operations:

$$(r_1,s_1) + (r_2,s_2) = (r_1 + r_2, s_1 + s_2)$$

and

$$(r_1,s_1)(r_2,s_2) = (r_1 r_2, s_1 s_2)$$

for all $r_1, r_2 \in R$, $s_1, s_2 \in S$. The discussion of direct products of groups in Section 13 carries over to show that $R \times S$ is a group with respect to addition: the additive identity is $(0,0) = (0_R, 0_S)$; and the negative of (r,s) is $(-r, -s)$. Here is verification of one of the distributive laws:

$$\begin{aligned}(r_1,s_1)((r_2,s_2)+(r_3,s_3)) &= (r_1,s_1)(r_2+r_3, s_2+s_3)\\ &= (r_1(r_2+r_3), s_1(s_2+s_3))\\ &= (r_1 r_2 + r_1 r_3, s_1 s_2 + s_1 s_3)\\ &= (r_1 r_2, s_1 s_2) + (r_1 r_3, s_1 s_3)\\ &= (r_1,s_1)(r_2,s_2) + (r_1,s_1)(r_3 s_3).\end{aligned}$$

The remaining details are left to Problem 18.7. This ring is called the *direct sum* of R and S.

When first working with any abstract concept such as *group* or *ring*, there is bound to be uncertainty over what is and is not allowed. For example, in a group written multiplicatively the left cancellation law holds: If $ab = ac$, then $b = c$ [Theorem 12.1(a)]. Because a ring is (among other things) a group with respect to addition, this can be translated into a statement about rings: If $a + b = a + c$, then $b = c$. But what about left cancellation for *multiplication* in a ring? In the ring of integers $ab = ac$ implies $b = c$ (for $a \neq 0$), but what about other rings? We shall come to see that sometimes it is safe to cancel and sometimes it is not. Only experience can guide us in such matters. Once it has been determined that there are enough important examples of a concept to make that concept worth studying, we set about trying to discover theorems about it. Not only do examples tell us whether the concept is worth studying, they are also the source of ideas for theorems. Given a property of a specific ring, for example, we can ask whether that property holds for all rings. Very often, the answer will be No, and it will be another example—a counterexample —which will give us that answer. But sometimes the answer will be Yes, and we shall then have another piece of the theory surrounding the general concept. In this sense the example of the integers is a good one to keep in mind when first studying rings. Even though not everything true about the integers is true about all rings, we can gradually improve our perspective by comparing each theorem and example with this familiar and important special case.

With these remarks and some examples behind us, we now turn to some elementary theorems. Because a ring is a group with respect to addition, the elementary properties of rings that involve only addition are obtained by simply translating the elementary properties of groups into additive notation. An example is the cancellation law mentioned above. And the law $(a^{-1})^{-1}=a$ becomes $-(-a)=a$. The following theorem gives a summary of such properties. The last part uses these conventions: If n is a positive integer, then $na=a+a+\cdots+a$ (n terms), and $(-n)a=-(na)$.

Theorem 18.1. *Let R be a ring and $a,b,c\in R$.*

(*a*) *The zero element of R is unique.*
(*b*) *Each element of R has a unique negative.*
(*c*) *If $a+b=a+c$, then $b=c$ (left cancellation law).*
(*d*) *If $b+a=c+a$, then $b=c$ (right cancellation law).*
(*e*) *Each of the equations $a+x=b$ and $x+a=b$ has a unique solution.*
(*f*) $-(-a)=a$ *and* $-(a+b)=(-a)+(-b)$.
(*g*) *If m and n are integers, then $(m+n)a=ma+na$, $m(a+b)=ma+mb$, and $m(na)=(mn)a$.*

The proof of Theorem 18.1 is left to Problem 18.8. The next theorem concerns properties involving multiplication in a ring. Here, and elsewhere, $a-b$ means $a+(-b)$.

Theorem 18.2. *Let R be a ring, 0 the zero of R, and $a,b,c\in R$.*

(*a*) $0a=a0=0$.
(*b*) $a(-b)=(-a)b=-(ab)$.
(*c*) $(-a)(-b)=ab$.
(*d*) $a(b-c)=ab-ac$ *and* $(a-b)c=ac-bc$.

PROOF. (a) $0a+0a=(0+0)a=0a=0a+0$, and therefore, by canceling $0a$ from the left of the first and last expressions, $0a=0$. The proof for $a0=0$ is similar.

(b) The equation $x+ab=0$ has $x=-(ab)$ as a solution. But $a(-b)+ab=a(-b+b)=a0=0$; hence $x=a(-b)$ is also a solution. By uniqueness of the solution [Theorem 18.1(e)], it follows that $-(ab)=a(-b)$. The proof for $-(ab)=(-a)b$ is similar.

(c) Using part (b) and Theorem 18.1(f), we have $(-a)(-b)=(-(-a))(b)=ab$.

(d) Using part (b), we can write $a(b-c)=a(b+(-c))=ab+a(-c)=ab+(-(ac))=ab-ac$. The proof for $(a-b)c=ac-bc$ is similar. □

In the definition of ring nothing is assumed about multiplication except associativity and the connection of multiplication and addition through the distributive laws. More assumptions will be made in the next section when we look at some special types of rings. We close this section by mentioning two of the assumptions that arise most frequently.

An element e in a ring R is called a *unity* (or *identity* or *unit element*) for the ring if $ea=ae=a$ for each $a\in R$. Thus a unity is just an identity for multiplication. The number 1 is a unity for the ring of integers. The ring of even integers has no unity.

A ring R is said to be *commutative* if $ab=ba$ for all $a,b\in R$. The rings in Examples 18.1, 18.2, 18.4, and 18.5 are commutative. If $ab\neq ba$ for some $a,b\in R$ the ring is *noncommutative*. The ring in Example 18.3 is noncommutative. The ring $R\times S$ in Example 18.6 will be commutative iff both R and S are commutative.

PROBLEMS

18.1. Prove that $([a]\oplus[b])\odot[c]=([a]\odot[c])\oplus([b]\odot[c])$ for all $[a],[b],[c]\in\mathbb{Z}_n$ (Example 18.2).

18.2. Prove that $M(2,\mathbb{Z})$ is a ring with the usual matrix operations (Example 18.3). Verify that this ring is noncommutative.

18.3. Prove that $\mathbb{Z}[\sqrt{2}\,]$ is a ring (Example 18.4).

18.4. Complete the verification that $F=M(\mathbb{R})$ is a ring (Example 18.5). Also verify that it is commutative and has a unity.

18.5. Show that if R is a ring and S is a nonempty set, then the set of all mappings from S into R can be made into a ring by using operations like those defined in (18.1) (Example 18.5).

18.6. Consider Example 18.5 with fg, as defined there, replaced by $f\circ g$. Verify that this does not define a ring.

18.7. Complete the verification that Example 18.6 is a ring.

18.8. Prove Theorem 18.1. (This can be done by referring to proofs that have already been carried out for groups.)

18.9. Prove that if R is a ring, then each of the following properties holds for all $a,b,c\in R$.

(a) $a0=0$ (This is part of Theorem 18.2.)

(b) $-(ab)=(-a)b$ (This is part of Theorem 18.2.)

(c) $(a-b)c=ac-bc$ (This is part of Theorem 18.2.)

(d) $-(a+b)=(-a)+(-b)$

(e) $(a-b)+(b-c)=a-c$.

18.10. Prove that if R is a ring, $a,b \in R$, and $ab=ba$, then $a(-b)=(-b)a$ and $(-a)(-b)=(-b)(-a)$. (Use Theorem 18.2.)

18.11. Prove that a ring has at most one unity.

18.12. Let E denote the set of even integers. Prove that with the usual addition, and with multiplication defined by $m*n=(\frac{1}{2})mn$, E is a ring. Is there a unity?

18.13. Prove that $a^2-b^2=(a+b)(a-b)$ for all a,b in a ring R iff R is commutative.

18.14. Verify that if A is an Abelian group, with addition as operation, and an operation $*$ is defined on A by $a*b=0$ for all $a,b \in A$, then A is a ring with respect to $+$ and $*$.

18.15. In the ring of integers, if $ab=ac$ and $a \neq 0$, then $b=c$. Is this true in all rings? (*Suggestion*: Look carefully at the rings in Examples 18.2, 18.3, 18.5, and 18.6.)

18.16. Which of the following properties hold in every ring R? What about every commutative ring R?

(a) $a^m a^n = a^{m+n}$ for all $a \in R$, $m,n \in \mathbb{N}$.

(b) $(a^m)^n = a^{mn}$ for all $a \in R$, $m,n \in \mathbb{N}$.

(c) $(ab)^m = a^m b^m$ for all $a,b \in R$, $m \in \mathbb{N}$.

(The powers occurring here are defined as in Section 12.)

18.17. Prove that $(a+b)^2=a^2+2ab+b^2$ for all a,b in a ring R iff R is commutative.

18.18. Verify that if R is a ring and $a,b \in R$, then

$$(a+b)^3 = a^3 + aba + ba^2 + b^2a + a^2b + ab^2 + bab + b^3.$$

Which ring axioms do you need?

18.19. Prove that if R is a commutative ring, $a,b \in R$, and n is a positive integer, then $(a+b)^n$ can be computed by the Binomial Theorem. (See Appendix C.)

18.20. For each set S, let $\mathcal{P}(S)$ denote the set of all subsets of S. For $A,B \in \mathcal{P}(S)$, define $A+B$ and AB by

$$A + B = (A \cup B) \setminus (A \cap B) = \{x \mid x \in A \cup B \text{ and } x \notin A \cap B\}$$

and

$$AB = A \cap B.$$

Verify that with these operations, $\mathcal{P}(S)$ is a ring. [$\mathcal{P}(S)$ is called the *power set* of A. See Appendix A for facts about sets. *Suggestion*: Make generous use of Venn diagrams in this problem.]

SECTION 19. INTEGRAL DOMAINS. SUBRINGS

In order to isolate what is unique about the integers—something we first promised in Section 15—we shall focus on the pertinent abstract ring properties. Commutativity and the existence of a unity are examples of these properties. After introducing one more such property we shall define a special class of rings known as *integral domains*. This class will contain the ring of integers, and will bring us very near to a characterization of that ring. The characterization will be completed in Section 23.

We begin by recalling a fact relating to zero in the ring of integers: If a and b are integers and $ab=0$, then either $a=0$ or $b=0$. This is not true in some rings. In $\mathbb{Z}_6$, for example, $[2]\odot[3]=[0]$ but $[2]\neq[0]$ and $[3]\neq[0]$. The following definition singles out such examples: An element $a\neq 0$ in a commutative ring R is called a *zero divisor* in R if there exists an element $b\neq 0$ in R such that $ab=0$. Thus the ring of integers has no zero divisors. But both [2] and [3] are zero divisors in $\mathbb{Z}_6$. Notice that the definition is restricted to elements in a commutative ring; for what can happen in a noncommutative ring see Problem 19.4. By the definition that follows, zero divisors are forbidden in an integral domain.

Definition. A commutative ring with unity $e\neq 0$ and no zero divisors is called an *integral domain.*

Example 19.1. The ring of integers, the ring of rational numbers, and the ring of real numbers are all integral domains. The ring of even integers is not an integral domain because it has no unity.

Example 19.2. The ring $\mathbb{Z}_6$ is not an integral domain because, as we have seen, it has zero divisors. This happens because 6 is not a prime. More generally, if n is not a prime, and $n=rs$ with r and s each greater than 1, then, in $\mathbb{Z}_n$, $[r]\odot[s]=[rs]=[n]=[0]$ with $[r]\neq[0]$ and $[s]\neq[0]$. Thus $\mathbb{Z}_n$ is not an integral domain if n is not a prime. On the other hand, it can be proved that if n is a prime, then $\mathbb{Z}_n$ is an integral domain (Problem 19.5). For the special case $n=5$ this can be seen from Table 11.2; every $\mathbb{Z}_n$ is commutative and has a unity, and Table 11.2 shows that the product of any two nonzero elements in $\mathbb{Z}_5$ is nonzero.

Of the other examples of rings in Section 18, $M(2, \mathbb{Z})$ is not an integral domain because it is noncommutative (Problem 18.2). The ring in Example 18.4 is an integral domain (Problem 19.6). The ring in Example 18.5 is not an integral domain (Problem 19.7). The ring $R\times S$ in Example 18.6 is an integral domain only if one of R or S is an integral domain and the other contains only a zero element (Problem 19.8).

The following theorem gives an easy-to-prove but important property of integral domains.

Theorem 19.1. *If D is an integral domain, $a,b,c\in D$, $a\neq 0$, and $ab=ac$, then $b=c$* (left cancellation property).

PROOF. From $ab=ac$ we have $ab-ac=0$ and thus $a(b-c)=0$. Since a is a nonzero element of the integral domain D, it cannot be a zero divisor, so that we must have $b-c=0$ and $b=c$. □

Because multiplication is commutative in an integral domain, the (left) cancellation property in Theorem 19.1 is equivalent to the *right cancellation property*: If $a\neq 0$ and $ba=ca$, then $b=c$. Moreover, for a commutative ring each of these cancellation properties implies that the ring has no zero divisors (Problem 19.9). Thus an alternative definition of an integral domain is this: *An integral domain is a commutative ring D with unity* $e\neq 0$ *such that if* $a,b,c\in D$, $ab=ac$, *and* $a\neq 0$, *then* $b=c$.

The notion of subring is the obvious analogue of the notion of subgroup.

Definition. A subset S of a ring R is a *subring* of R if S is itself a ring with respect to the operations on R.

Example 19.3. The ring of even integers is a subring of the ring of all integers. The ring of integers is a subring of the ring of rational numbers.

Theorem 7.1 gave a criterion for determining when a subset of a group is a subgroup. Here is the corresponding criterion for subrings.

Theorem 19.2. *A subset S of a ring R is a subring of R iff S is nonempty, S is closed under both the addition and the multiplication of R, and S contains the negative of each of its elements.*

PROOF. Problem 19.10. □

Example 19.4. Let $F=M(\mathbb{R})$ denote the ring of all functions $f:\mathbb{R}\to\mathbb{R}$, introduced in Example 18.5. Let S denote the set of all $f\in F$ such that $f(1)=0$. Then S is a subring of F. For certainly $0_F\in S$, since $0_F(x)=0$ for all $x\in\mathbb{R}$ so that in particular $0_F(1)=0$. And if f and g are in S, then

$$(f+g)(1) = f(1)+g(1) = 0+0 = 0$$

and

$$(fg)(1) = f(1)g(1) = 0\cdot 0 = 0$$

so that $f+g$ and fg are in S. Finally, if $f(1)=0$ then $(-f)(1)=-f(1)=-0=0$; therefore, the negative of each element of S is also in S.

PROBLEMS

19.1. Which elements of $\mathbb{Z}_4$ are zero divisors?

19.2. Which elements of $\mathbb{Z}_{10}$ are zero divisors?

19.3. Which elements of $\mathbb{Z}\times\mathbb{Z}$ are zero divisors? (See Example 18.6.)

19.4. Compute $\begin{bmatrix} 0 & 1 \\ 0 & 0 \end{bmatrix}\begin{bmatrix} 1 & 0 \\ 0 & 0 \end{bmatrix}$ and $\begin{bmatrix} 1 & 0 \\ 0 & 0 \end{bmatrix}\begin{bmatrix} 0 & 1 \\ 0 & 0 \end{bmatrix}$ in $M(2,\mathbb{Z})$.

19.5. Prove that if p is a prime, then $\mathbb{Z}_p$ is an integral domain.

19.6. Explain why $\mathbb{Z}[\sqrt{2}\,]$ (Example 18.4) is an integral domain. (Assume that $\mathbb{R}$ is an integral domain.)

19.7. Prove that the ring $F=M(\mathbb{R})$ in Example 18.5 is not an integral domain.

19.8. Prove that $R\times S$ (Example 18.6) is an integral domain iff one of R or S is an integral domain and the other contains only a zero element.

19.9. Prove that for a commutative ring the cancellation property of Theorem 19.1 implies that there are no zero divisors.

19.10. Prove Theorem 19.2.

19.11. Let $C(\mathbb{R})$ denote the set of all continuous functions from $\mathbb{R}$ into $\mathbb{R}$. Prove that $C(\mathbb{R})$ is a subring of the ring $F=M(\mathbb{R})$ in Example 18.5. What properties of continuous functions are required? What happens if $C(\mathbb{R})$ is replaced by the set of differentiable functions from $\mathbb{R}$ into $\mathbb{R}$?

19.12. Prove that if $\mathcal{C}$ denotes any collection of subrings of a ring R, then the intersection of all rings in $\mathcal{C}$ is also a subring of R. (Compare Theorem 7.2.)

19.13. State and prove a theorem for rings that is analogous to Theorem 7.3 for groups. (Use Problem 19.12.)

19.14. What is the smallest subring of $\mathbb{Z}$ containing 3?

19.15. What is the smallest subring of $\mathbb{R}$ containing $\frac{1}{2}$?

19.16. Prove or disprove that if $\mathcal{C}$ denotes a collection of subrings of a ring R, and each ring in $\mathcal{C}$ is an integral domain, then the intersection of all rings in $\mathcal{C}$ is also an integral domain.

19.17. For which $n\in\mathbb{Z}$ is the smallest subring containing n an integral domain?

19.18. Verify that $\{a+b\sqrt[3]{2}+c\sqrt[3]{4}\,|\,a,b,c\in\mathbb{Z}\}$ is a subring of $\mathbb{R}$. Is this subring an integral domain?

19.19. The *center* of a ring R is defined to be $\{c\in R\,|\,cr=rc$ for every $r\in R\}$. Prove that the center of a ring is a subring. What is the center of a commutative ring?

19.20. What is the center of $M(2,\mathbb{Z})$? (See Example 18.3 and Problem 19.19.)

19.21. What is the center of $M(\mathbb{R})$? (See Example 18.5 and Problem 19.19.)

19.22. Which of the following are subrings of $M(2,\mathbb{Z})$ (Example 18.3)?

(a) $\left\{\begin{bmatrix} a & b \\ 0 & 0 \end{bmatrix}\middle|\, a,b\in\mathbb{Z}\right\}$ (b) $\left\{\begin{bmatrix} a & b \\ 0 & c \end{bmatrix}\middle|\, a,b,c\in\mathbb{Z}\right\}$.

19.23. State and prove a theorem giving a necessary and sufficient condition for a subset of an integral domain to be an integral domain. (Compare Theorem 19.2.)

SECTION 20. FIELDS

Because an integral domain has no zero divisors, its set of nonzero elements is closed with respect to multiplication. Therefore, multiplication is an operation on this set of nonzero elements, and it is natural to ask if it yields a group. Because multiplication is required to be associative in a ring, the operation is associative. Because an integral domain has a unity, the operation has an identity element. Thus, with respect to multiplication, the set of nonzero elements of an integral domain can fail to be a group only because of the absence of inverse elements. In the integral domain of integers, for instance, the only nonzero elements with inverses relative to multiplication are 1 and -1. In the integral domain of rational numbers, however, each nonzero element has an inverse relative to multiplication. Such integral domains are called fields:

Definition. A commutative ring in which the set of nonzero elements forms a group with respect to multiplication is called a *field*.

Alternatively, a field can be defined as an integral domain in which each nonzero element has an inverse relative to multiplication (Problem 20.2). Another example of a field, besides the ring of rational numbers, is the ring of real numbers. Fields are indispensable to much of mathematics —for example, the field of real numbers is automatically being used whenever we apply calculus. In Chapter XVI we shall see that there are also useful finite fields.

Here are some relationships between classes of rings:

$$\text{fields} \subset \text{integral domains} \subset \text{commutative rings} \subset \text{rings}.$$

Each class is contained in, but different from, the class that follows it. If we restrict attention to rings having only finitely many elements, however, then the first two classes are the same, because of the following theorem.

Theorem 20.1. *Every finite integral domain is a field.*

PROOF. Let D be a finite integral domain. We must show that each nonzero element $a \in D$ has an inverse relative to multiplication; that is, if $a \in D$ and $a \neq 0$, then there is an element $b \in D$ such that $ab = e$. This means we must show that e is among the set of elements ax for $x \in D$. To show this, assume that $a \neq 0$ and consider the mapping $\lambda_a : D \to D$ defined by $\lambda_a(x) = ax$ for each $x \in D$. If this mapping is onto, then in particular $\lambda_a(x) = e$ for some $x \in D$, say $x = b$, and then $\lambda_a(b) = ab = e$. Thus it suffices to show that λ_a is onto.

Because λ_a is a mapping of a finite set to itself, it suffices to establish that λ_a is one-to-one, for it will then necessarily be onto. To show that λ_a is one-to-one, assume that $\lambda_a(x_1) = \lambda_a(x_2)$. Then $ax_1 = ax_2$, and therefore, by Theorem 19.1, $x_1 = x_2$. Thus λ_a is one-to-one, as required. □

Corollary. *$\mathbb{Z}_n$ is a field iff n is a prime.*

PROOF. From the remarks in Example 19.2 we know that $\mathbb{Z}_n$ is an integral domain iff n is a prime. That makes this corollary an immediate consequence of Theorem 20.1. □

Example 20.1. Tables 20.1 and 20.2 show operations on $\{0, e, a, b\}$ that produce a field.

Table 20.1

$+$	0	e	a	b
0	0	e	a	b
e	e	0	b	a
a	a	b	0	e
b	b	a	e	0

Table 20.2

$\cdot$	0	e	a	b
0	0	0	0	0
e	0	e	a	b
a	0	a	b	e
b	0	b	e	a

Once 0 has been chosen for the zero element, and e for the unity, there is no choice about how to complete the table for multiplication (Problem 20.6). The table for addition must produce a group of order 4; Problem 20.7 asks for verification that the additive group here is isomorphic to $\mathbb{Z}_2 \times \mathbb{Z}_2$ (with operation $\oplus$ on each $\mathbb{Z}_2$). This example shows that there is a field of order 4, even though $\mathbb{Z}_4$ (with $\oplus$ and $\odot$) is not a field by the corollary of Theorem 20.1. It can be proved that there is a finite field of order n iff n is a power of a prime (see Section 49).

Definition. A subset K of a field F is a *subfield* of F if K is itself a field with respect to the operations on F.

Theorem 20.2. *A subset K of a field F is a subfield of F iff*

(*a*) *K contains the zero and unity of F,*
(*b*) *if* $a,b\in K$, *then* $a+b\in K$ *and* $ab\in K$,
(*c*) *if* $a\in K$, *then* $-a\in K$, *and*
(*d*) *if* $a\in K$ *and* $a\neq 0$, *then* $a^{-1}\in K$.

PROOF. Problem 20.10. □

Example 20.2. The ring $\mathbb{Z}[\sqrt{2}\,]=\{a+b\sqrt{2}\,|a,b\in\mathbb{Z}\}$, considered in Example 18.4, is a subring of $\mathbb{R}$. Although $\mathbb{Z}[\sqrt{2}\,]$ is an integral domain (see Problem 19.6), it is not a field. For instance, $-2+\sqrt{2}\in\mathbb{Z}[\sqrt{2}\,]$, but $(-2+\sqrt{2}\,)^{-1}=-1-(\frac{1}{2})\sqrt{2}\notin\mathbb{Z}[\sqrt{2}\,]$. If, however, $\mathbb{Z}$ is replaced by $\mathbb{Q}$, then we do get a field: $\mathbb{Q}[\sqrt{2}\,]=\{a+b\sqrt{2}\,|a,b\in\mathbb{Q}\}$ is a subfield of $\mathbb{R}$ (Problem 20.5).

If the requirement of commutativity is dropped from the definition of a field, what is left is the definition of a *division ring*: a ring in which the set of nonzero elements forms a group with respect to multiplication. Thus a commutative division ring is a field. Problem 26.16 gives an example of a division ring that is not commutative.

PROBLEMS

20.1. Give an example of each of the following.
(a) An integral domain that is not a field.
(b) A commutative ring without zero divisors that is not an integral domain.
(c) A commutative ring with a zero divisor.
(d) A noncommutative ring.

20.2. Prove that an integral domain is a field iff each nonzero element has an inverse relative to multiplication.

20.3. Prove that an integral domain D is a field iff each equation $ax=b$ ($a,b\in D$ and $a\neq 0$) has a unique solution in D.

20.4. Let $\mathbb{Z}_n^{\#}$ denote the nonzero elements of $\mathbb{Z}_n$. For which n is $\mathbb{Z}_n^{\#}$ a group with respect to $\odot$? (Compare Examples 11.4 and 11.5.)

20.5. Verify that $\mathbb{Q}[\sqrt{2}\,]=\{a+b\sqrt{2}\,|a,b\in\mathbb{Q}\}$ is a subfield of the field of real numbers (Example 20.2).

20.6. Prove that if $\{0,e,a,b\}$ is to be a field with 0 as zero element and e as unity, then the multiplication must be as defined in Example 20.1.

20.7. Prove that the additive group of the field in Example 20.1 is isomorphic to $\mathbb{Z}_2\times\mathbb{Z}_2$.

20.8. Show that the ring $\mathbb{Z}_2\times\mathbb{Z}_2$ is not a field. Why is this not in conflict with Problem 20.7?

20.9. Prove that a direct sum of two or more fields is never a field.

20.10. Prove Theorem 20.2.

20.11. Prove that if $\mathcal{C}$ denotes any collection of subfields of a field F, then the intersection of all fields in $\mathcal{C}$ is also a subfield of F. (Compare Problem 19.12.)

20.12. State and prove a theorem for fields that is analagous to Theorem 7.3 for groups. (Use Problem 20.11, and compare Problem 19.13.)

20.13. What is the smallest subfield of $\mathbb{R}$ containing $\mathbb{Z}$? (There is such a subfield by Problem 20.12.)

20.14. (a) An element a in a commutative ring R with unity e is said to be *invertible* if there is an element $b \in R$ such that $ab = e$. Prove that if R is a commutative ring with unity, then the invertible elements form a group with respect to the multiplication of the ring.
(b) What is the group of invertible elements in a field?
(c) What is the group of invertible elements in $\mathbb{Z}_4$?
(d) What is the group of invertible elements in $\mathbb{Z}_n$?

20.15. Prove that a zero divisor in a commutative ring with unity cannot be invertible. (See Problem 20.14.)

SECTION 21. ISOMORPHISM. CHARACTERISTIC

In Section 15 we met the idea of isomorphism for groups and learned that isomorphic groups are essentially the same—they differ at most in the names of the elements and the operations. A similar idea applies to rings.

Definition. Let R and S be rings. An *isomorphism* of R onto S is a mapping $\theta: R \to S$ that is one-to-one and onto and satisfies

$$\theta(a+b) = \theta(a) + \theta(b)$$

and

$$\theta(ab) = \theta(a)\theta(b)$$

for all $a, b \in R$. If there is an isomorphism of R onto S, then R and S are said to be *isomorphic*, and we write $R \approx S$.

In the conditions $\theta(a+b) = \theta(a) + \theta(b)$ and $\theta(ab) = \theta(a)\theta(b)$, the operations on the left in each equation are, of course, those of R, and the operations on the right are those of S. Notice that because of the first of these conditions, a ring isomorphism is necessarily an isomorphism of the additive groups of R and S. It follows that $\theta(0) = 0$ and $\theta(-a) = -\theta(a)$ for each $a \in R$, by translation into additive notation of parts (a) and (b) of Theorem 15.2.

The following example shows that an isomorphism between the additive groups of two rings is not necessarily a ring isomorphism.

Example 21.1. Let θ be the mapping from the ring of integers to the ring of even integers defined by $\theta(n)=2n$ for each n. We verified in Example 15.3 that this is an isomorphism between additive groups. But it is not a ring isomorphism because it does not preserve multiplication: $\theta(mn)=2mn$ but $\theta(m)\theta(n)=(2m)(2n)=4mn$.

Although this mapping θ is not a ring isomorphism, we might ask whether some other mapping from the ring of integers to the ring of even integers can be a ring isomorphism. The answer is No. For example, the ring of integers has a unity but the ring of even integers does not, and if one of two isomorphic rings has a unity, then the other must as well (Problem 21.1). Notice, then, that although the integers and even integers cannot be distinguished as groups (Example 15.3), they can be distinguished as rings.

Example 21.2. Consider the ring $\mathbb{Z}[\sqrt{2}\,]$ from Example 18.4, and define $\theta:\mathbb{Z}[\sqrt{2}\,]\to\mathbb{Z}[\sqrt{2}\,]$ by $\theta(a+b\sqrt{2}\,)=a-b\sqrt{2}$. This mapping is clearly one-to-one and onto. It also preserves both ring operations: For addition,

$$\begin{aligned}\theta((a+b\sqrt{2}\,)+(c+d\sqrt{2}\,)) &= \theta((a+c)+(b+d)\sqrt{2}\,)\\ &= (a+c)-(b+d)\sqrt{2}\end{aligned}$$

and also

$$\begin{aligned}\theta(a+b\sqrt{2}\,)+\theta(c+d\sqrt{2}\,) &= (a-b\sqrt{2}\,)+(c-d\sqrt{2}\,)\\ &= (a+c)-(b+d)\sqrt{2}\,.\end{aligned}$$

For multiplication,

$$\begin{aligned}\theta((a+b\sqrt{2}\,)(c+d\sqrt{2}\,)) &= \theta((ac+2bd)+(bc+ad)\sqrt{2}\,)\\ &= (ac+2bd)-(bc+ad)\sqrt{2}\end{aligned}$$

and also

$$\begin{aligned}\theta(a+b\sqrt{2}\,)\theta(c+d\sqrt{2}\,) &= (a-b\sqrt{2}\,)(c-d\sqrt{2}\,)\\ &= (ac+2bd)-(bc+ad)\sqrt{2}\,.\end{aligned}$$

Thus θ is an isomorphism of $\mathbb{Z}[\sqrt{2}\,]$ onto $\mathbb{Z}[\sqrt{2}\,]$. An isomorphism like this, of a ring onto itself, is called an *automorphism*.

Example 21.3. We can prove that $\mathbb{Z}_6\approx\mathbb{Z}_2\times\mathbb{Z}_3$ by using the mapping $\theta:\mathbb{Z}_6\to\mathbb{Z}_2\times\mathbb{Z}_3$ defined by $\theta([a]_6)=([a]_2,[a]_3)$. The following string of equivalent statements shows that θ is both well-defined and one-to-one:

$$\begin{aligned}[a]_6=[b]_6 \quad &\text{iff} \quad 6\,|\,(a-b)\\ &\text{iff} \quad 2\,|\,(a-b) \text{ and } 3\,|\,(a-b)\\ &\text{iff} \quad [a]_2=[b]_2 \text{ and } [a]_3=[b]_3\\ &\text{iff} \quad ([a]_2,[a]_3)=([b]_2,[b]_3)\end{aligned}$$

(the "only if" portion shows that θ is well-defined; the "if" portion shows that it is one-to-one). Because $|\mathbb{Z}_6| = |\mathbb{Z}_2 \times \mathbb{Z}_3|$, θ is onto since it is one-to-one. Next, θ preserves addition:

$$\begin{aligned}
\theta([a]_6 \oplus [b]_6) &= \theta([a+b]_6) \\
&= ([a+b]_2, [a+b]_3) \\
&= ([a]_2 \oplus [b]_2, [a]_3 \oplus [b]_3) \\
&= ([a]_2, [a]_3) + ([b]_2, [b]_3) \\
&= \theta([a]_6) + \theta([b]_6).
\end{aligned}$$

Similarly, θ preserves multiplication (Problem 21.14). Therefore, $\mathbb{Z}_6 \approx \mathbb{Z}_2 \times \mathbb{Z}_3$, as claimed. (Problem 21.13 suggests how to show that $\mathbb{Z}_4 \not\approx \mathbb{Z}_2 \times \mathbb{Z}_2$. Problem 39.15 will ask you to show that $\mathbb{Z}_{mn} \approx \mathbb{Z}_m \times \mathbb{Z}_n$ iff m and n have no common divisor greater than 1.)

If one of two isomorphic groups is Abelian, then the other must also be Abelian (Theorem 15.1). In the same way, if one of two isomorphic rings is commutative, then the other must also be commutative (Problem 21.2). Other properties shared by isomorphic rings include the existence of a unity, existence of a zero divisor, that of being an integral domain, and that of being a field (see the problems). The most common method of showing that two rings are not isomorphic is by finding some such property that one of the rings has but the other does not.

The next concept will help in determining what is unique about the ring of integers. It is also especially useful in the study of fields. Recall that if n is a positive integer and a is a ring element, then $na = a + a + \cdots + a$ (n terms).

Definition. Let R be a ring. If there is a positive integer n such that $na = 0$ for each $a \in R$, then the least such integer is called the *characteristic* of R. If there is no such positive integer, then R is said to have *characteristic* 0 (zero).

If a ring has a unity e and characteristic $n \neq 0$, then in particular $ne = 0$. On the other hand, if $ne = 0$ and $a \in R$, then $na = n(ea) = (ne)a = 0a = 0$. *Thus, for a ring with unity e, the characteristic can be defined alternatively as the least positive integer n such that $ne = 0$, if there is such an integer; otherwise the ring has characteristic* 0.

Example 21.4. (a) The ring of integers has characteristic 0, for there is no positive integer n such that $n \cdot 1 = 0$. For the same reason, the ring of rational numbers and the ring of real numbers also have characteristic 0.

(b) The characteristic of $\mathbb{Z}_n$ is n, because $n[1]=[n]=[0]$ but $k[1]=[k]\neq[0]$ for $0<k<n$.

In Example 19.2 we observed that if a ring $\mathbb{Z}_n$ is an integral domain, then n is a prime. Thus, in view of the last example, if a ring $\mathbb{Z}_n$ is an integral domain, then its characteristic is a prime. Here is a more general statement.

Theorem 21.1. *If D is an integral domain, then the characteristic of D is either* 0 *or a prime.*

PROOF. Assume that D is an integral domain with characteristic $n\neq 0$. We shall prove that n must be a prime. Assume otherwise. Then $n=rs$ for some integers r and s with $1<r<n$ and $1<s<n$. If e is the unity of D, we have $ne=0$ so that $(rs)e=0$ and $(rs)(ee)=(re)(se)=0$. But D, being an integral domain, has no zero divisors, and so this implies that $re=0$ or $se=0$. Since $1<r<n$ and $1<s<n$, either possibility contradicts the fact that n is the characteristic of D. Thus n must be a prime. □

Theorem 21.2. *If D is an integral domain of characteristic* 0, *then D contains a subring isomorphic to $\mathbb{Z}$.*

PROOF. Let e denote the unity of D, and define $\theta:\mathbb{Z}\to D$ by $\theta(n)=ne$ for each $n\in\mathbb{Z}$. We shall prove that θ is one-to-one and that it preserves both ring operations.

If $\theta(m)=\theta(n)$, then $me=ne$, $me-ne=0$, and $(m-n)e=0$ so that $m=n$ because D has characteristic 0. Therefore, θ is one-to-one.

If $m,n\in\mathbb{Z}$, then

$$\theta(m+n) = (m+n)e = me + ne = \theta(m) + \theta(n)$$

and

$$\theta(mn) = (mn)e = (me)(ne) = \theta(m)\theta(n).$$

The image of θ is a subring of D (Problem 21.6), and θ is an isomorphism of $\mathbb{Z}$ onto that subring. This completes the proof. □

Theorem 21.3. *If D is an integral domain of prime characteristic p, then D contains a subring isomorphic to $\mathbb{Z}_p$.*

PROOF. The proof of this theorem is similar to that of Theorem 21.2. The relevant mapping is $\theta:\mathbb{Z}_p\rightarrow D$ defined by $\theta([k])=ke$ for each $[k]\in\mathbb{Z}_p$. The details are left as an exercise (Problem 21.7). □

If a ring R contains a subring isomorphic to a ring S, then it is said that S can be *embedded* in R. Using this terminology, Theorem 21.2 becomes: *The ring of integers can be embedded in every integral domain of characteristic* 0. And Theorem 21.3 becomes: *The ring* $\mathbb{Z}_p$ *can be embedded in every integral domain of prime characteristic* p.

PROBLEMS

21.1. Prove that if R and S are rings, $\theta:R\rightarrow S$ is an isomorphism, and e is a unity of R, then $\theta(e)$ is a unity of S.

21.2. Prove that if R and S are isomorphic rings and R is commutative, then S is commutative.

21.3. Prove that if R and S are isomorphic rings and R is an integral domain, then S is an integral domain.

21.4. Prove that if R and S are isomorphic rings and R is a field, then S is a field.

21.5. Prove that if E and F are fields, $\theta:E\rightarrow F$ is an isomorphism, and $a\in E$, $a\neq 0$, then $\theta(a^{-1})=\theta(a)^{-1}$.

21.6. Prove that if R and S are rings, $\theta:R\rightarrow S$, and θ preserves both ring operations, then $\theta(R)$ is a subring of S.

21.7. Complete the proof of Theorem 21.3. Prove, in particular, that the mapping θ is well-defined.

21.8. Prove that isomorphism is an equivalence relation on the class of all rings.

21.9. List five ring properties that hold for each ring isomorphic to $\mathbb{Z}$ but not for every ring. (By a *ring property* is meant a property of a ring R that is shared by every ring isomorphic to R.)

21.10. Prove that if R is a ring with unity e, and $a,b\in R$ and $m,n\in\mathbb{Z}$, then each of the following is true.

(a) $m(ab)=(ma)b=a(mb)$

(b) $(mn)e=(me)(ne)$

(c) $(mn)a=m(na)$

(d) $m(ea)=(me)a$

21.11. What can be said about the characteristic of a ring R in which $x=-x$ for each $x\in R$?

21.12. Prove that if R and S are isomorphic rings, then their characteristics are equal.

21.13. Use Problem 21.12 to explain why $\mathbb{Z}_4\not\approx\mathbb{Z}_2\times\mathbb{Z}_2$.

21.14. Verify that the mapping θ in Example 21.3 preserves multiplication.

21.15. A ring R is called a *Boolean ring* if $x^2=x$ for each $x\in R$. Prove that every Boolean ring R is commutative and satisfies $2x=0$ for each $x\in R$. Give an example of such a ring.

21.16. Prove that if R is a finite ring, then the characteristic of R is a divisor of $|R|$. (Section 14 is relevant.)

21.17. Let R denote the subfield $\{a+b\sqrt{2} \mid a,b\in\mathbb{Q}\}$ of $\mathbb{R}$, and let S denote the subfield $\{a+b\sqrt{3} \mid a,b\in\mathbb{Q}\}$ of $\mathbb{R}$. Verify that $\theta: R\to S$ defined by $\theta(a+b\sqrt{2})=a+b\sqrt{3}$ is not a ring isomorphism.

21.18. (a) Prove that if R and S are rings with unities e and f, respectively, $R\approx S$, and $x^2=e+e$ has a solution in R, then $x^2=f+f$ has a solution in S. (See Problem 21.1.)

(b) Verify that $x^2=2$ has a solution in the ring R of Problem 21.17.

(c) The rings R and S in Problem 21.17 are not isomorphic. Give a reason. (Notice that Problem 21.17 shows that a *particular* mapping from R to S is not an isomorphism. This problem shows that *no* mapping from R to S is an isomorphism.)

21.19. Every ring R can be embedded in a ring with a unity. Prove this by verifying the following steps.

(a) The set $\mathbb{Z}\times R$ is a ring with respect to the operations

$$(m,a)+(n,b) = (m+n,a+b)$$

and

$$(m,a)(n,b) = (mn,na+mb+ab).$$

(Notice that this is not the direct sum of the rings $\mathbb{Z}$ and R.)

(b) The ring in part (a) has unity $(1,0)$.

(c) $R'=\{(0,a) \mid a\in R\}$ is a subring of the ring in part (a), and $R\approx R'$.

NOTES ON CHAPTER V

We shall return to rings in a number of later chapters. References [2] and [3] are standard sources that go beyond what is in this book. Chapter 49 of [1] contains remarks on the history of ring theory.

1. Kline, M., *Mathematical Thought from Ancient to Modern Times*, Oxford University Press, London, 1972.
2. McCoy, N. H., *Rings and Ideals*, Carus Monograph Series, No. 8, Mathematical Association of America (1948).
3. Zariski, O., and P. Samuel, *Commutative Algebra*, Vol. I, D. Van Nostrand, Princeton, N.J., 1958.

CHAPTER VI

The Familiar Number Systems

This chapter will show what distinguishes each of the familiar number systems—integers, rational numbers, real numbers, complex numbers—in terms of its special properties as a ring or field. This will require that we introduce ideas relating to order ($<$) and to the existence of solutions of polynomial equations. Complete proofs will be given or sketched in the case of the integers and rational numbers, but not in the case of the real or complex numbers. This chapter gives part of what is necessary to replace an intuitive understanding of the familiar number systems by an understanding based on a more solid logical foundation.

SECTION 22. ORDERED INTEGRAL DOMAINS

In this section and the one that follows we shall take the final steps in characterizing the ring of integers. The first definition given, that of an *ordered* integral domain, applies to the integers as well as to many other integral domains. It will lead to the ideas of *positive, negative, greater than*, and *less than*. In order to read the definition with the integers in mind as an example, think of D^p as being the set of positive integers.

Definition. An integral domain D is said to be *ordered* if there is a subset D^p of D such that:

closure under addition

if $a, b \in D^p$, then $a + b \in D^p$,

closure under multiplication

if $a, b \in D^p$, then $ab \in D^p$,

law of trichotomy

if $a \in D$, then exactly one of the following is true:

$$a = 0, \quad a \in D^p, \quad \text{or} \quad -a \in D^p.$$

The elements of D^p are called the *positive* elements of D. Elements that are neither zero nor positive are said to be *negative*.

Besides the integers, other ordered integral domains include the rational numbers and the real numbers, with the set of positive elements being the set of positive numbers in each case. We shall see that the integral domains $\mathbb{Z}_p$ are not ordered (regardless of what one tries to use for the set of positive elements).

Assume in the remainder of this section that D is an ordered integral domain with unity e.

Lemma 22.1. *If $a \in D$ and $a \neq 0$, then $a^2 \in D^p$.*

PROOF. By definition of D^p, since $a \neq 0$ we have either $a \in D^p$ or $-a \in D^p$. If $a \in D^p$, then $a \cdot a = a^2 \in D^p$ by closure of D^p under multiplication. If $-a \in D^p$, then $(-a)(-a) = (-a)^2 \in D^p$ by closure of D^p under multiplication; but $(-a)^2 = a^2$ in any ring, so that again $a^2 \in D^p$. □

Corollary. $e \in D^p$.

PROOF. Since $e \in D$ and $e \neq 0$, $e^2 \in D^p$ by Lemma 22.1. But $e^2 = e$. □

Lemma 22.2. *If $a \in D^p$ and n is a positive integer, then $na \in D^p$.*

PROOF. The proof is by induction on n (which is reviewed in Appendix C). We are given $1a = a \in D^p$. Assume that $na \in D^p$. Then $(n+1)a = na + 1a \in D^p$ by closure of D^p under addition. □

Theorem 22.1. *If D is an ordered integral domain, then D has characteristic* 0.

PROOF. By the corollary of Lemma 22.1, $e \in D^p$ so that $ne \in D^p$ for each positive integer n by Lemma 22.2. Thus $ne \neq 0$ for each positive integer n, since a positive element cannot be 0. Therefore the characteristic cannot be $n \neq 0$. □

Corollary. *If D is an ordered integral domain, then D contains a subring isomorphic to $\mathbb{Z}$.*

PROOF. Apply Theorems 22.1 and 21.2. □

Corollary. *A finite integral domain cannot be ordered. In particular, $\mathbb{Z}_p$ (p a prime) cannot be ordered.*

PROOF. This is a direct consequence of the preceding corollary. □

Definition. For D an ordered integral domain and $a,b \in D, a > b$ will mean that $a - b \in D^p$. If $a > b$ we say that a is *greater than* b, and that b is *less than* a.

For the ring of integers (or rational numbers or real numbers), this is the usual meaning of $>$. As usual, $b < a$ means $a > b$, $a \geq b$ means $a > b$ or $a = b$, and $a \leq b$ means $a < b$ or $a = b$. The following theorem brings together many of the properties of the relation $>$.

Theorem 22.2. *Let D be an ordered integral domain and let $a,b,c \in D$.*

(a) If $a > 0$ and $b > 0$, then $a + b > 0$.
(b) If $a > 0$ and $b > 0$, then $ab > 0$.
(c) Exactly one of the following is true: $a = b$, $a > b$, or $b > a$.
(d) If $a > b$, then $a + c > b + c$.
(e) If $a > b$ and $c > 0$, then $ac > bc$.
(f) If $a \neq 0$, then $a^2 > 0$.
(g) If $a > b$ and $b > c$, then $a > c$.

PROOF. Each property follows from the definitions or other properties already given in this section. To prove (e), for instance, suppose that $a > b$ and $c > 0$. Then $a - b \in D^p$ and $c \in D^p$ by the definition of $>$. Therefore $(a - b)c \in D^p$ by the closure of D^p under multiplication. But $(a - b)c = ac - bc$ so that $ac - bc \in D^p$. Therefore, applying the definition of $>$ once more, $ac > bc$.

The proofs of the other properties are left as exercises (Problem 22.3). □

PROBLEMS

Throughout this set of problems D denotes an ordered integral domain.

22.1. List as many integral domains as possible, and indicate which are ordered and which are not.

22.2. Prove or disprove that if E is a subring of an ordered integral domain D, and E is also an integral domain, then E is ordered.

22.3. Complete the proof of Theorem 22.2.

22.4. If e is the unity of D, then $x^2+e=0$ has no solution in D. Why?

22.5. Prove that if $a,b,c \in D$, $a>b$, and $c<0$, then $ac<bc$.

22.6. Prove that if $a,b,c \in D$, $ac>bc$, and $c>0$, then $a>b$.

22.7. Prove that if $a,b \in D$, and $a<0$ and $b<0$, then $ab>0$.

22.8. Prove that if $a,b \in D$ and $a>b$, then $-a<-b$.

22.9. Prove or disprove that if $a,b \in D$ and $a>b$, then $a^2>b^2$.

22.10. Prove or disprove that if $a,b \in D$ and $a>b$, then $a^3>b^3$.

22.11. For $a \in D$, define $|a|$ by

$$|a| = \begin{cases} a & \text{if} \quad a \in D^p \\ 0 & \text{if} \quad a=0 \\ -a & \text{if} \quad -a \in D^p. \end{cases}$$

Prove that if $a,b \in D$, then:

(a) $|ab|=|a|\cdot|b|$

(b) $|a| \geq a \geq -|a|$ and $|b| \geq b \geq -|b|$

(c) $|a|+|b| \geq |a+b|$ [*Suggestion*: Add inequalities from part (b).]

(d) $|a-b| \geq ||a|-|b||$

SECTION 23. THE INTEGERS

Definition. An element a in a subset S of an ordered integral domain D is a *least element* of S if $x>a$ for each $x \in S$ such that $x \neq a$.

Definition. An ordered integral domain D is *well-ordered* if every nonempty subset of D^p has a least element.

The Least Integer Principle (Section 10) states that the integral domain of integers is well-ordered. (What we have called the Least Integer Principle is sometimes even called the *Well-Ordering Principle*.) The integral domain of rational numbers is not well-ordered, because the set of positive rational numbers has no least element (Problem 23.1). In fact, the integers form the "only" well-ordered integral domain. The following theorem makes this precise.

Theorem 23.1. *If D is a well-ordered integral domain, then D is isomorphic to the ring of integers.*

The proof of the theorem will be easier to grasp if the following fact is proved separately.

Lemma 23.1. *If D is a well-ordered integral domain with unity e, then e is the least element of D^p.*

PROOF. Because D is well-ordered, D^p must have a least element; assume it to be $a \neq e$ (this will lead to a contradiction). Since $e \in D^p$ by the corollary of Lemma 22.1, and a is the least element of D^p by our assumption, we must have $e > a$. Now $e > a$ and $a > 0$ imply $a > a^2$, by Theorem 22.2(e). However, $a^2 \in D^p$ by Lemma 22.1. Thus we have $a^2 \in D^p$ and $a > a^2$, which contradicts the assumption that a is the least element of D^p. Thus the least element of D^p must be e. □

PROOF OF THEOREM 23.1. Assume that D is a well-ordered integral domain with unity e, and define $\theta: \mathbb{Z} \to D$ by $\theta(n) = ne$ for each $n \in \mathbb{Z}$. We showed in proving Theorem 21.2 that this θ is an isomorphism of $\mathbb{Z}$ onto $\theta(Z)$; therefore it suffices to prove that $\theta(\mathbb{Z}) = D$. We shall use an indirect proof for this.

Assume that θ is not onto, and let d denote an element such that $d \in D$ but $d \notin \theta(\mathbb{Z})$. Then also $-d \in D$ and $-d \notin \theta(\mathbb{Z})$: for if $-d \in \theta(\mathbb{Z})$, say $\theta(m) = -d$, then $\theta(m) = me = -d$ so that $\theta(-m) = (-m)e = -(me) = d$, implying $d \in \theta(\mathbb{Z})$, which is false. Since $d \notin \theta(\mathbb{Z})$ and $-d \notin \theta(\mathbb{Z})$, and either $d \in D^p$ or $-d \in D^p$, we conclude that there is a positive element in D that is not in $\theta(\mathbb{Z})$. Therefore the set of elements that are in D^p but not in $\theta(\mathbb{Z})$ is nonempty, and so since D is well-ordered there is a least such element—call it s.

Thus s is the least element of D^p that is not in $\theta(\mathbb{Z})$. Since $\theta(1) = 1e = e$, we have $e \in \theta(\mathbb{Z})$ so that $e \neq s$. Therefore, since e is the least element of D^p (Lemma 23.1), we must have $s > e$, $s - e > 0$, and $s - e \in D^p$. But $s > s - e$ (because $s - (s - e) = e \in D^p$), so that since s is the least element of D^p not in $\theta(\mathbb{Z})$, we must have $s - e \in \theta(\mathbb{Z})$. But if $\theta(k) = ke = s - e$ for $k \in \mathbb{Z}$, then $\theta(k+1) = (k+1)e = ke + e = (s - e) + e = s$, and hence $s \in \theta(\mathbb{Z})$, which is a contradiction. Therefore $\theta(\mathbb{Z}) = D$. □

If we do not distinguish between isomorphic rings, we now have a characterization of the ring of integers: *$\mathbb{Z}$ is the unique well-ordered integral domain.*

PROBLEMS

23.1. Prove (without Theorem 23.1) that $\mathbb{Q}$ is not well-ordered.

23.2. Prove that if D is an ordered integral domain with unity e, and $a \in D$, then $a > a - e$.

23.3. Prove that if n is an integer, then $n+1$ is the least integer greater than n. (Everyone "knows" this, but prove it.)

23.4. Prove that the integral domain in Example 18.4 is not well-ordered.

23.5. Show that if $\theta: \mathbb{Z} \to \mathbb{Z}$ is a ring isomorphism, then θ must be the identity mapping. Is there an additive group isomorphism $\theta: \mathbb{Z} \to \mathbb{Z}$ other than the identity mapping?

23.6. Prove the Principle of Mathematical Induction (Appendix C) from the Least Integer Principle, that is, from the fact that $\mathbb{Z}$ is well-ordered. [*Suggestion*: Let $S = \{k \mid P(k) \text{ is false}\}$ and show that S must be empty.]

SECTION 24. FIELD OF QUOTIENTS. THE FIELD OF RATIONAL NUMBERS

If a and b are integers with $a \neq 0$, then the equation $ax = b$ may not have a solution in the integral domain of integers. The equation does have a solution in the field of rational numbers, however. Moreover, the field of rational numbers is just large enough to contain all such solutions, because every rational number has the form $a^{-1}b$ for a and b integers with $a \neq 0$, so that every rational number is a solution of an equation $ax = b$. We shall prove in this section that if D is *any* integral domain, then there is a "unique smallest" field "containing" D such that each equation $ax = b$, with $a, b \in D$ and $a \neq 0$, has a solution in that field. This field will be called the *field of quotients* of D. The field of quotients of the integral domain of integers will be the field of rational numbers. Since we have already characterized the integers, the uniqueness of its field of quotients will give us a characterization of the rational numbers.

Before starting through the formalities needed to construct a field of quotients, here is the basic idea as it applies to the integers. A fraction a/b gives us an ordered pair of integers (a, b) with the second component nonzero. Instead of thinking of the fraction, think of the ordered pair. To account for the fact that different fractions (such as $\frac{2}{3}$, $\frac{16}{24}$, and $\frac{-20}{-30}$) can represent the same rational number, we agree not to distinguish between ordered pairs if they correspond to such fractions. This is done with an equivalence relation: pairs (a, b) and (c, d) will be equivalent if the corresponding fractions are equal [thus $(2, 3)$ and $(-20, -30)$ will be equivalent]. The equivalence classes for this equivalence relation form a set, and we define two operations on this set that make it a field—the field of rational numbers. Next, we prove that this field contains an integral domain isomorphic to the integral domain of integers. Finally, we prove that any field containing an integral domain isomorphic to the integers must contain a field isomorphic to the rational numbers.

Why do all this if we already know about the rational numbers? First, it will tell us what is unique about the rational numbers in terms of the appropriate abstract ring properties. Second, the procedure used will apply to any integral domain, not just the integers.

Throughout the following discussion D will denote an integral domain with unity e, and D' will denote the set of all nonzero elements of D. The Cartesian product of D and D' is

$$D \times D' = \{(a,b) | a,b \in D, b \neq 0\}.$$

For elements (a,b) and (c,d) in $D \times D'$ we write

$$(a,b) \sim (c,d) \text{ iff } ad = bc.$$

Lemma 24.1. *The relation $\sim$ is an equivalence relation on $D \times D'$.*

PROOF. The verification of the reflexive and symmetric properties is Problem 24.1. To prove that $\sim$ is transitive, assume that $(a,b) \sim (c,d)$ and $(c,d) \sim (f,g)$, with each pair in $D \times D'$. Then $ad = bc$ and $cg = df$. By the first of these equations $adg = bcg$, and by the second $bcg = bdf$. From these last equations we conclude that $adg = bdf$, or $(ag)d = (bf)d$ (remember that D is commutative). But D is an integral domain and $d \neq 0$, so that by cancellation (Theorem 19.1) $ag = bf$. This proves that $(a,b) \sim (f,g)$, which establishes transitivity. □

If $(a,b) \in D \times D'$, we shall denote the equivalence class to which (a,b) belongs relative to $\sim$ by $[a,b]$. Thus

$$[a,b] = \{(x,y) \in D \times D' | (a,b) \sim (x,y)\}.$$

The set of all such equivalence classes will be denoted by F_D. If we recall how fractions are added and multiplied $[(a/b) + (c/d) = (ad + bc)/bd$ and $(a/b)(c/d) = ac/bd]$, then we are led to define two operations on F_D as follows:

$$[a,b] + [c,d] = [ad + bc, bd]$$

and (24.1)

$$[a,b] \cdot [c,d] = [ac, bd]$$

for all $[a,b], [c,d] \in F_D$. (Notice that each second component, bd, is in D' because $b \in D'$ and $d \in D'$ and D has no zero divisors.) The next lemma shows that these operations are well-defined.

Lemma 24.2. *If $[a_1,b_1] = [a_2,b_2]$ and $[c_1,d_1] = [c_2,d_2]$, then*

$$[a_1d_1 + b_1c_1, b_1d_1] = [a_2d_2 + b_2c_2, b_2d_2]$$

and

$$[a_1c_1, b_1d_1] = [a_2c_2, b_2d_2].$$

Proof. The proof for multiplication is left to Problem 24.2. Here is the proof for addition. From $[a_1,b_1]=[a_2,b_2]$ we know that $(a_1,b_1)\sim(a_2,b_2)$, and so

$$a_1b_2 = b_1a_2. \tag{24.2}$$

Similarly, from $[c_1,d_1]=[c_2,d_2]$ we know that

$$c_1d_2 = d_1c_2. \tag{24.3}$$

We must show that

$$[a_1d_1+b_1c_1,b_1d_1] = [a_2d_2+b_2c_2,b_2d_2],$$

that is,

$$(a_1d_1+b_1c_1)(b_2d_2) = (b_1d_1)(a_2d_2+b_2c_2). \tag{24.4}$$

Using (24.2) and (24.3), and the commutativity of D, we can deduce (24.4) as follows:

$$\begin{aligned}(a_1d_1+b_1c_1)(b_2d_2) &= a_1d_1b_2d_2 + b_1c_1b_2d_2\\ &= a_1b_2d_1d_2 + b_1b_2c_1d_2\\ &= b_1a_2d_1d_2 + b_1b_2d_1c_2\\ &= b_1d_1a_2d_2 + b_1d_1b_2c_2\\ &= b_1d_1(a_2d_2+b_2c_2). \quad \square\end{aligned}$$

It is easy to verify that

if $c \in D'$ and $[a,b] \in F_D$, then $[a,b]=[ac,bc]=[ca,cb]$.

This will be used without explicit reference.

Lemma 24.3. *F_D with the operations defined in (24.1) is a field. The zero is $[0,e]$; the negative of $[a,b]$ is $[-a,b]$; the unity is $[e,e]$; and the inverse of $[a,b]\neq[0,e]$ is $[b,a]$.*

Proof. We shall verify that $[0,e]$ is a zero and that one of the two distributive laws is satisfied. The remainder of the proof is left to Problem 24.3.

As to zero: if $[a,b]\in F_D$, then $[a,b]+[0,e]=[ae+b0,be]=[ae,be]=[a,b]$, and $[0,e]+[a,b]=[0b+ea,eb]=[ea,eb]=[a,b]$.

If $[a,b],[c,d],[f,g]\in F_D$, then

$$\begin{aligned}[a,b]([c,d]+[f,g]) &= [a,b][cg+df,dg]\\ &= [acg+adf,bdg]\\ &= [b(acg+adf),b(bdg)]\\ &= [(ac)(bg)+(bd)(af),(bd)(bg)]\\ &= [ac,bd]+[af,bg]\\ &= [a,b][c,d]+[a,b][f,g].\end{aligned}$$

This establishes one of the two distributive laws. $\square$

Lemma 24.4. *Let D_1 denote the subset of F_D consisting of all $[a,e]$ for $a \in D$. Then D_1 is a subring of F_D and $D \approx D_1$.*

PROOF. Define $\theta : D \to F_D$ by $\theta(a) = [a,e]$ for each $a \in D$. The image of θ is clearly D_1. We shall prove that θ is one-to-one and preserves sums and products. It will follow from this that D_1 is a subring and that $D \approx D_1$. [The proof that D_1 is a subring is analogous to the proof of Theorem 15.2(c).]

If $\theta(a_1) = \theta(a_2)$, then $[a_1, e] = [a_2, e]$ so that $a_1 e = e a_2$ and $a_1 = a_2$. Thus θ is one-to-one. If $a, b \in D$, then $\theta(a+b) = [a+b, e] = [ae+eb, ee] = [a,e] + [b,e] = \theta(a) + \theta(b)$, and $\theta(ab) = [ab, e] = [ab, ee] = [a,e][b,e] = \theta(a)\theta(b)$. □

The field F_D, constructed from the integral domain D in this way, is called the *field of quotients* of D. Recall that to say that a ring S can be embedded in a ring R means that R contains a subring isomorphic to S (Section 21). Using this terminology, what we have shown is that *any integral domain can be embedded in a field*—its field of quotients. But in general an integral domain can be embedded in more than one field. For example, the integral domain of integers can be embedded in the field of rational numbers and also the field of real numbers. However, the field of quotients is the smallest such field, in that it can be embedded in any field in which the given integral domain can be embedded. We put all of this together in the following theorem.

Theorem 24.1. *If D is an integral domain, then there exists a field F_D, the field of quotients of D, such that*

(*a*) *F_D contains an integral domain isomorphic to D, and*
(*b*) *if K is any field containing an integral domain isomorphic to D, then K contains a field isomorphic to F_D.*

PROOF. Lemma 24.3 proves that F_D is a field. Lemma 24.4 proves property (a).

In proving property (b), we shall assume that D is actually a subring of K; this amounts to identifying D with an integral domain to which it is isomorphic, and allows us to get at the main idea without having it obscured by distractive notation. Because of the way in which the field of quotients F_D is constructed from the elements of D, we have a natural correspondence between elements $ab^{-1} \in K$, with $a, b \in D$ and $b \neq 0$, and elements $[a,b] \in F_D$. If we identify each element of F_D with the element to which it corresponds ($[a,b] \leftrightarrow ab^{-1}$), then we can think of F_D as a subring of K, which is what property (b) asserts. [If you prefer, the map $\phi : F_D \to K$ defined by $\phi([a,b]) = ab^{-1}$ is one-to-one and preserves both addition and multiplication, and thus is an isomorphism of F_D onto a subfield of K.] □

We can now characterize the field of rational numbers among all sets with two operations: *The field of rational numbers is the* (*unique*) *field of quotients of the* (*unique*) *well-ordered integral domain*. Again, this assumes that we do not distinguish between isomorphic rings or fields.

If K is any field of characteristic 0, then K must contain a subring isomorphic to $\mathbb{Z}$, by Theorem 21.2. But then K must also contain a subfield isomorphic to $\mathbb{Q}$, by Theorem 24.1. This gives the following corollary.

Corollary. *If K is any field of characteristic* 0, *then K contains a subfield isomorphic to $\mathbb{Q}$.*

PROBLEMS

24.1. Complete the proof of Lemma 24.1.
24.2. Complete the proof of Lemma 24.2.
24.3. Complete the proof of Lemma 24.3.
24.4. The ring $\mathbb{Z}_6$ cannot be embedded in a field. Why?
24.5. Verify that the field of quotients of the integral domain $\mathbb{Z}[\sqrt{2}\,]$ (Example 18.4) is isomorphic to $\mathbb{Q}[\sqrt{2}\,]=\{a+b\sqrt{2}\,|a,b\in\mathbb{Q}\}$.
24.6. The field of quotients of any field D is isomorphic to D. Why?
24.7. Assume that D, D_1, and F_D are as in Lemma 24.4. Show that each element of F_D is a solution of some equation $ax=b$ with $a,b\in D_1$.
24.8. State and prove the analogue of the corollary of Theorem 24.1 with characteristic p in place of characteristic 0.

SECTION 25. ORDERED FIELDS. THE FIELD OF REAL NUMBERS

By moving from the integral domain of integers to the field of rational numbers, we have obtained solutions to all equations $ax=b$ (a and b integers, $a\neq 0$). But other deficiencies remain. As we shall prove in Theorem 25.1, for instance, there is no rational number x such that $x^2=2$. In other words, $\sqrt{2}$ is *irrational*, that is, not rational. This was first discovered by the Pythagoreans, in the fifth century B.C., in its geometric form: there is no rational number that will measure the hypotenuse of a right triangle with each leg of unit length (see Figure 25.1). In terms of a

Figure 25.1

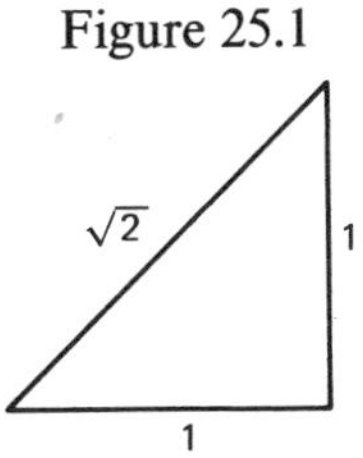

number line, this means that if two points are chosen on a straight line and labeled 0 and 1, and if other points are then made to correspond to the rational numbers in the obvious way, there will be no number corresponding to the point "$\sqrt{2}$ units" from 0 in the positive direction. In fact, there will be many points not corresponding to rational numbers. The basic assumption of coordinate geometry is that this problem can be overcome by using real numbers in place of (just) rational numbers. This important use of real numbers emphasizes the fact that the deficiency of the rational numbers that the real numbers corrects has more to do with order than with the solution of equations (such as $x^2=2$). This should become clearer with the discussion of real numbers in this section and complex numbers in the next.

We now prove the irrationality of $\sqrt{2}$. The usual notation a/b will be used in place of the ordered pair notation (a,b) of Section 24, and also to denote the equivalence class $[a,b]$.

Theorem 25.1. *There is no rational number x such that $x^2=2$.*

PROOF. The proof is by contradiction, and so we begin by assuming the theorem to be false. Thus assume that there are integers a and b, with $b\neq 0$, such that $(a/b)^2=2$. We also assume a/b reduced to lowest terms, so that a and b have no common factor except ± 1. From $(a/b)^2=2$ we have $a^2=2b^2$. The right side of this equation, $2b^2$, is even; therefore the left side, a^2, must also be even. But if a^2 is even then a must be even, and hence $a=2k$ for some integer k. Substituting $a=2k$ in $a^2=2b^2$, we obtain $(2k)^2=2b^2$, or $2k^2=b^2$. Since $2k^2$ is even, b^2 is even, so that b must be even. Thus we have deduced that a and b are both even, and they therefore have 2 as a common factor. This contradicts the assumption that a/b was reduced to lowest terms, and completes the proof. $\square$

The next theorem shows that the rational numbers form an *ordered field*, that is, an ordered integral domain that is also a field. (The proof of this theorem should be considered optional.)

Theorem 25.2. *Let $\mathbb{Q}$ denote the field of rational numbers, and let $\mathbb{Q}^p$ denote the set of all elements of $\mathbb{Q}$ with representations a/b such that $ab>0$. Then $\mathbb{Q}$ is an ordered field with $\mathbb{Q}^p$ as its set of positive elements.*

PROOF. We must first verify that the condition ($ab>0$) for an element of $\mathbb{Q}$ to be in $\mathbb{Q}^p$ is independent of the particular fraction chosen to represent it. To this end, assume that $a/b=c/d$. Then $ad=bc$. Making use of this, and $b^2>0$ and $d^2>0$, together with properties of the relation $>$, we can deduce that if $ab>0$ then $abd^2>0$, $adbd>0$, $bcbd>0$, $cdb^2>0$, and finally $cd>0$. Similarly, if $cd>0$ then $ab>0$. Thus $ab>0$ iff $cd>0$, as required.

It remains to verify the three properties in the definition of ordered integral domain (Section 22).

If a/b and c/d represent elements of $\mathbb{Q}^p$, then $ab>0$ and $cd>0$, and thus $abd^2>0$ and $cdb^2>0$. This implies that $(ad+bc)bd=abd^2+cdb^2>0$, so that $(a/b)+(c/d)=(ad+bc)/bd\in\mathbb{Q}^p$.

As to closure of $\mathbb{Q}^p$ under multiplication, if $a/b\in\mathbb{Q}^p$ and $c/d\in\mathbb{Q}^p$, then $ab>0$ and $cd>0$, so that $(ac)(bd)=(ab)(cd)>0$ and $(a/b)(c/d)=(ac)/(bd)\in\mathbb{Q}^p$.

Finally, if a/b represents a nonzero element of $\mathbb{Q}$, then $a\neq 0$ and $b\neq 0$; therefore $ab\neq 0$. Therefore $ab>0$ or $0>ab$, and, correspondingly, $a/b\in\mathbb{Q}^p$ or $-(a/b)=(-a)/b\in\mathbb{Q}^p$. This establishes the law of trichotomy. □

An element u of an ordered field F is said to be an *upper bound* for a subset S of F if $u\geq x$ for each $x\in S$. For example, any positive rational number is an upper bound for the set of all negative rational numbers. The set of integers has no upper bound in the field of rational numbers (or in any other field). An element u of an ordered field F is said to be a *least upper bound* for a subset S of F provided that

1. u is an upper bound for S, and
2. if $v\in F$ is an upper bound for S, then $v\geq u$.

Thus 0 is a least upper bound for the set of negative rational numbers.

If S denotes the set of all rational numbers r such that $r^2<2$, then S has an upper bound in the field of rational numbers (1.5, for instance); but it does not have a *least* upper bound in the field of rational numbers. However, S does have a least upper bound in the field of real numbers—namely $\sqrt{2}$. This leads to the following definition, which isolates the property distinguishing the field of real numbers from the field of rational numbers.

Definition. An ordered field F is said to be *complete* if every nonempty subset of F having an upper bound in F has a least upper bound in F.

Theorem 25.3. *There exists a complete ordered field. Any two such fields are isomorphic, and any such field contains a subfield isomorphic to the field of rational numbers.*

The field of real numbers is a complete ordered field. Theorem 25.3 shows that such a field exists and is essentially unique. The fact that it must contain a subfield isomorphic to $\mathbb{Q}$ follows quickly from results in Sections 22 and 24 (Problem 25.3). We shall not prove the other parts of the theorem. (For details of the construction of the real numbers, see, for example, the book by Landau referred to in the notes at the end of this chapter.)

In applications the real numbers are usually thought of as the numbers having decimal representations. Examples are

$$\tfrac{1}{2}=0.5 \qquad \tfrac{1}{3}=0.\overline{3}$$

$$-12.138 \qquad \tfrac{11}{7}=1.\overline{571428}$$

$$\sqrt{2}=1.414213\cdots \qquad \pi=3.141592\cdots,$$

where the lines over 3 and 571428 mean that they repeat without end. It can be shown that the decimal numbers representing rational numbers are precisely those that either terminate or become periodic (Problems 25.6 and 25.8). The number $0.1010010001\cdots$, where the number of 0's between 1's increases each time, is irrational. Each number whose decimal representation terminates (such as 0.5 or -12.138) also has a representation with 9 repeating on the end. For example, $1.0=0.\overline{9}$ (Problem 25.4). The numbers π and e (the base for natural logarithms) are both irrational, but the proofs are more difficult than that for $\sqrt{2}$.

PROBLEMS

25.1. Prove that if p is a prime, then $\sqrt{p}$ is irrational. (Problem 39.7 gives a more general statement.)

25.2. (a) Write definitions of *lower bound* and *greatest lower bound* for a subset S of an ordered field F.

(b) Prove that if F is a complete ordered field, then every nonempty subset of F having a lower bound in F has a greatest lower bound in F.

(c) Is the converse of the statement in part (b) true?

25.3. Prove the last part of Theorem 25.3, that is, prove that each complete ordered field contains a subfield isomorphic to $\mathbb{Q}$.

25.4. Explain why $0.\overline{9}=1$. (*Suggestion*: Let $x=0.\overline{9}$, compute $10x-x$, and solve for x. Or sum the geometric series $0.9+0.09+0.009+\cdots$; Appendix C shows how.)

25.5. Express $1.9\overline{35}$ as a fraction. (*Suggestion*: Let $x=1.9\overline{35}$, compute $1000x-10x$, and solve for x. Or sum a geometric series, as suggested in Problem 25.4.)

25.6. Prove that a decimal number that becomes periodic represents a rational number. (See Problems 25.4 and 25.5.)

25.7. Determine the decimal representations of each of the following numbers.

(a) $\frac{3}{11}$ (b) $\frac{1984}{7}$

(c) $\frac{9}{8}$ (d) $\frac{8}{9}$

25.8. Explain why the decimal representation of a rational number must terminate or become periodic. (*Suggestion*: In computing the decimal representation a/b by long division, there are only b possible remainders. Look at what happens in a special case.)

25.9. Prove that if F is an ordered field, and $a,b \in F$, then
(a) $b > a > 0$ implies $1/a > 1/b > 0$, and
(b) $0 > b > a$ implies $0 > 1/a > 1/b$.

25.10. Prove that if a and b are two distinct positive real numbers, then $(a+b)/2 > \sqrt{ab}$. ("The arithmetic mean is greater than the geometric mean." *Suggestion*: First explain why $(\sqrt{a} - \sqrt{b})^2 > 0$.)

25.11. Prove that if $a,b \in \mathbb{Q}$, and $a > b$, then there are infinitely many $x \in \mathbb{Q}$ such that $a > x > b$. [*Suggestion*: If c and d are rational, then so is $(c+d)/2$.]

25.12. Prove that the statement in Problem 25.11 is true if $\mathbb{Q}$ is replaced by any other ordered field.

25.13. Prove that if a is rational and b is irrational, then $a+b$ is irrational.

25.14. Prove that if a is rational, $a \neq 0$, and b is irrational, then ab is irrational.

25.15. True or false: If a is irrational, then a^{-1} is irrational.

25.16. Give examples to show that if a and b are irrational, then ab may be either rational or irrational, depending on a and b.

25.17. Prove that if a and b are positive real numbers, then there exists an integer n such that $na > b$. (This is called the *Archimedean Property* of $\mathbb{R}$. Suppose that the statement is false, so that b is an upper bound for $\{na \mid n \in \mathbb{Z}\}$, and then deduce a contradiction by using the completeness of $\mathbb{R}$.)

25.18. Prove that if $a,b \in \mathbb{R}$, and $a > b$, then there exists a rational number m/n such that $a > m/n > b$. [*Suggestion*: By Problem 25.17 there is a positive integer n such that $n(a-b) > 1$, or $(a-b) > 1/n$. Let m be the least integer such that $m > nb$. Then $(m-1)/n \leq b$, and so $m/n = (m-1)/n + 1/n < b + (a-b) = a$.]

25.19. Prove that every real number is a least upper bound of some set of rational numbers. (For $u \in \mathbb{R}$, let $S = \{r \in \mathbb{Q} \mid r \leq u\}$. Use Problem 25.18 to explain why u must be a least upper bound for S.)

25.20. Prove that a subset of an ordered field has at most one least upper bound in the field.

25.21. Prove that if u is a least upper bound for a subset S of $\mathbb{R}$, then $2u$ is a least upper bound for $\{2x \mid x \in S\}$.

25.22. Prove that if u is a least upper bound for a subset S of $\mathbb{R}$, then $3+u$ is a least upper bound for $\{3+x \mid x \in S\}$.

25.23. Prove that the order on $\mathbb{Q}$ given by Theorem 25.2 is the only one that will make $\mathbb{Q}$ an ordered field.

SECTION 26. THE FIELD OF COMPLEX NUMBERS

There is no real number x such that $x^2 = -1$, because the square of any nonzero element in an ordered integral domain must be positive (Lemma 22.1). The field of complex numbers, which contains the field of real numbers as a subfield, overcomes this deficiency. It does much more than that, in fact, as can be seen from the following theorem.

Fundamental Theorem of Algebra. *Every polynomial equation*

$$a_n x^n + a_{n-1} x^{n-1} + \cdots + a_1 x + a_0 = 0, \tag{26.1}$$

which is of degree at least one and whose coefficients $a_n, a_{n-1}, \ldots, a_1, a_0$ *are complex numbers, has at least one solution in the field of complex numbers.*

Notice the implications of this theorem. To have solutions for all equations $ax = b$ (coefficients integers), we extended the integers to the rational numbers. To have a solution for $x^2 = 2$, we went outside the rational numbers to the real numbers. To have a solution for $x^2 = -1$, we are extending the real numbers to the complex numbers. The Fundamental Theorem of Algebra asserts that in looking for solutions to polynomial equations there will be no need to extend further, because any such equation with complex numbers as coefficients will have a solution *in* the field of complex numbers.

We shall not prove the Fundamental Theorem of Algebra, but we shall say more about polynomial equations in Chapter XIV. We now give a description of the complex numbers. The rational numbers were constructed using equivalence classes of ordered pairs of integers. The complex numbers will be constructed using ordered pairs of real numbers. (*Suggestion*: After studying the statement of Theorem 26.1, pass over the proof and read through Example 26.2; then return to the proof. This should make the operations in the theorem seem more natural.)

Theorem 26.1. *Let* $\mathbb{C}$ *denote the set of all ordered pairs* (a,b) *with* $a, b \in \mathbb{R}$. *Define addition and multiplication of these pairs by*

$$(a,b) + (c,d) = (a+c, b+d)$$

and

$$(a,b)(c,d) = (ac - bd, ad + bc)$$

for all $a,b,c,d \in \mathbb{R}$. *With these operations,* $\mathbb{C}$ *is a field. The subset of* $\mathbb{C}$ *consisting of all* $(a,0)$ *with* $a \in \mathbb{R}$ *forms a subfield of* $\mathbb{C}$, *isomorphic to* $\mathbb{R}$.

PROOF. Most of the details will be left as an exercise, including the verification that $(0,0)$ serves as an identity element for addition and $(-a, -b)$ serves as a negative of (a,b).

To prove that multiplication is associative, assume that $(a,b), (c,d), (e,f) \in \mathbb{C}$. Using the definition of multiplication, and properties of the real numbers, we can write

$$\begin{aligned}(a,b)[(c,d)(e,f)] &= (a,b)(ce-df,cf+de)\\ &= (a(ce-df)-b(cf+de),a(cf+de)+b(ce-df))\\ &= (ace-adf-bcf-bde,acf+ade+bce-bdf)\\ &= ((ac-bd)e-(ad+bc)f,(ac-bd)f+(ad+bc)e)\\ &= (ac-bd,ad+bc)(e,f)\\ &= [(a,b)(c,d)](e,f),\end{aligned}$$

as required.

To prove that multiplication is commutative, write

$$\begin{aligned}(a,b)(c,d) &= (ac-bd,ad+bc)\\ &= (ca-db,cb+da)\\ &= (c,d)(a,b).\end{aligned}$$

The unity is (1,0):

$$(a,b)(1,0) = (a\cdot 1-b\cdot 0,a\cdot 0+b\cdot 1) = (a,b).$$

Because of the commutativity of multiplication we need not check separately that $(1,0)(a,b)=(a,b)$. A similar remark applies to verification of inverse elements and the distributive laws.

Assume that (a,b) is different from (0,0), the zero of $\mathbb{C}$. Then $a\neq 0$ or $b\neq 0$, so that $a^2>0$ or $b^2>0$, and $a^2+b^2>0$. Thus $(a/(a^2+b^2), -b/(a^2+b^2))$ is an element of $\mathbb{C}$, and it is the inverse of (a,b) relative to multiplication:

$$(a,b)\left(\frac{a}{(a^2+b^2)},\frac{-b}{(a^2+b^2)}\right) =$$

$$\left(\frac{(a^2+b^2)}{(a^2+b^2)},\frac{(-ab+ba)}{(a^2+b^2)}\right) = (1,0).$$

The remainder of the proof that $\mathbb{C}$ is a field is left to Problem 26.1.

To prove that $\{(a,0)\colon a\in\mathbb{R}\}$ is a subfield isomorphic to $\mathbb{R}$, consider the mapping $\theta:\mathbb{R}\to\mathbb{C}$ defined by $\theta(a)=(a,0)$ for each $a\in\mathbb{R}$. The mapping θ preserves both operations:

$$\theta(a+b) = (a+b,0) = (a,0)+(b,0) = \theta(a)+\theta(b)$$

and

$$\theta(ab) = (ab,0) = (a,0)(b,0) = \theta(a)\theta(b)$$

for all $a,b\in\mathbb{R}$. Also, θ is one-to-one because if $\theta(a)=\theta(b)$, then $(a,0)=(b,0)$ so that $a=b$. Thus θ is an isomorphism of $\mathbb{R}$ onto $\{(a,0)|a\in\mathbb{R}\}$; the proof that the latter is a field is left to Problem 26.2. □

In light of the last part of this theorem it is natural to identify $a \in \mathbb{R}$ with $(a,0) \in \mathbb{C}$. In this way $\mathbb{R}$ actually becomes a subset of $\mathbb{C}$, so that every real number is a complex number. The element $(0,1)$ of $\mathbb{C}$ is customarily denoted by i, and then each element $(0,b)$ by bi. This leads to the notation $a+bi$ for the element (a, b) of $\mathbb{C}$. For $a,b,c,d \in \mathbb{R}$,

$$a+bi = c+di \quad \text{iff} \quad a=c \quad \text{and} \quad b=d.$$

The rules for addition and multiplication become

$$(a+bi)+(c+di) = (a+c)+(b+d)i$$

and

$$(a+bi)(c+di) = (ac-bd)+(ad+bc)i.$$

In particular, $i^2 = (0+1i)(0+1i) = -1+0i = -1$, and hence i is a solution in $\mathbb{C}$ to the equation $x^2 = -1$.

To compute with elements of $\mathbb{C}$, simply apply the various associative, commutative, and distributive laws, and replace i^2 by -1 wherever it occurs. In this way any expression involving complex numbers can be reduced to the form $a+bi$ with $a,b \in \mathbb{R}$. When such a number has $b \neq 0$ it is said to be *imaginary*.

Example 26.1. (a) $(1+i)^2 = 1+2i+i^2 = 1+2i-1 = 2i$.

(b) $i^4 = (i^2)^2 = (-1)^2 = 1$.

(c) $(-i)^2 = (-1)^2(i)^2 = i^2 = -1$.

(d) $i(1-i)+2(3+i) = i-i^2+6+2i = i-(-1)+6+2i = i+1+6+2i = 7+3i$.

The number $a-bi$ is called the *conjugate* of $a+bi$. To simplify a fraction with an imaginary number $a+bi$ in the denominator, multiply both numerator and denominator by this conjugate, making use of $(a+bi)(a-bi) = a^2-(bi)^2 = a^2+b^2$.

Example 26.2. (a) $\dfrac{1}{1+i} = \dfrac{1}{1+i} \cdot \dfrac{1-i}{1-i} = \dfrac{1-i}{2} = \dfrac{1}{2} - \dfrac{1}{2}i.$

(b) $\dfrac{2+i}{2-i} = \dfrac{2+i}{2-i} \cdot \dfrac{2+i}{2+i} = \dfrac{4+4i-1}{4+1} = \dfrac{3}{5} + \dfrac{4}{5}i.$

More will be said about calculations with complex numbers in the next section. Following now is a concise discussion of the ideas needed to characterize the field $\mathbb{C}$.

If E and F are fields, then E is said to be an *extension* of F if E contains a subfield isomorphic to F; for convenience, F is often thought of as actually being a subfield of E. Thus $\mathbb{R}$ is an extension of $\mathbb{Q}$, and $\mathbb{C}$ is an extension of both $\mathbb{R}$ and $\mathbb{Q}$. Any field is an extension of itself.

Assume that E is an extension of F. An element $a \in E$ is said to be *algebraic* over F if a is a solution of some polynomial equation (26.1) with coefficients in F. For example, $\sqrt{2}$ is algebraic over $\mathbb{Q}$ because it is a solution of $x^2 - 2 = 0$. Neither π nor e is algebraic over $\mathbb{Q}$, but these facts are not easy to prove. A field E is an *algebraic extension* of F if every $a \in E$ is algebraic over F. The remarks about π and e show that $\mathbb{R}$ is not an algebraic extension of $\mathbb{Q}$. However, $\mathbb{C}$ is an algebraic extension of $\mathbb{R}$ (Problem 26.5).

A field F is *algebraically closed* if every polynomial equation (26.1) with coefficients in F has a solution in F. By the Fundamental Theorem of Algebra $\mathbb{C}$ is algebraically closed. But neither $\mathbb{Q}$ nor $\mathbb{R}$ is algebraically closed. (Why?) A field E is an *algebraic closure* of a field F if

1. E is an algebraic extension of F, and
2. E is algebraically closed.

It can be proved that every field has an algebraic closure. Moreover, this algebraic closure is essentially unique: If E_1 and E_2 are algebraic closures of F, then E_1 must be isomorphic to E_2.

Because $\mathbb{C}$ is an algebraic extension of $\mathbb{R}$, and $\mathbb{C}$ is also algebraically closed, we see that $\mathbb{C}$ is an algebraic closure of $\mathbb{R}$. If we put all of this together with what we know about $\mathbb{R}$, and agree not to distinguish between isomorphic fields, we arrive at the following characterization: *The field of complex numbers is the (unique) algebraic closure of the (unique) complete ordered field.*

In the introductory remarks of Section 25 it was stated that the necessity for extending $\mathbb{Q}$ to $\mathbb{R}$ had more to do with order than with algebra. We can now put this in better focus. The question of order was covered in Section 25. Regarding algebra, the algebraic closure of $\mathbb{Q}$ cannot be $\mathbb{R}$ because $\mathbb{R}$ is not algebraically closed. Also, the algebraic closure of $\mathbb{Q}$ cannot be $\mathbb{C}$ because $\mathbb{C}$ is not an algebraic extension of $\mathbb{Q}$. The algebraic closure of $\mathbb{Q}$ is, in fact, a subfield of $\mathbb{C}$ known as the *field of algebraic numbers*. This field consists precisely of those elements of $\mathbb{C}$ that are algebraic over $\mathbb{Q}$. If we were to begin with $\mathbb{Q}$, and concern ourselves only with finding solutions for polynomial equations, we could work wholly within the field of algebraic numbers—we would not need all of $\mathbb{C}$, and, although we would need some elements outside $\mathbb{R}$, we would not need all of $\mathbb{R}$. Questions about algebraic numbers have been important throughout the history of modern algebra; more will be said about this in Chapter XII.

Further questions about polynomial equations and field extensions will be dealt with in Chapters XIII, XIV, and XV.

PROBLEMS

26.1. Complete the proof that $\mathbb{C}$ is a field, in the proof of Theorem 26.1.

26.2. Prove that $\{(a,0)|a\in\mathbb{R}\}$ is a subfield of $\mathbb{C}$ (Theorem 26.1).

26.3. (a) Verify that in $\mathbb{C}$, thought of as $\{(a,b)|a,b\in\mathbb{R}\}$,

$$(0,1)(0,1) = (-1,0).$$

(b) What is $(0,1)^4$ in $\mathbb{C}$?

26.4. Express each of the following in the form $a+bi$, with $a,b\in\mathbb{R}$.

(a) $(2-i)(1+i)$ (b) i^3

(c) $(-i)^3$ (d) $\dfrac{1}{1+2i}$

(e) $(1+i)^3$ (f) $\dfrac{1+i}{2-3i}$

26.5. Prove that $\mathbb{C}$ is an algebraic extension of $\mathbb{R}$. [*Suggestion*: Consider the equation $[x-(a+bi)][x-(a-bi)]=0$.]

26.6. Explain why neither $\mathbb{Q}$ nor $\mathbb{R}$ is algebraically closed.

26.7. Explain why $\mathbb{C}$ cannot be an ordered field.

26.8. Prove or disprove that the mapping $\theta:\mathbb{C}\to\mathbb{C}$ defined by $\theta(a+bi)=a-bi$ is a ring isomorphism. (Compare Problem 23.5.)

26.9. Prove that $\mathbb{Z}_2$ is not an algebraically closed field.

26.10. Prove that if p is a prime, then the field $\mathbb{Z}_p$ is not algebraically closed.

26.11. Find two complex numbers that are solutions of $x^2=-4$.

26.12. Determine a pair of complex numbers $z=a+bi$ and $w=c+di$ giving a solution of the system

$$\begin{aligned} 3z-2w &= -i \\ iz+2iw &= -5. \end{aligned}$$

(The usual methods of solving systems of equations over $\mathbb{R}$ also work over $\mathbb{C}$. Why?)

26.13. Repeat Problem 26.12 for the system

$$\begin{aligned} z+iw &= 1 \\ -2iz+w &= -1. \end{aligned}$$

26.14. Let z^* denote the conjugate of the complex number z, that is, $(a+bi)^*=a-bi$. Prove that each of the following is true for each $z\in\mathbb{C}$.

(a) $(z^*)^*=z$ (b) $z+z^*\in\mathbb{R}$

(c) $z=z^*$ iff $z\in\mathbb{R}$ (d) $(z^{-1})^*=(z^*)^{-1}$

26.15. Verify that

$$\theta(a+bi)=\begin{bmatrix} a & b \\ -b & a \end{bmatrix}$$

defines an isomorphism of $\mathbb{C}$ onto a subring of $M(2,\mathbb{R})$, the ring of 2×2 matrices over $\mathbb{R}$.

26.16. Let z^* denote the conjugate of z, as in Problem 26.14. Let Q denote the set of all matrices in $M(2,\mathbb{C})$ that have the form

$$\begin{bmatrix} z & w \\ -w^* & z^* \end{bmatrix}. \tag{26.2}$$

For example,

$$\begin{bmatrix} 1+2i & 2+i \\ -2+i & 1-2i \end{bmatrix}$$

is in Q. Prove that Q is a division ring, that is, a ring with a unity in which each nonzero element has an inverse relative to multiplication. [*Suggestion*: Assume that $z=a+bi$ and $w=c+di$ in (26.2), and let $k=a^2+b^2+c^2+d^2$. Then

$$\frac{1}{k}\begin{bmatrix} z^* & -w \\ w^* & z \end{bmatrix}$$

is an inverse in Q for the matrix in (26.2).] Also prove that Q is noncommutative, so that it is not a field. The division ring Q is called the ring of *Hamilton's quaternions*, after the British mathematician W. R. Hamilton (1805–1865).

SECTION 27. COMPLEX ROOTS OF UNITY

The other sections of this chapter have been concerned primarily with general properties and abstractions. This section has to do with computation. We shall look at some useful ways of representing complex numbers and show how they can be used to determine the complex roots of unity—the solutions of equations of the form $x^n=1$. These roots of unity will be useful for examples; they also arise often enough in other areas of mathematics to make their inclusion here worthwhile.

Just as the points on a line can be used to represent real numbers geometrically, the points in a plane can be used to represent complex numbers geometrically. A rectangular coordinate system is chosen for the plane, and then each complex number $a+bi$ is represented by the points with coordinates (a,b). Because

$$a+bi=c+di \quad \text{iff} \quad a=c \quad \text{and} \quad b=d$$

(for $a,b,c,d \in \mathbb{R}$), the correspondence

$$a+bi \leftrightarrow (a,b)$$

is one-to-one between complex numbers and points of the plane. Figure 27.1 shows some examples.

Figure 27.1

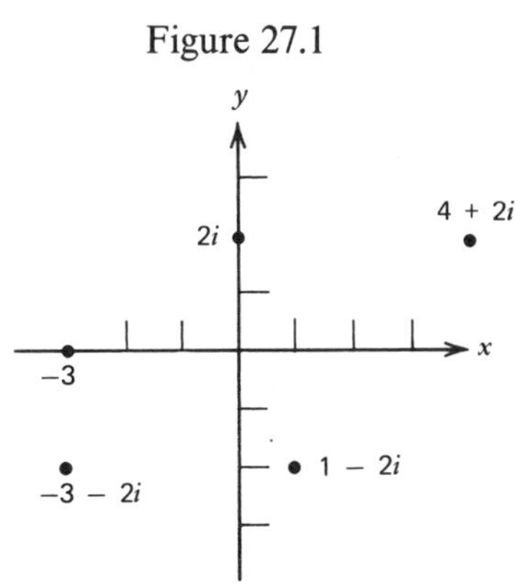

Figure 27.2

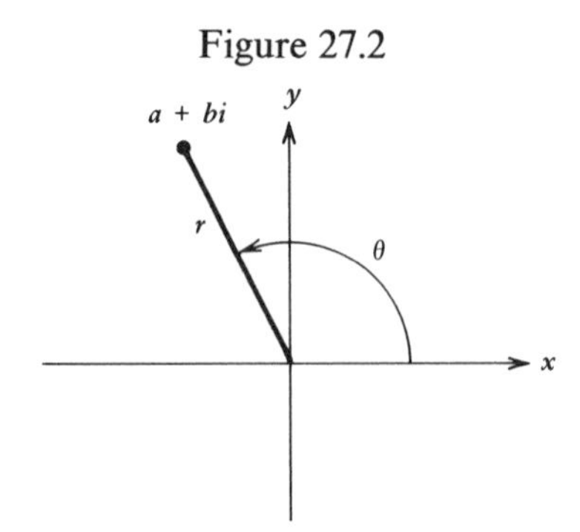

Addition of complex numbers corresponds to vector addition of points in the plane:

$$(a+bi)+(c+di) = (a+c)+(b+d)i \leftrightarrow (a+c,b+d) = (a,b)+(c,d).$$

To describe multiplication of complex numbers geometrically, we turn to polar coordinates. Recall that the polar representation of a point with rectangular coordinates (a,b) is (r,θ), where r denotes the distance between the origin and the given point, and θ denotes the angle from the positive x-axis to the ray from the origin through the given point, with the positive direction taken counterclockwise (Figure 27.2).

Thus $r=\sqrt{a^2+b^2}$, $a=r\cos\theta$, $b=r\sin\theta$, and

$$a+bi = r(\cos\theta + i\sin\theta).$$

The latter is called the *polar* (or *trigonometric*) *form* of $a+bi$. The nonnegative number r appearing here is called the *absolute value* (or *modulus*) of $a+bi$ and is denoted by $|a+bi|$. The angle θ is called the *argument* (or *amplitude*) of $a+bi$. The absolute value of $a+bi$ is unique, but the argument is not: if θ is an argument of $a+bi$, then so is $\theta+2n\pi$ for any integer n. If θ is restricted so that $0\le\theta<2\pi$, then each nonzero complex number does have a unique argument.

Example 27.1. The absolute value of $-2+2i$ is $|-2+2i|=\sqrt{(-2)^2+2^2}=2\sqrt{2}$. The argument is $\theta=135°=3\pi/4$. Thus the polar form of $-2+2i$ is $2\sqrt{2}\ [\cos(3\pi/4)+i\sin(3\pi/4)]$.

The following theorem shows that polar form is especially well-suited for computing products.

Theorem 27.1. *If $z=r(\cos\theta+i\sin\theta)$ and $w=s(\cos\phi+i\sin\phi)$, then*

$$zw = rs\left[\cos(\theta+\phi)+i\sin(\theta+\phi)\right].$$

Thus the absolute value of a product is the product of the absolute values, and the argument of a product is the sum of the arguments.

PROOF. Recall the following two addition formulas from trigonometry:

$$\cos(\theta+\phi) = \cos\theta\cos\phi - \sin\theta\sin\phi$$

and

$$\sin(\theta+\phi) = \sin\theta\cos\phi + \cos\theta\sin\phi.$$

Using these, we can write

$$\begin{aligned} zw &= r(\cos\theta + i\sin\theta)\cdot s(\cos\phi + i\sin\phi) \\ &= rs\big[(\cos\theta\cos\phi - \sin\theta\sin\phi) \\ &\quad + i(\sin\theta\cos\phi + \cos\theta\sin\phi)\big] \\ &= rs\big[\cos(\theta+\phi) + i\sin(\theta+\phi)\big]. \quad \square \end{aligned}$$

De Moivre's Theorem. *If n is a positive integer and $z = r(\cos\theta + i\sin\theta)$, then $z^n = r^n(\cos n\theta + i\sin n\theta)$.*

PROOF. Use induction on n. For $n=1$ the result is obvious. Assuming the theorem true for $n=k$, and using Theorem 27.1 with $w=z^k$, we have

$$\begin{aligned} z^{k+1} = zz^k &= r(\cos\theta + i\sin\theta)\cdot r^k(\cos k\theta + i\sin k\theta) \\ &= r^{k+1}\big[\cos(k+1)\theta + i\sin(k+1)\theta\big], \end{aligned}$$

as required. $\square$

Example 27.2. To compute $(-2+2i)^5$, begin with

$$-2 + 2i = 2\sqrt{2}\ \big[\cos(3\pi/4) + i\sin(3\pi/4)\big],$$

from Example 27.1. Now apply DeMoivre's Theorem. We have $(2\sqrt{2}\,)^5 = 128\sqrt{2}$ and $5(3\pi/4) = 15\pi/4 = 2\pi + 7\pi/4$. Therefore

$$\begin{aligned} (-2+2i)^5 &= 128\sqrt{2}\ \big[\cos(7\pi/4) + i\sin(7\pi/4)\big] \\ &= 128\sqrt{2}\ (\sqrt{2}\,/2 - i\sqrt{2}\,/2) \\ &= 128 - 128i. \end{aligned}$$

For each integer $n \geq 1$, there can be at most n distinct complex numbers that are solutions of $x^n = 1$. (This will be a corollary of Theorem 50.1.) There are, in fact, exactly n solutions, called the complex nth *roots of unity*. They can be determined by using DeMoivre's Theorem.

Theorem 27.2. *For each integer $n \geq 1$, the n complex nth roots of unity are*

$$\cos\frac{2k\pi}{n} + i\sin\frac{2k\pi}{n}, \qquad k = 0, 1, \ldots, n-1. \tag{27.1}$$

Figure 27.3

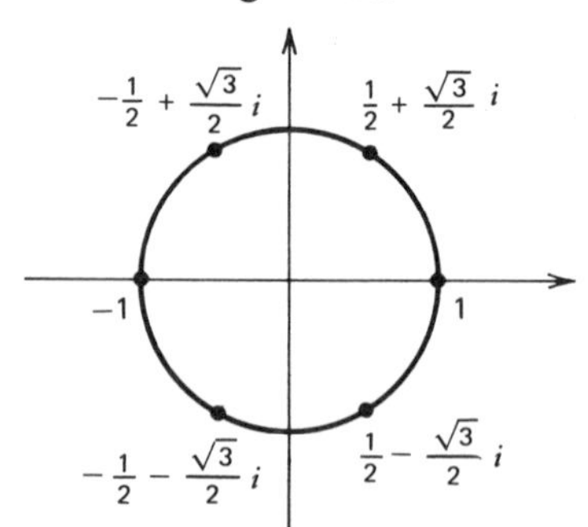

PROOF. By DeMoivre's Theorem, the argument of the nth power of each number in (27.1) is $n(2k\pi/n)=2k\pi$, and the absolute value is $1^n=1$. Thus each of the numbers is an nth root of unity. The n numbers in (27.1) are distinct because the numbers $2k\pi/n$ are distinct and $0\leq 2k\pi/n<2\pi$ for $k=0,1,\ldots,n-1$. Thus the numbers in (27.1) represent all the nth roots of unity. □

The nth roots of unity are represented geometrically by n equally-spaced points on the circle with center at the origin and radius 1, with one of the points being 1. Figure 27.3 shows the case $n=6$.

Theorem 27.2 can be extended to give a formula for the nth roots of any complete number (see Problems 27.14 and 27.15).

PROBLEMS

27.1. Write each of the following complex numbers in the form $a+bi$.
(a) The number with absolute value 2 and argument $\pi/6$.
(b) The number with absolute value $1/2$ and argument $5\pi/3$.
(c) The number with absolute value 5 and argument $9\pi/4$.
(d) The number with absolute value 3 and argument π.

27.2. Use DeMoivre's Theorem to write each of the following complex numbers in the form $a+bi$.
(a) $(1+i)^{10}$ (b) $(\sqrt{3}+i)^5$
(c) $(1-i)^6$ (d) $(-i)^{10}$
(e) $(\frac{1}{2}-\frac{1}{2}i)^4$ (f) $(-\sqrt{2}-\sqrt{2}\,i)^{12}$

27.3. Determine all complex 8th roots of unity, and represent them geometrically.

27.4. Determine all complex 5th roots of unity, and represent them geometrically.

27.5. Prove that if $z=r(\cos\theta+i\sin\theta)$, then

$$z^{-1}=r^{-1}[\cos(-\theta)+i\sin(-\theta)].$$

27.6. State and prove DeMoivre's Theorem for negative integers. (For $n=-1$ see Problem 27.5.)

27.7. Prove that if n is a positive integer, then the set of all nth roots of unity forms a cyclic group of order n with respect to multiplication. [Any generator of this group is called a *primitive* root of unity. One such root is $\cos(2\pi/n)+i\sin(2\pi/n)$.]

27.8. Prove that for each positive integer n the sum of the nth roots of unity is 0. (*Suggestion*: They form a geometric progression, which can be summed by a formula in Appendix C.)

27.9. Let n be a positive integer. What is the product of all the nth roots of unity?

27.10. Prove that the set of all roots of unity forms a group with respect to multiplication.

27.11. Prove that with respect to multiplication, the set of all complex numbers of absolute value 1 forms a group that is isomorphic to the group of all rotations of the plane about a fixed point p (Example 5.7).

27.12. Let z^* denote the conjugate of the complex number z, that is, $(a+bi)^*=a-bi$. Prove that the following are true for each $z\in\mathbb{C}$.

(a) $|z^*|=|z|$

(b) $zz^*=|z|^2$

(c) $z^{-1}=z^*/|z|^2$ if $z\neq 0$.

27.13. Express each of the following complex numbers in polar form.

(a) $1+i$ (b) $\sqrt{3}-i$

(c) -5 (d) $-2i$

(e) $2-2i$ (f) $2i+2\sqrt{3}$

27.14. Prove that for each integer $n\geq 1$, the n complex nth roots of

$$z = r(\cos\theta+i\sin\theta)$$

are

$$r^{1/n}\left(\cos\frac{\theta+2k\pi}{n}+i\sin\frac{\theta+2k\pi}{n}\right), \qquad k=0,1,\ldots,n-1,$$

where $r^{1/n}$ is the positive real nth root of r.

27.15. Prove that for each integer $n\geq 1$, the n complex nth roots of

$$z=r(\cos\theta+i\sin\theta)$$

are

$$v, vw, vw^2,\ldots, vw^{n-1},$$

where

$$v = r^{1/n}\left[\cos\frac{\theta}{n}+i\sin\frac{\theta}{n}\right]$$

and

$$w = \cos\frac{2\pi}{n}+i\sin\frac{2\pi}{n}.$$

27.16 (a) to (f). Use either Problem 27.14 or 27.15 to determine the complex cube roots of each of the numbers in Problem 27.13.

27.17. (a) to (f). Use either Problem 27.14 or 27.15 to determine the complex fourth roots of each of the numbers in Problem 27.13.

NOTES ON CHAPTER VI

References [1], [2], [3], and [5] each discuss the history of the ideas in this chapter; the remarks in [5] are the most extensive. Reference [4] begins with the Peano Postulates—a set of five postulates for the system of natural numbers— and then works through a careful development of each of the systems covered in this chapter.

1. Bell, E. T., *Development of Mathematics*, 2nd ed., McGraw-Hill, New York, 1945.
2. Boyer, C. B., *A History of Mathematics*, John Wiley, New York, 1968.
3. Kline, M., *Mathematical Thought from Ancient to Modern Times*, Oxford University Press, London, 1972.
4. Landau, E., *Foundations of Analysis*, Chelsea, New York, 1960.
5. Nový, L., *Origins of Modern Algebra*, Noordhoff, Leyden, The Netherlands, 1973.

CHAPTER VII

Group Homomorphisms

One way to study a relatively large and complicated group is to study its smaller and less complicated subgroups. But it would also be useful to be able to study the group as a whole, much as a globe allows us to study the earth's surface—brought down to manageable size in a way that preserves as many essential features as possible. Homomorphisms, which are more general than isomorphisms, can help us do just that. A homomorphism is a mapping from one group to another that is not necessarily one-to-one, but does preserve the operations. Thus the image of a homomorphism can be smaller than the domain, but it will generally reflect some essential features of the domain. We shall see by the end of this chapter how subgroups and images of homomorphisms can even be used together to show that most groups are built up from smaller component groups. The concept of homomorphism also extends to rings (Chapter XII) and other algebraic structures; it is unquestionably one of the most important concepts in algebra.

SECTION 28. HOMOMORPHISMS OF GROUPS. KERNELS

Definition. If G is a group with operation $*$, and H is a group with operation $\#$, then a mapping $\theta: G \to H$ is a *homomorphism* if

$$\theta(a*b) = \theta(a)\#\theta(b)$$

for all $a,b \in G$.

Every isomorphism is a homomorphism. But a homomorphism need not be one-to-one, and it need not be onto.

Example 28.1. For any positive integer n, define $\theta: \mathbb{Z} \to \mathbb{Z}_n$ by $\theta(a) = [a]$ for each $a \in \mathbb{Z}$. Then $\theta(a+b) = [a+b] = [a] \oplus [b] = \theta(a) \oplus \theta(b)$ for all $a, b \in \mathbb{Z}$, so that θ is a homomorphism. It is onto, but not one-to-one.

Example 28.2. Define $\theta: \mathbb{Z} \to \mathbb{Z}$ by $\theta(a) = 2a$ for each $a \in \mathbb{Z}$. Then $\theta(a+b) = 2(a+b) = 2a + 2b = \theta(a) + \theta(b)$ for all $a, b \in \mathbb{Z}$. Thus θ is a homomorphism. It is one-to-one, but not onto.

Example 28.3. Let R be a ring, and let $r \in R$. Define $\rho_r: R \to R$ by $\rho_r(a) = ar$ for each $a \in R$. Then ρ_r is a homomorphism from the additive group of R to itself: $\rho_r(a+b) = (a+b)r = ar + br = \rho_r(a) + \rho_r(b)$ for all $a, b \in R$. Notice that ρ_r is a homomorphism precisely because of the distributive law $(a+b)r = ar + br$.

Example 28.4. Let A and B be groups, and $A \times B$ their direct product (defined in Section 13). Then $\pi_1: A \times B \to A$ defined by $\pi_1((a,b)) = a$ is a homomorphism from $A \times B$ onto A:

$$\pi_1((a_1,b_1)(a_2,b_2)) = \pi_1((a_1a_2,b_1b_2)) = a_1a_2 = \pi_1((a_1,b_1))\pi_1((a_2,b_2)).$$

$\pi_2: A \times B \to B$ defined by $\pi_2((a,b)) = b$ is also a homomorphism (Problem 28.17).

We can see now that the basic assumption of Theorem 15.2 is that the mapping θ there is a homomorphism. Thus that theorem tells us that if $\theta: G \to H$ is a homomorphism, then

1. $\theta(e_G) = e_H$,
2. $\theta(a^{-1}) = \theta(a)^{-1}$ for each $a \in G$, and
3. $\theta(G)$, the image of θ, is a subgroup of H.

The subgroup $\theta(G)$ is called a *homomorphic image* of G. We shall see that nearly everything about a homomorphic image is determined by the domain of the homomorphism and the following subset of the domain.

Definition. If $\theta: G \to H$ is a homomorphism, then the *kernel* of θ is the set of all elements $a \in G$ such that $\theta(a) = e_H$. This set will be denoted by *Ker* θ.

The kernel of a homomorphism is always a subgroup of the domain. Before proving that, however, let us look at some examples.

Example 28.5. For the homomorphism $\theta: \mathbb{Z} \to \mathbb{Z}_n$ in Example 28.1, $a \in \operatorname{Ker}\theta$ iff $\theta(a) = [a] = [0]$. Therefore $\operatorname{Ker}\theta$ consists of the set of all integral multiples of n.

Example 28.6. In Example 28.2, $\operatorname{Ker}\theta=\{0\}$.

Example 28.7. If R is a commutative ring, $r\in R$, and ρ_r is the homomorphism in Example 28.3, then $\operatorname{Ker}\rho_r\neq\{0\}$ iff $r=0$ or r is a zero divisor in R. If R is a field and $r\neq 0$, then ρ_r will be an isomorphism of the additive group of R onto itself (Problem 28.1).

Example 28.8. In Example 28.4, $\operatorname{Ker}\pi_1=\{e_A\}\times B$.

Theorem 28.1 *If $\theta: G\to H$ is a homomorphism, then* $\operatorname{Ker}\theta$ *is a subgroup of G. Moreover, θ is one-to-one iff* $\operatorname{Ker}\theta=\{e_G\}$.

PROOF. Let the operations of G and H be $*$ and $\#$, respectively. To show that $\operatorname{Ker}\theta$ is a subgroup of G, we use Theorem 7.1. Theorem 15.2(a) shows that $e_G\in\operatorname{Ker}\theta$. If $a,b\in\operatorname{Ker}\theta$, then $\theta(a*b)=\theta(a)\#\theta(b)=e_H\#e_H=e_H$, so that $a*b\in\operatorname{Ker}\theta$. Theorem 15.2(b) shows that $\theta(a^{-1})=\theta(a)^{-1}$; therefore, if $a\in\operatorname{Ker}\theta$, then $\theta(a^{-1})=\theta(a)^{-1}=e_H^{-1}=e_H$, and $a^{-1}\in\operatorname{Ker}\theta$. Thus $\operatorname{Ker}\theta$ is a subgroup.

Since $e_G\in\operatorname{Ker}\theta$, it is clear that if θ is one-to-one then $\operatorname{Ker}\theta=\{e_G\}$. Assume, on the other hand, that $\operatorname{Ker}\theta=\{e_G\}$. If $a,b\in G$ and $\theta(a)=\theta(b)$, then $\theta(a)\#\theta(b)^{-1}=e_H$, $\theta(a)\#\theta(b^{-1})=e_H$, $\theta(a*b^{-1})=e_H$, $a*b^{-1}\in\operatorname{Ker}\theta=\{e_G\}$; hence $a*b^{-1}=e_G$ and $a=b$. This proves that θ is one-to-one. □

Kernels have one more property in common—they are all *normal*, in the sense of the next definition.

Definition. A subgroup N of a group G is a *normal subgroup* of G if $gng^{-1}\in N$ for all $n\in N$ and all $g\in G$. If N is a normal subgroup, we shall write $N\triangleleft G$.

If N is a subgroup of an Abelian group G, and $n\in N$ and $g\in G$, then $gng^{-1}=n\in N$; thus every subgroup of an Abelian group is a normal subgroup. The subgroup $\langle(1\ 2)\rangle=\{(1),(1\ 2)\}$ of S_3 is not normal, since, for example, $(1\ \ 2\ \ 3)(1\ \ 2)(1\ \ 2\ \ 3)^{-1}=(2\ \ 3)\notin\langle(1\ \ 2)\rangle$.

It is often helpful to realize that $gng^{-1}\in N$ for all $n\in N$ and $g\in G$ iff $g^{-1}ng\in N$ for all $n\in N$ and $g\in G$. Thus $N\triangleleft G$ iff $g^{-1}ng\in N$ for all $n\in N$ and $g\in G$ (Problem 28.13).

A large collection of examples of normal subgroups—in fact, all examples, as we shall see in the next section—is given by the following theorem.

Theorem 28.2. *If G and H are groups and $\theta: G\to H$ is a homomorphism, then* $\operatorname{Ker}\theta\triangleleft G$.

PROOF. We know from Theorem 28.1 that $\operatorname{Ker}\theta$ is a subgroup of G; thus it suffices to show that it is normal. Let $n \in \operatorname{Ker}\theta$ and $g \in G$. Then $\theta(n) = e$ so that $\theta(gng^{-1}) = \theta(g)\theta(n)\theta(g^{-1}) = \theta(g)e\theta(g)^{-1} = e$. Therefore $gng^{-1} \in \operatorname{Ker}\theta$, as required. □

A homomorphism that is one-to-one is sometimes called a *monomorphism*; a homomorphism that is onto is sometimes called an *epimorphism*. We shall not use either of these terms.

PROBLEMS

28.1. Prove the last statement in Example 28.7.

28.2. Let n and k denote positive integers, and define $\theta: \mathbb{Z} \to \mathbb{Z}_n$ by

$$\theta(a) = [ka] \qquad \text{for each } a \in \mathbb{Z}.$$

(a) Prove that θ is a homomorphism.
Determine $\operatorname{Ker}\theta$ in each of the following cases.
(b) $n=6$, $k=5$.
(c) $n=6$, $k=3$.
(d) General n and k.

28.3. Define $\theta: \mathbb{Z}_6 \to \mathbb{Z}_3$ by $\theta([a]_6) = [a]_3$ for each $[a]_6 \in \mathbb{Z}_6$.
(a) Prove that θ is well-defined.
(b) Prove that θ is a homomorphism.
(c) What is $\operatorname{Ker}\theta$?
(d) Show that $\alpha: \mathbb{Z}_3 \to \mathbb{Z}_6$ by $\alpha([a]_3) = [a]_6$ is not well-defined.
(e) For which pairs m, n is $\beta: \mathbb{Z}_m \to \mathbb{Z}_n$ by $\beta([a]_m) = [a]_n$ well-defined?

28.4. Let G denote the subgroup $\{1, -1, i, -i\}$ of complex numbers (operation multiplication). Define $\theta: \mathbb{Z} \to G$ by $\theta(n) = i^n$ for each $n \in \mathbb{Z}$. Verify that θ is a homomorphism and determine $\operatorname{Ker}\theta$.

28.5. If R is a ring, $r \in R$, and $\lambda_r: R \to R$ is defined by $\lambda_r(a) = ra$ for each $a \in R$, then λ_r is a homomorphism from the additive group of R to itself. Which property of R makes this true?

28.6. Prove that every homomorphic image of an Abelian group is Abelian.

28.7. Prove that every homomorphic image of a cyclic group is cyclic.

28.8. Prove that if $\theta: G \to H$ is a homomorphism, and $a \in G$, then $o(\theta(a)) | o(a)$.

28.9. Consider Example 28.3 for the special case $R = M(2, \mathbb{Z})$,

$$r = \begin{bmatrix} 1 & 0 \\ 0 & 0 \end{bmatrix}.$$

Determine $\operatorname{Ker}\rho_r$. [See Example 18.3 for $M(2, \mathbb{Z})$.]

28.10. Repeat Problem 28.9, except let

$$r = \begin{bmatrix} 1 & 1 \\ 0 & 0 \end{bmatrix}.$$

28.11. There is a unique homomorphism $\theta: \mathbb{Z}_6 \to S_3$ such that $\theta([1]) = (1\ 2\ 3)$. Determine $\theta([k])$ for each $[k] \in \mathbb{Z}_6$. Which elements are in $\operatorname{Ker}\theta$?

28.12. True or false: If $N \lhd G$, then $gng^{-1} = n$ for all $n \in N$ and $g \in G$. Justify your answer.

28.13. Prove that $N \triangleleft G$ iff $g^{-1}ng \in N$ for all $n \in N$ and $g \in G$. (In other words, prove that $gng^{-1} \in N$ for all $n \in N$ and $g \in G$ iff $g^{-1}ng \in N$ for all $n \in N$ and $g \in G$.)

28.14. By choosing a rectangular coordinate system, the points of a plane can be identified with the elements of $\mathbb{R} \times \mathbb{R}$. What is the geometric interpretation of $\pi_1 : \mathbb{R} \times \mathbb{R} \to \mathbb{R}$ defined by $\pi_1((a,b)) = a$ (as in Example 28.4)?

28.15. Determine all normal subgroups of the group of symmetries of a square. (See Example 14.2.)

28.16. Prove that if θ is a homomorphism from G onto H, and $N \triangleleft G$, then $\theta(N) \triangleleft H$.

28.17. Prove that $\pi_2 : A \times B \to B$, as defined in Example 28.4, is a homomorphism.

28.18. If A and B are groups, then $\{e\} \times B \triangleleft A \times B$. Give two different proofs of this, one using the definition of normal subgroup, and the other using Theorem 28.2 and Example 28.4.

28.19. Prove that if $\mathcal{C}$ denotes any collection of normal subgroups of a group G, then the intersection of all the groups in $\mathcal{C}$ is also a normal subgroup of G. (See Theorem 7.2.)

28.20. Prove that if N is a subgroup of G, then $N \triangleleft G$ iff $Ng = gN$ for each $g \in G$.

28.21. Prove that if N is a subgroup of G and $[G:N] = 2$, then $N \triangleleft G$. (*Suggestion*: Use Problem 28.20.)

28.22. Determine all normal subgroups of S_3. (See Problem 14.4.)

28.23. If $G = \langle a \rangle$, and $\theta : G \to H$ is a homomorphism, then θ is completely determined by $\theta(a)$. Explain.

28.24. There is only one homomorphism from $\mathbb{Z}_2$ to $\mathbb{Z}_3$. Why?

28.25. Verify that if G and H are any groups, and $\theta : G \to H$ is defined by $\theta(g) = e_H$ for each $g \in G$, then θ is a homomorphism.

28.26. Let $\mathbb{C}^{\#}$ denote the multiplicative group of nonzero complex numbers and $\mathbb{R}^p$ the multiplicative group of positive real numbers. Why is $\theta : \mathbb{C}^{\#} \to \mathbb{R}^p$, defined by $\theta(z) = |z|$, a homomorphism? What is $\operatorname{Ker}\theta$?

28.27. Define $\theta : \mathbb{Z} \times \mathbb{Z} \to \mathbb{Z}$ by $\theta((a,b)) = a + b$. Verify that θ is a homomorphism and determine $\operatorname{Ker}\theta$.

28.28. Determine all homomorphisms of $\mathbb{Z}$ onto $\mathbb{Z}$.

28.29. Let A be an Abelian group, written additively. If α and β are homomorphisms from A to A, define $\alpha + \beta$ and $\alpha\beta$ by

$$(\alpha + \beta)(a) = \alpha(a) + \beta(a)$$

and

$$(\alpha\beta)(a) = \alpha(\beta(a))$$

for all $a \in A$. Prove that $\alpha + \beta$ and $\alpha\beta$ are also homomorphisms from A to A, and with these operations the set of all homomorphisms from A to A is a ring. (A homomorphism from a group to itself is called an *endomorphism*. The ring in this problem is called the *ring of endomorphisms* of the additive group A. Compare Problem 18.6.)

28.30. Let R denote the ring of endomorphisms of $\mathbb{Z}$ (see Problem 28.29). Prove that $R \approx \mathbb{Z}$ (ring isomorphism). (*Suggestion*: For each $k \in \mathbb{Z}$, define $\alpha_k : \mathbb{Z} \to \mathbb{Z}$ by $\alpha_k(n) = kn$ for all $n \in \mathbb{Z}$. Verify that $\alpha_k \in R$ for each $k \in \mathbb{Z}$ and that $\theta : \mathbb{Z} \to R$ defined by $\theta(k) = \alpha_k$ is a ring isomorphism.)

SECTION 29. QUOTIENT GROUPS

It will not be obvious at the outset, but quotient groups, which we are about to introduce, are essentially the same as homomorphic images. The proof that they are essentially the same will come with Theorem 29.2 and the Fundamental Homomorphism Theorem (in Section 30).

Each group $\mathbb{Z}_n$ is constructed in a simple way from the group of integers. The set of all multiples of the integer n forms a subgroup, $\langle n \rangle$, of $\mathbb{Z}$, and the elements of $\mathbb{Z}_n$ are the right cosets of that subgroup (Section 13). Moreover, the operation $\oplus$ of $\mathbb{Z}_n$ depends in a natural way on the operation $+$ of the integers: $[a]\oplus[b]=[a+b]$. We shall now see how this idea can be used to construct new groups in much more general circumstances. Indeed, the following theorem shows that $\mathbb{Z}$ can be replaced by any group G, and $\langle n \rangle$ by any normal subgroup N of G. (Notice that $\langle n \rangle \lhd \mathbb{Z}$ because $\mathbb{Z}$ is Abelian.) It may help to review Section 13, especially Theorem 13.1 and Lemma 13.1, before reading this section. We shall continue to use juxtaposition to denote unspecified group operations.

Theorem 29.1. *Let N be a normal subgroup of G, and let G/N denote the set of all right cosets of N in G. For $Na \in G/N$ and $Nb \in G/N$, let $(Na)(Nb)=N(ab)$. With this operation G/N is a group, called the* quotient group (*or* factor group) *of G by N.*

Proof. We must first prove that the operation on G/N is well-defined:

$$\text{if} \quad Na_1 = Na_2 \quad \text{and} \quad Nb_1 = Nb_2, \quad \text{then} \quad N(a_1b_1) = N(a_2b_2).$$

From $Na_1=Na_2$ we have $a_1=n_1a_2$ for some $n_1 \in N$; from $Nb_1=Nb_2$ we have $b_1=n_2b_2$ for some $n_2 \in N$. Therefore, $a_1b_1=n_1a_2n_2b_2$. But $a_2n_2a_2^{-1}=n_3$ for some $n_3 \in N$ because $N \lhd G$. This gives $a_2n_2=n_3a_2$, so that $a_1b_1=n_1n_3a_2b_2$ with $n_1n_3 \in N$. This proves that $N(a_1b_1)=N(a_2b_2)$, as required.

If $a, b, c \in G$, then $Na(NbNc)=Na(N(bc))=N(a(bc))=N((ab)c)=N(ab)Nc=(NaNb)Nc$, so that the operation on G/N is associative. From $(Na)(Ne)=N(ae)=Na$ and $(Ne)(Na)=N(ea)=Na$, we see that Ne is an identity element for the operation. Finally, $(Na)(Na^{-1})=N(aa^{-1})=Ne$ and $(Na^{-1})(Na)=N(a^{-1}a)=Ne$, and so Na^{-1} is an inverse for Na. This proves that G/N is a group. □

To emphasize: the *elements* of G/N are *subsets* of G. If G is finite, then the order of G/N is the number of right cosets of N in G; this is $[G:N]$, the index of N in G (Section 14). From the remarks after Lagrange's

Theorem, $[G:N]=|G|/|N|$. Therefore,

$$|G/N| = |G|/|N|.$$

Example 29.1. Consider G/N for $G=S_3$ and $N=\langle(1\ \ 2\ \ 3)\rangle$. We have $\langle(1\ \ 2\ \ 3)\rangle=\{(1),\ (1\ \ 2\ \ 3),\ (1\ \ 3\ \ 2)\}\lhd S_3$, and $|S_3/\langle(1\ \ 2\ \ 3)\rangle|=6/3=2$. The elements of $S_3/\langle(1\ \ 2\ \ 3)\rangle$ are

$$\langle(1\ \ 2\ \ 3)\rangle = \{(1), (1\ \ 2\ \ 3), (1\ \ 3\ \ 2)\}$$

and

$$\langle(1\ \ 2\ \ 3)\rangle(1\ \ 2) = \{(1\ \ 2), (1\ \ 3), (2\ \ 3)\}.$$

The element (coset) $\langle(1\ \ 2\ \ 3)\rangle$ is the identity, and

$$\begin{aligned}\langle(1\ \ 2\ \ 3)\rangle(1\ \ 2)\cdot\langle(1\ \ 2\ \ 3)\rangle(1\ \ 2) &= \langle(1\ \ 2\ \ 3)\rangle(1\ \ 2)^2\\ &= \langle(1\ \ 2\ \ 3)\rangle(1)\\ &= \langle(1\ \ 2\ \ 3)\rangle.\end{aligned}$$

Example 29.2. Let G be the group of symmetries of a square (Example 8.3), and let $N=\langle\mu_3\rangle$. Then $N=\{\mu_1,\mu_3\}$ and $N\lhd G$. (In fact, if μ is any element of G, then $\mu\circ\mu_1\circ\mu^{-1}=\mu_1$ and $\mu\circ\mu_3\circ\mu^{-1}=\mu_3$.) The elements of G/N are $\{\mu_1,\mu_3\}$, $\{\mu_2,\mu_4\}$, $\{\mu_5,\mu_6\}$, and $\{\mu_7,\mu_8\}$. If we denote these cosets by $[\mu_1]$, $[\mu_2]$, $[\mu_5]$, and $[\mu_7]$, respectively, then the Cayley table for G/N is as shown in Table 29.1. For instance, $[\mu_2][\mu_7]=[\mu_5]$ because $[\mu_2][\mu_7]=N\mu_2N\mu_7=N(\mu_2\circ\mu_7)=N\mu_6$, and $N\mu_6=N\mu_5$.

Example 29.3. Let $G=\mathbb{Z}_{12}\times\mathbb{Z}_4$ and $N=\langle[3]\rangle\times\langle[2]\rangle$, with the notation interpreted in the following natural way. The first factor, $\langle[3]\rangle$, is a subgroup of $\mathbb{Z}_{12}$:

$$\langle[3]\rangle = \{[0],[3],[6],[9]\}\subseteq\mathbb{Z}_{12}.$$

The second factor is

$$\langle[2]\rangle = \{[0],[2]\}\subseteq\mathbb{Z}_4.$$

Table 29.1

	$[\mu_1]$	$[\mu_2]$	$[\mu_5]$	$[\mu_7]$
$[\mu_1]$	$[\mu_1]$	$[\mu_2]$	$[\mu_5]$	$[\mu_7]$
$[\mu_2]$	$[\mu_2]$	$[\mu_1]$	$[\mu_7]$	$[\mu_5]$
$[\mu_5]$	$[\mu_5]$	$[\mu_7]$	$[\mu_1]$	$[\mu_2]$
$[\mu_7]$	$[\mu_7]$	$[\mu_5]$	$[\mu_2]$	$[\mu_1]$

Because G is Abelian, $N \lhd G$. And $|G/N| = 48/8 = 6$. A complete list of right coset representatives of N in G is

$$([0],[0]), ([0],[1]), ([1],[0]), ([1],[1]), ([2],[0]), ([2],[1]).$$

Denote the corresponding cosets of N by

$$N_{0,0}, N_{0,1}, N_{1,0}, N_{1,1}, N_{2,0}, N_{2,1},$$

respectively. Then, for example,

$$\begin{aligned} N_{2,1}N_{1,0} &= N([2],[1])N([1],[0]) \\ &= N([2]\oplus[1],[1]\oplus[0]) \\ &= N([3],[1]) \\ &= N([0],[1]) \\ &= N_{0,1}. \end{aligned}$$

Problem 29.4 asks for the Cayley table of this quotient group.

It can be proved that if N is not normal, then the operation on G/N from Theorem 29.1 will not be well-defined (Problem 29.11). Thus the concepts of quotient group and normal subgroup are inseparable. Also, a subgroup N of a group G is normal iff the right coset of N determined by each element is the same as the left coset of N determined by that same element (Problem 28.20). Thus in working with cosets of a normal subgroup N, it is immaterial whether we use Na or aN; we shall always use Na.

By Theorem 28.2, a kernel of a homomorphism is necessarily a normal subgroup. The next theorem shows that, conversely, every normal subgroup is a kernel of some homomorphism.

Theorem 29.2. *If G is a group and $N \lhd G$, then the mapping $\eta: G \to G/N$ defined by*

$$\eta(a) = Na \qquad \textit{for each} \qquad a \in G$$

is a homomorphism of G onto G/N, and $\operatorname{Ker}\eta = N$.

PROOF. The mapping η is clearly well-defined and onto. Also, if $a, b \in G$, then $\eta(ab) = N(ab) = (Na)(Nb) = \eta(a)\eta(b)$, and so η is a homomorphism. Finally, if $a \in G$, then $a \in \operatorname{Ker}\eta$ iff $\eta(a) = Na = Ne$, because Ne is the identity element of G/N. Thus $a \in \operatorname{Ker}\eta$ iff $a \in N$, and so $\operatorname{Ker}\eta = N$. $\square$

The mapping η in Theorem 29.2 is called the *natural homomorphism* of G onto G/N.

Example 29.4. If n is a positive integer, then $\mathbb{Z}/\langle n\rangle=\mathbb{Z}_n$. The natural homomorphism $\eta:\mathbb{Z}\to\mathbb{Z}/\langle n\rangle$ is given by $\eta(a)=\langle n\rangle+a=[a]$. And $\operatorname{Ker}\eta=\langle n\rangle$. This η is the same as the mapping θ in Example 28.1, whose kernel we computed in Example 28.5.

PROBLEMS

29.1. Determine the order of each of the following quotient groups.
(a) $\mathbb{Z}_8/\langle[4]\rangle$ (b) $\mathbb{Z}_8/\langle[2]\rangle$
(c) $\mathbb{Z}_8/\langle[3]\rangle$ (d) $\langle 3\rangle/\langle 6\rangle$, where $\langle 6\rangle\subseteq\langle 3\rangle\subseteq\mathbb{Z}$.

29.2. Determine the order of $\mathbb{Z}_{12}\times\mathbb{Z}_4/\langle([3],[2])\rangle$. Explain the difference between this quotient group and the one in Example 29.3.

29.3. Construct the Cayley table for $\mathbb{Z}_{12}/\langle[4]\rangle$. [*Suggestion*: Use $[[k]]$ to denote the coset of $[k]$.]

29.4. Construct the Cayley table for the group in Example 29.3.

29.5. Prove that every quotient group of an Abelian group is Abelian.

29.6. Prove that every quotient group of a cyclic group is cyclic.

29.7. Prove that if $N\lhd G$ and $a\in G$, then $o(Na)|o(a)$. (Here $o(Na)$ denotes the order of Na as an element of G/N.) How is this problem related to Problem 28.8?

29.8 Prove that every element of $\mathbb{Q}/\mathbb{Z}$ has finite order.

29.9. The elements of finite order in an Abelian group form a subgroup (Problem 12.9). Show that the subgroup of elements of finite order in $\mathbb{R}/\mathbb{Z}$ is $\mathbb{Q}/\mathbb{Z}$. (See Problem 29.8.)

29.10. Prove that G/N is Abelian iff $aba^{-1}b^{-1}\in N$ for all $a,b\in G$.

29.11. Prove that if N is a subgroup of G, and the operation $(Na)(Nb)=N(ab)$ is well-defined on the set G/N of all right cosets of N in G, then $N\lhd G$.

29.12. Verify that $\alpha:\mathbb{R}\to\mathbb{C}$ defined by $\alpha(x)=\cos(2\pi x)+i\sin(2\pi x)$ is a homomorphism of the additive group of $\mathbb{R}$ onto the group of all complex numbers of absolute value 1. What is $\operatorname{Ker}\alpha$? Interpret α geometrically. (See Problem 27.11.)

SECTION 30. THE FUNDAMENTAL HOMOMORPHISM THEOREM

The natural homomorphism $\eta:G\to G/N$ shows that each quotient group of a group G is a homomorphic image of G (Theorem 29.2). The next theorem shows that the converse is also true: each homomorphic image of G is (isomorphic to) a quotient group of G. Thus the claim made at the beginning of Section 29 is justified: quotient groups are essentially the same as homomorphic images.

Fundamental Homomorphism Theorem. *Let G and H be groups, and let $\theta:G\to H$ be a homomorphism from G onto H with $\operatorname{Ker}\theta=K$. Then the* *mapping* $\phi: G/K \to H$ *defined by*

$$\phi(Ka) = \theta(a) \qquad \text{for each} \qquad Ka \in G/K$$

is an isomorphism of G/K *onto* H. *Therefore*

$$G/K \approx H.$$

PROOF. We must first verify that ϕ is well-defined. If $Ka_1 = Ka_2$, then $ka_1 = a_2$ for some $k \in K = \operatorname{Ker}\theta$, and so $\theta(ka_1) = \theta(a_2)$. But $\theta(ka_1) = \theta(k)\theta(a_1) = e\theta(a_1) = \theta(a_1)$, and so $\theta(a_1) = \theta(a_2)$. Therefore, $\theta(a)$ is determined solely by the coset of K to which a belongs, and so ϕ is well-defined.

To prove that ϕ preserves the operations, assume that $Ka \in G/K$ and $Kb \in G/K$. Then $\phi((Ka)(Kb)) = \phi(K(ab)) = \theta(ab) = \theta(a)\theta(b) = \phi(Ka)\phi(Kb)$, as required. Clearly ϕ is onto, because θ is onto. It remains only to prove that ϕ is one-to-one; or, equivalently, by Theorem 28.1, that $\operatorname{Ker}\phi$ contains only the identity element, Ke, of G/K. This is true because if $Ka \in Ker\ \phi$, then $\theta(a) = \phi(Ka) = e$, and therefore $a \in \operatorname{Ker}\theta = K$, or $Ka = Ke$. □

If a homomorphism $\theta: G \to H$ is not onto, then H should be replaced by $\theta(G)$ in the last two sentences of the theorem. Then the last statement of the theorem becomes $G/K \approx \theta(G)$. In any case, with θ, ϕ, and K as in the theorem, and $\eta: G \to G/K$ the natural homomorphism, it can be verified that $\phi \circ \eta = \theta$. Schematically, the two ways (θ and $\phi \circ \eta$) of getting from G to H in Figure 30.1 give the same result for every element of G (Problem 30.1).

Example 30.1. Let G denote the group of all rotations of the plane about a fixed point p (Example 5.7). For each real number r, let $\theta(r)$ in G denote clockwise rotation through r radians. Then θ is a homomorphism of $\mathbb{R}$ onto G, and

$$\operatorname{Ker}\theta = \langle 2\pi \rangle = \{2k\pi \mid k \in \mathbb{Z}\}.$$

The Fundamental Homomorphism Theorem shows that $\mathbb{R}/\langle 2\pi \rangle \approx G$.

Figure 30.1

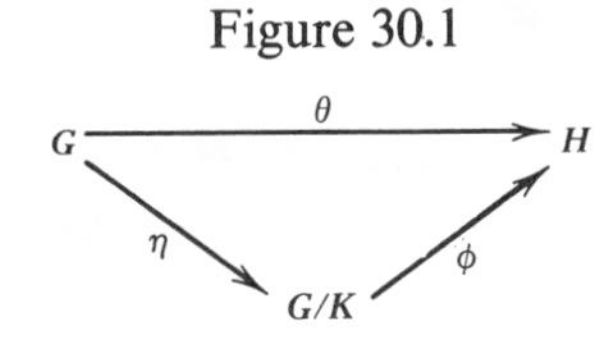

Example 30.2. For $a \in \mathbb{Z}$, let $[a]_{12}$ and $[a]_4$ denote the congruence classes determined by a in $\mathbb{Z}_{12}$ and $\mathbb{Z}_4$, respectively. Define

$$\theta : \mathbb{Z}_{12} \to \mathbb{Z}_4 \text{ by } \theta([a]_{12}) = [a]_4.$$

θ is well-defined because if $[a]_{12} = [b]_{12}$, then $12|(a-b)$, and therefore $4|(a-b)$ and $[a]_4 = [b]_4$. Also, θ is a homomorphism:

$$\begin{aligned}\theta([a]_{12} \oplus [b]_{12}) &= \theta([a+b]_{12}) = [a+b]_4 = [a]_4 \oplus [b]_4 \\ &= \theta([a]_{12}) \oplus \theta([b]_{12}).\end{aligned}$$

Clearly θ is onto, and

$$\operatorname{Ker}\theta = \{[0]_{12}, [4]_{12}, [8]_{12}\} = \langle [4]_{12} \rangle.$$

Therefore, by the Fundamental Homomorphism Theorem,

$$\frac{\mathbb{Z}_{12}}{\langle [4]_{12} \rangle} \approx \mathbb{Z}_4.$$

[Notice that from $|\operatorname{Ker}\theta| = 3$ alone we could deduce that $|\mathbb{Z}_{12}/(\operatorname{Ker}\theta)| = 4$; but this would still leave the possibility that $\mathbb{Z}_{12}/(\operatorname{Ker}\theta) \approx \mathbb{Z}_2 \times \mathbb{Z}_2$.]

Example 30.3. Let G be the group of symmetries of a square (Example 8.3); we shall determine all homomorphic images of G. By the Fundamental Homomorphism Theorem, this is equivalent to determining all quotient groups of G. To do that, we must determine all normal subgroups of G. There are six in all (Problem 28.15); we can ignore two of the three of order 4 here, and work with these:

$$G, \qquad \langle \mu_2 \rangle, \qquad \langle \mu_3 \rangle, \qquad \langle \mu_1 \rangle.$$

These are of order 8, 4, 2, 1, respectively. Therefore,

$$\left|\frac{G}{G}\right| = 1, \qquad \left|\frac{G}{\langle \mu_2 \rangle}\right| = 2, \qquad \left|\frac{G}{\langle \mu_3 \rangle}\right| = 4, \qquad \left|\frac{G}{\langle \mu_1 \rangle}\right| = 8.$$

Any group of order 1 is isomorphic to $\mathbb{Z}_1$, and any group of order 2 is isomorphic to $\mathbb{Z}_2$. Therefore,

$$\frac{G}{G} \approx \mathbb{Z}_1, \qquad \frac{G}{\langle \mu_2 \rangle} \approx \mathbb{Z}_2.$$

There are two isomorphism classes of groups of order 4; namely, those determined by $\mathbb{Z}_4$ and $\mathbb{Z}_2 \times \mathbb{Z}_2$ (Section 16). Problem 30.2 suggests how to show that $G/\langle \mu_3 \rangle$ is isomorphic to the latter. Finally, $G/\langle \mu_1 \rangle \approx G$ (see Problem 30.3).

Summarizing, we see that any homomorphic image of this group G is isomorphic either to $\mathbb{Z}_1$, $\mathbb{Z}_2$, $\mathbb{Z}_2 \times \mathbb{Z}_2$, or G.

We shall now see how the ideas in this chapter can be used to construct many groups from smaller component groups. *The groups in the remainder of this section are assumed to be finite* (even though some of the statements are also true for infinite groups).

A group G is said to be an *extension* of a group A by a group B if G contains a subgroup N such that

$$A \approx N \lhd G \quad \text{and} \quad G/N \approx B.$$

When this is so, we can think of G as being built up from component groups A and B. Because $|A|=|N|$ and $|G/N|=|G|/|N|=|B|$, we must have $|G|=|A|\cdot|B|$.

Example 30.4. We saw in Example 29.1 that S_3 has a normal subgroup of order 3, namely $\langle(1\ \ 2\ \ 3)\rangle$. Also, $|S_3/\langle(1\ \ 2\ \ 3)\rangle|=2$. Every group of prime order p is isomorphic to $\mathbb{Z}_p$ (Theorem 16.2). Therefore,

$$\mathbb{Z}_3 \approx \langle(1\ \ 2\ \ 3)\rangle \lhd S_3 \quad \text{and} \quad \frac{S_3}{\langle(1\ \ 2\ \ 3)\rangle} \approx \mathbb{Z}_2.$$

Thus S_3 is an extension of $\mathbb{Z}_3$ by $\mathbb{Z}_2$.

Example 30.5. There is always at least one extension of A by B, whatever A and B; namely $A \times B$. This is because

$$A \approx A \times \{e\} \lhd A \times B \quad \text{and} \quad \frac{A \times B}{A \times \{e\}} \approx B$$

(Problem 30.4).

If we apply Example 30.5 with $A=\mathbb{Z}_3$ and $B=\mathbb{Z}_2$, we see that $\mathbb{Z}_3 \times \mathbb{Z}_2$ is an extension of $\mathbb{Z}_3$ by $\mathbb{Z}_2$. But Example 30.4 showed that S_3 is also an extension of $\mathbb{Z}_3$ by $\mathbb{Z}_2$, and $\mathbb{Z}_3 \times \mathbb{Z}_2 \not\approx S_3$ ($\mathbb{Z}_3 \times \mathbb{Z}_2$ is Abelian but S_3 is non-Abelian). Thus there are nonisomorphic extensions of $\mathbb{Z}_3$ by $\mathbb{Z}_2$. This leads to the following general problem: *Given groups A and B, determine all (isomorphism classes of) extensions of A by B.* This problem was solved in part by Otto Hölder in the 1890s, and more completely by Otto Schreier in the 1920s. Their results show how to construct a collection of groups (Cayley tables if you like), from A and B, with the property that any extension of A by B will be isomorphic to some group in that collection. The details are complicated and will not be given here. The important point is that, in theory, there is a procedure for determining all extensions of A by B.

Some groups can be constructed from smaller groups by extension, and some cannot. Each group $\mathbb{Z}_p$ with p a prime, for example, has only the two obvious (normal) subgroups of orders 1 and p; therefore, $\mathbb{Z}_p$ can be thought of as an extension in only a trivial way—either as $\mathbb{Z}_1$ by $\mathbb{Z}_p$, or as $\mathbb{Z}_p$ by $\mathbb{Z}_1$. A group G having no normal subgroup other than $\{e\}$ and G itself is called a *simple* group. An Abelian simple group must be isomorphic to $\mathbb{Z}_p$ for some prime p (Problem 30.5).

All finite groups can be constructed from simple groups by forming repeated extensions. This can be seen as follows. Assume that G is finite. Let G_1 denote a maximal proper normal subgroup of G: $G_1 \neq G$ (proper), $G_1 \triangleleft G$ (normal), and G has no normal subgroup strictly between G_1 and G (maximal normal). It can be shown that G/G_1 is a simple group (Problem 30.11). Now let G_2 denote a maximal proper normal subgroup of G_1. Then G_1/G_2 is simple also (Problem 30.11 again). Continuing in this way, with G denoted by G_0, we eventually arrive at a series of subgroups

$$G = G_0 \triangleright G_1 \triangleright G_2 \triangleright \cdots \triangleright G_{k-2} \triangleright G_{k-1} \triangleright G_k = \{e\}$$

such that each factor group G_{i-1}/G_i is simple ($1 \leq i \leq k$). Such a series is called a *composition series* of G. We see from this that G can be constructed by a succession of extensions by simple groups: extend G_{k-1} by G_{k-2}/G_{k-1} to get G_{k-2}; extend G_{k-2} by G_{k-3}/G_{k-2} to get G_{k-3}; and so on until we arrive at G. Moreover, it can be proved that, although G may have more than one composition series, the factor groups arising from any two composition series can be paired (after rearrangement, perhaps) in such a way that corresponding factor groups are isomorphic (*Jordan-Hölder Theorem*).

These remarks about composition series show that, with the problem of how to determine extensions effectively solved, the place to direct attention is the class of simple groups. If all finite simple groups could be determined, then, in theory, all finite groups would be determined, just by constructing all successive extensions by simple groups. In recent years the problem of determining all finite simple groups has received more attention from group theorists than any other single problem. We shall close our discussion with some general remarks about this problem.

We have already observed that groups of prime order are simple; they are the only Abelian simple groups. These groups of prime order are simple in the technical sense, defined above, and they are also simple in the sense of being uncomplicated. Other simple groups (technical sense) can be quite complicated. The smallest non-Abelian simple group is of order 60, and is isomorphic to a subgroup of index 2 in S_5. In general, each symmetric group S_n ($n \geq 5$) contains a subgroup A_n, called the *alternating group* of degree n, such that $[S_n : A_n] = 2$ and A_n is simple. (A_n consists of all elements of S_n that can be represented as products of an even number of 2-cycles.) Thus there is a simple group of order $\frac{1}{2}(n!)$ for each $n \geq 5$. These groups account for 5 of the 56 non-Abelian simple groups of order less than 1,000,000.

The most comprehensive easy-to-state theorem about non-Abelian simple groups is a celebrated theorem proved in the 1960s by the American group theorists Walter Feit and John Thompson: *There are no non-Abelian simple groups of odd order* [1]. The fact that the paper proving this theorem is over 250 pages long suggests how complicated this area of group theory has become. For a survey of what was known about finite simple groups by the mid-1970s, see either of the survey papers [2] or [3].

PROBLEMS

30.1. With θ, ϕ, and η as in the paragraph following the proof of the Fundamental Homomorphism Theorem, prove that $\phi \circ \eta = \theta$.

30.2. Verify the isomorphism $G/\langle \mu_3 \rangle \approx \mathbb{Z}_2 \times \mathbb{Z}_2$, from Example 30.3. (It is sufficient to check that the factor group on the left has no element of order 4. Why?)

30.3. Use the Fundamental Homomorphism Theorem to prove that if G is any group with identity e, then $G/\{e\} \approx G$.

30.4. Verify the isomorphisms in Example 30.5. (*Suggestion*: Use the Fundamental Homomorphism Theorem and the homomorphism $\pi_2 : A \times B \to B$ from Problem 28.17.)

30.5. Prove that if G is a simple Abelian group, then $G \approx \mathbb{Z}_p$ for some prime p.

30.6. Verify that both $\mathbb{Z}_4$ and $\mathbb{Z}_2 \times \mathbb{Z}_2$ are extensions of $\mathbb{Z}_2$ by $\mathbb{Z}_2$.

30.7. Prove that $\mathbb{Z}_{18}/\langle [3] \rangle \approx \mathbb{Z}_3$.

30.8. Prove that if k and n are positive integers and k is a divisor of n, then $\mathbb{Z}_n/\langle [k] \rangle \approx \mathbb{Z}_k$.

30.9. Determine all homomorphic images of each of the following groups.
(a) $\mathbb{Z}_6$ (b) $\mathbb{Z}_5$ (c) $\mathbb{Z}_n$ (d) S_3

30.10. Prove that if θ is a homomorphism of G onto H, $B \lhd H$, and $A = \{ g \in G | \theta(g) \in B \}$, then $A \lhd G$.

30.11. Prove that if $N \lhd G$, $N \neq G$, and G has no normal subgroup strictly between N and G, then G/N is simple. (*Suggestion*: Use the natural homomorphism $\eta : G \to G/N$ and Problem 30.10.)

30.12. If G is a simple group, then any homomorphic image of G is either isomorphic to G or of order one. Why?

30.13. Let A denote the group of all mappings $\alpha_{a,b} : \mathbb{R} \to \mathbb{R}$, defined as in Example 5.8. Let $B = \{ \alpha_{1,b} | b \in \mathbb{R} \}$, and let $\mathbb{R}^{\#}$ denote the multiplicative group of $\mathbb{R}$. Define $\theta : A \to \mathbb{R}^{\#}$ by $\theta(\alpha_{a,b}) = a$ for each $\alpha_{a,b} \in A$.
(a) Verify that θ is a homomorphism with $\operatorname{Ker} \theta = B$.
(b) Explain why $B \lhd A$ and $A/B \approx \mathbb{R}^{\#}$.
(c) Explain why A is an extension of $\mathbb{R}$ (operation addition) by $\mathbb{R}^{\#}$ (operation multiplication).
(d) Give an example of a group that is an extension of $\mathbb{R}$ (operation addition) by $\mathbb{R}^{\#}$ (operation multiplication) and is not isomorphic to A.

30.14. (First Isomorphism Theorem) Assume that H and K are subgroups of a group G and that $K \lhd G$.
(a) Prove that $HK = \{ hk | h \in H \text{ and } k \in K \}$ is a subgroup of G, and that $K \lhd HK$. (Compare Problem 30.16.)

(b) Define $\theta: H \to HK/K$ by $\theta(h) = Kh$ for each $h \in H$. Verify that θ is a homomorphism of H onto HK/K.

(c) Verify that $\operatorname{Ker}\theta = H \cap K$. Now explain why

$$\frac{H}{H \cap K} \approx \frac{HK}{K}.$$

30.15. (Second Isomorphism Theorem) Assume that H and K are subgroups of a group G, and that $K \lhd H$, $K \lhd G$, and $H \lhd G$. Prove that $H/K \lhd G/K$ and that $(G/K)/(H/K) \approx G/H$. [*Suggestion*: Consider $\theta: G/K \to G/H$ defined by $\theta(Kg) = Hg$. Verify that θ is well-defined and is a homomorphism, and apply the Fundamental Homomorphism Theorem.]

30.16. Give an example to show that if A and B are subgroups of a group G, then $AB = \{ab \mid a \in A \text{ and } b \in B\}$ need not be a subgroup of G. [*Suggestion*: Try $G = S_3$. If $A \lhd G$ or $B \lhd G$, then AB is a subgroup; see Problem 30.14(a).]

NOTES ON CHAPTER VII

1. Feit, W., and J. G. Thompson, *Solvability of groups of odd order*, Pacific Journal of Mathematics **13** (1963) 755–1029.

2. Hurley, J. F., and A. Rudvalis, *Finite simple groups*, American Mathematical Monthly **84** (1977) 693–714.

3. Gallian, J. A., *The search for finite simple groups*, Mathematics Magazine **49** (1976) 163–179.

CHAPTER VIII

Applications of Permutation Groups

This chapter gives an introduction to the use of permutation groups for the solution of counting problems. This is one of the applications of modern algebra to the increasingly important branch of mathematics known as *combinatorics* or *combinatorial analysis*. Section 31 introduces some basic terminology, which is then used in Section 32 in the proof and application of Burnside's Counting Theorem. In Section 33 the ideas in Section 31 are applied to prove a theorem within group theory itself.

SECTION 31. GROUPS ACTING ON SETS

We begin by recalling the group of symmetries of a square, from Example 8.3. For convenience, here are the definitions of the elements and also the accompanying figure, Figure 31.1.

Group of symmetries of a square (Figure 31.1)

$$\begin{aligned}
\mu_1 &= \text{identity permutation} \\
\mu_2 &= \text{rotation } 90^\circ \text{ clockwise around } p \\
\mu_3 &= \text{rotation } 180^\circ \text{ clockwise around } p \\
\mu_4 &= \text{rotation } 270^\circ \text{ clockwise around } p \\
\mu_5 &= \text{reflection through } H \\
\mu_6 &= \text{reflection through } V \\
\mu_7 &= \text{reflection through } D_1 \\
\mu_8 &= \text{reflection through } D_2
\end{aligned}$$

Figure 31.1

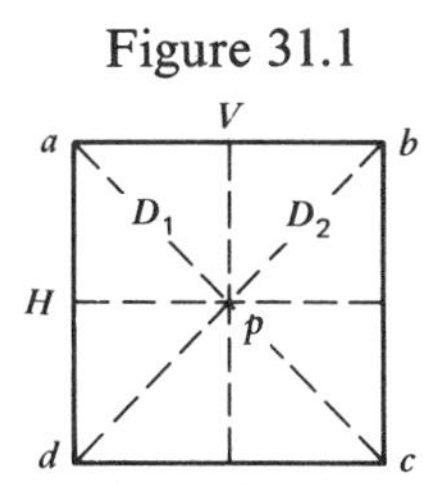

Denote the group by G. The elements of G are isometries of the plane; therefore, they are permutations of the set of all points of the plane. But these isometries also induce permutations of the set of vertices of the square in a natural way. If we assign to each isometry the corresponding permutation of $\{a,b,c,d\}$, then we have a mapping from G to $\mathrm{Sym}\{a,b,c,d\}$. The mapping with the induced permutations written in cycle notation is

$$\begin{array}{ll} \mu_1 \mapsto (a)(b)(c)(d) & \mu_5 \mapsto (a\ \ d)(b\ \ c) \\ \mu_2 \mapsto (a\ \ b\ \ c\ \ d) & \mu_6 \mapsto (a\ \ b)(c\ \ d) \\ \mu_3 \mapsto (a\ \ c)(b\ \ d) & \mu_7 \mapsto (a)(c)(b\ \ d) \\ \mu_4 \mapsto (a\ \ d\ \ c\ \ b) & \mu_8 \mapsto (a\ \ c)(b)(d) \end{array}$$

Because the operations on both G and $\mathrm{Sym}\{a,b,c,d\}$ are composition, this mapping is a homomorphism.

The isometries in G also induce permutations of the set of diagonals $\{D_1,D_2\}$ in a natural way. For instance, $\mu_3(D_1)=D_1$ and $\mu_3(D_2)=D_2$ (μ_3 interchanges the ends of each of the diagonals, but that is not important here). In this case we can assign to each isometry the corresponding permutation of $\{D_1,D_2\}$. This gives a mapping from G to $\mathrm{Sym}\{D_1,D_2\}$. The induced permutations are

$$\begin{array}{ll} \mu_1 \mapsto (D_1)(D_2) & \mu_5 \mapsto (D_1\ \ D_2) \\ \mu_2 \mapsto (D_1\ \ D_2) & \mu_6 \mapsto (D_1\ \ D_2) \\ \mu_3 \mapsto (D_1)(D_2) & \mu_7 \mapsto (D_1)(D_2) \\ \mu_4 \mapsto (D_1\ \ D_2) & \mu_8 \mapsto (D_1)(D_2) \end{array}$$

Again, the mapping is a homomorphism.

These two examples are special cases of the following generalization of the notion of a permutation group on a set. (Here, and throughout this chapter, the notation $\beta\circ\alpha$ for composition will be shortened to $\beta\alpha$.)

Definition. A group G *acts on a set* S if to each $g\in G$ there is assigned a permutation π_g of S in such a way that $\pi_{ab}=\pi_a\pi_b$ for all $a,b\in G$.

In other words, G acts on S if there is a homomorphism $g \mapsto \pi_g$ of G into Sym(S). [If $g \in G$, so that $\pi_g \in \mathrm{Sym}(S)$, and $s \in S$, then $\pi_g(s) \in S$; we are writing π_g rather than $\pi(g)$ because $\pi_g(s)$ seems preferable to $\pi(g)(s)$.] Any subgroup of Sym(S) acts on S. The idea of a group's acting on S is more general than that of a permutation group on S because, in the former case, unequal group elements can give rise to equal permutations; that is, the mapping $g \mapsto \pi_g$ need not be one-to-one. If the mapping $g \mapsto \pi_g$ is one-to-one, then G is said to act *faithfully* on S. In the first example above, G acts faithfully—unequal group elements give rise to unequal permutations of $\{a,b,c,d\}$. In the second example, G does not act faithfully—considered only as permutations of $\{D_1, D_2\}$, $\mu_1 = \mu_3 = \mu_7 = \mu_8$ and $\mu_2 = \mu_4 = \mu_5 = \mu_6$.

Because $g \mapsto \pi_g$ is a homomorphism, necessarily $\pi_e = \iota_S$ and $\pi_{g^{-1}} = \pi_g^{-1}$ for all $g \in G$. This is used in the proof of the following theorem.

Theorem 31.1. *Assume that G is a group acting on a set S. Define the relation $\sim$ on S by*

$$s_1 \sim s_2 \quad \text{iff} \quad \pi_g(s_1) = s_2 \quad \text{for some } g \in G.$$

Then $\sim$ is an equivalence relation on S.

PROOF. *Reflexive*: if $s \in S$, then $\pi_e(s) = \iota_S(s) = s$ so that $s \sim s$. *Symmetric*: if $s_1 \sim s_2$, with $\pi_g(s_1) = s_2$, then $\pi_{g^{-1}}(s_2) = \pi_g^{-1}(s_2) = s_1$ so that $s_2 \sim s_1$. *Transitive*: if $s_1 \sim s_2$ and $s_2 \sim s_3$, with $\pi_g(s_1) = s_2$ and $\pi_h(s_2) = s_3$, then $\pi_{hg}(s_1) = \pi_h \pi_g(s_1) = \pi_h(s_2) = s_3$ so that $s_1 \sim s_3$. □

The equivalence classes relative to $\sim$ in Theorem 31.1 are called *orbits*. The term *G-orbit* can be used if it is necessary to specify the group. Also, $\sim_G$ can be written in place of $\sim$.

Example 31.1. Let G denote the group of all rotations of a plane about a point p in the plane (Example 5.7). Then G acts on the set of points in the plane, and the orbits are the circles with centers at p.

Example 31.2. If S is any nonempty set, then Sym(S) acts on S, and there is only one orbit, namely S.

Example 31.3. The subgroup $\{\mu_1, \mu_3\}$ of the symmetry group of a square acts on $\{a,b,c,d\}$ (Figure 31.1). The orbits are $\{a,c\}$ and $\{b,d\}$.

For G acting on S and $s \in S$, let Orb(s) denote the orbit of s:

$$\mathrm{Orb}(s) = \{\pi_g(s) \mid g \in G\}.$$

Then $|\mathrm{Orb}(s)|$ is the number of elements in the orbit of s. Let G_s denote the set of all $g \in G$ such that $\pi_g(s) = s$. Problem 31.5 asks for verification that G_s is a subgroup of G.

Example 31.4. The group $G = \{(1), (1\ \ 2), (3\ \ 4), (1\ \ 2)(3\ \ 4)\}$ acts on $\{1,2,3,4\}$. In this case

$$\mathrm{Orb}(1) = \mathrm{Orb}(2) = \{1,2\}, \qquad \mathrm{Orb}(3) = \mathrm{Orb}(4) = \{3,4\},$$
$$G_1 = G_2 = \{(1), (3\ \ 4)\}, \qquad G_3 = G_4 = \{(1), (1\ \ 2)\}.$$

The following theorem is the key to the connection between group theory and combinatorics.

Theorem 31.2. *If a finite group G acts on a set S, and $s \in S$, then*

$$|\mathrm{Orb}(s)| = [G:G_s] = \frac{|G|}{|G_s|}.$$

PROOF. To compute $|\mathrm{Orb}(s)|$, we must compute the number of *distinct* elements $\pi_g(s)$ with $g \in G$. Thus to prove that $|\mathrm{Orb}(s)| = [G:G_s]$, it suffices to prove that

$$\pi_g(s) = \pi_h(s) \quad \text{iff} \quad gG_s = hG_s \quad (\text{for } g,h \in G), \tag{31.1}$$

for this will show that $\pi_g(s) \neq \pi_h(s)$ iff g and h are in distinct left cosets of G_s in G, from which it will follow that the number of distinct elements $\pi_g(s)$ is the same as the number of distinct left cosets of G_s in G, which is $[G:G_s]$. The relation in (31.1) holds because of the following sequence of equivalent statements (for $g,h \in G$):

$$\pi_g(s) = \pi_h(s) \quad \text{iff} \quad (\pi_h^{-1}\pi_g)(s) = s \quad \text{iff} \quad \pi_{h^{-1}}\pi_g(s) = s$$
$$\text{iff} \quad \pi_{h^{-1}g}(s) = s \quad \text{iff} \quad h^{-1}g \in G_s \quad \text{iff} \quad gG_s = hG_s.$$

The equality $[G:G_s] = |G|/|G_s|$ in the theorem follows from Lagrange's Theorem. □

PROBLEMS

31.1. Let E_1, E_2, E_3, and E_4 denote, respectively, the edges ab, bc, cd, and da of the square in Figure 31.1. Write the permutation induced on $\{E_1, E_2, E_3, E_4\}$ by each isometry (symmetry) of the square. [*Example*: $\mu_5 \mapsto (E_1\ \ E_3)$.] Does the symmetry group of the square act faithfully on $\{E_1, E_2, E_3, E_4\}$?

31.2. Determine the orbits for each of the following subgroups of the symmetry group of the square, acting on $\{a,b,c,d\}$. (Compare Example 31.3.)
(a) $\langle \mu_2 \rangle$ (b) $\langle \mu_5 \rangle$ (c) $\langle \mu_7 \rangle$

31.3. (a) to (c). Repeat Problem 31.2, except replace $\{a,b,c,d\}$ by the set of diagonals, $\{D_1,D_2\}$.

31.4. The paragraph preceding Theorem 31.1 states that because $g \mapsto \pi_g$ is a homomorphism if G acts on a set, necessarily $\pi_e = \iota_S$ and $\pi_{g^{-1}} = \pi_g^{-1}$. Give a precise reference to justify this. (See Section 28.)

31.5. Prove that if G acts on S by $g \mapsto \pi_g$, $s \in S$, and

$$G_s = \{ g \in G \mid \pi_g(s) = s \},$$

then G_s is a subgroup of G.

31.6. The group $G = \langle (1 \quad 2 \quad 3)(4 \quad 5) \rangle$ is of order 6 and acts on $\{1,2,3,4,5\}$.
(a) Determine Orb(k) for $1 \le k \le 5$.
(b) Determine G_k for $1 \le k \le 5$.
(c) Use parts (a) and (b) to verify that $|\text{Orb}(k)| = |G|/|G_k|$ for $1 \le k \le 5$. (Compare Theorem 31.2.)

31.7. Assume that G is a group, and for each $a \in G$ define $\pi_a : G \to G$ by $\pi_a(x) = axa^{-1}$ for each $x \in G$.
(a) Verify that with this definition G acts on G. (The orbits in this case are known as the *conjugate classes* of G; elements $s,t \in G$ are *conjugate* if $t = asa^{-1}$ for some $a \in G$.)
(b) For $G = S_3$, determine Orb(s) for each $s \in G$.
(c) Verify that G is Abelian iff $|\text{Orb}(s)| = 1$ for each $s \in G$.
(d) Verify that, in general, $|\text{Orb}(s)| = 1$ iff $sa = as$ for each $a \in S$.
(e) Verify that the set of all elements satisfying the two equivalent conditions in part (d) is a subgroup of G. This subgroup is called the *center* of G; denote it by $Z(G)$.
(f) Verify that $Z(S_3) = \{(1)\}$.
(g) Verify that the center of the symmetry group of a square is $\{\mu_1, \mu_3\}$.

31.8. Let G be a group, and let Aut(G) denote the automorphism group of G (Problem 16.12).
(a) Verify that if π_a is defined as in Problem 31.7, then $\pi_a \in \text{Aut}(G)$. (An automorphism of the form π_a is called an *inner automorphism* of G.)
(b) Define $\theta : G \to \text{Aut}(G)$ by $\theta(a) = \pi_a$ for each $a \in G$. Verify that θ is a homomorphism.
(c) With θ as in part (b), $\theta(G)$, being the image of a homomorphism, must be a subgroup of Aut(G). Denote this subgroup by Inn(G) (it is called the *group of inner automorphisms* of G). Verify that $|\text{Inn}(G)| = 1$ iff G is Abelian.
(d) Explain why $G/Z(G) \approx \text{Inn}(G)$. [See Problem 31.7(e) for the definition of $Z(G)$.]

31.9. Verify that $\mathbb{Z}_3$ has an automorphism that is not an inner automorphism. [Compare Problem 31.8(c).]

31.10. Let S denote the collection of all subgroups of a finite group G. For $a \in G$ and $H \in S$, let $\pi_a(H) = aHa^{-1}$.
(a) Verify that with this definition G acts on S. (Each subgroup aHa^{-1} is called a *conjugate* of H.)
(b) For $G = S_3$, determine Orb($\langle (1 \quad 2) \rangle$).
(c) For $G = S_3$, determine $G_{\langle (1 \quad 2) \rangle}$.

(d) Use parts (b) and (c) to verify that

$$|\text{Orb}(\langle(1\quad 2)\rangle)| = \frac{|G|}{|G_{\langle(1\quad 2)\rangle}|}.$$

(Compare Theorem 31.2.)

(e) For $H \in S$, the *normalizer of H in G* is defined by $N_G(H) = \{a \in G \mid aHa^{-1} = H\}$. Using results from this section, explain why $N_G(H)$ is a subgroup of G, and the number of conjugates of H is $[G : N_G(H)]$.

31.11. Prove that if G acts on S and $\pi_a(s) = t$, then $G_s = a^{-1}G_t a$.

31.12. If G acts on S and H is a subgroup of G, then H acts on S.

(a) Verify that $s \sim_H t$ implies $s \sim_G t$.

(b) Give an example (specific S, G, H, s, t) showing that $s \sim_G t$ does not imply $s \sim_H t$.

31.13. Assume that H acts on S and that $\theta : G \to H$ is a homomorphism from G onto H.

(a) Verify that G acts on S if π_a is defined to be $\pi_{\theta(a)}$ for each $a \in G$.

(b) Verify that if $s, t \in S$, then $s \sim_H t$ iff $s \sim_G t$.

31.14. Explain Cayley's Theorem (Section 17) in terms of a group acting on a set. Is the action faithful?

31.15. (Generalization of Cayley's Theorem) Let G be a group, H a subgroup of G, and S the set of left cosets of H in G. For $a \in G$, define $\pi_a : S \to S$ by $\pi_a(gH) = (ag)H$.

(a) Verify that in this way G acts on S, that is, $a \mapsto \pi_a$ defines a homomorphism from G into $\text{Sym}(S)$.

(b) Prove that $\text{Ker}\,\pi = \cap\{gHg^{-1} \mid g \in G\}$.

(c) Explain and prove the statement "$\text{Ker}\,\pi$ is the largest normal subgroup of G that is contained in H."

31.16. Prove that if a finite group G contains a subgroup $H \neq G$ such that $|G| \nmid [G:H]!$ then H contains a nontrivial normal subgroup of G. (Use Problem 31.15, Lagrange's Theorem, and facts about homomorphisms.)

31.17. (a) Prove that if a finite group G contains a subgroup $H \neq G$ such that $|G| \nmid [G:H]!$ then G is not simple. (Use Problem 31.16. Simple groups, and their importance, were discussed in Section 30.)

(b) Assume that each group of order 100 contains a subgroup of order 25 (this will be proved in Section 33). Prove that there is no simple group of order 100.

SECTION 32. BURNSIDE-PÓLYA COUNTING

One of the central problems of combinatorics is to compute the number of distinguishable ways in which something can be done. A simple example is to compute the number of distinguishable ways the three edges of an equilateral triangle can be painted so that one edge is red (R), one is white (W), and one is blue (B). The six possibilities are shown in the top row of Figure 32.1.

Figure 32.1

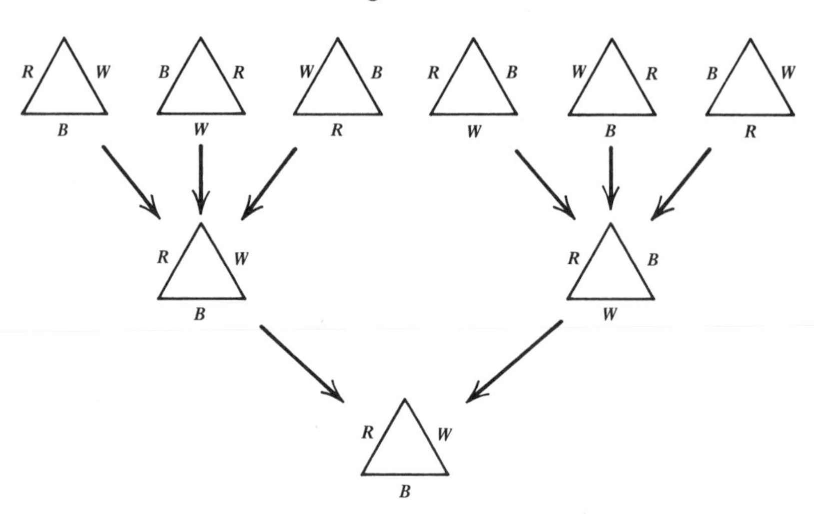

If we permit rotation of the triangle in the plane, then the first three possibilities become indistinguishable—they collapse into the first possibility in the second row of Figure 32.1. The last three possibilities in the top row collapse into the second possibility in the second row.

If we permit reflections through lines, as well as rotation in the plane, then the only possibility is that shown in the third row of Figure 32.1.

In passing from one row to the next in the example, we have treated different ways of painting the triangle as being equivalent, and we have shown one representative from each equivalence class. The problem at each step is to compute the number of equivalence classes. In the terminology used most often in combinatorics, possibilities in the same equivalence class are indistinguishable, while possibilities in different equivalence classes are distinguishable. So the problem of computing the number of distinguishable ways in which something can be done is the same as that of computing the number of equivalence classes under an appropriate equivalence relation. The link between group theory and combinatorics rests on the fact that the equivalence classes are very often orbits under the action of an appropriate group. In the example, the appropriate group for passing to the second row of Figure 32.1 is the group of rotations of the triangle; the appropriate group for passing to the third row contains these three rotations and also three reflections, each through a line connecting a vertex with the midpoint of the opposite side (this group is D_3 in the notation of Section 34).

To solve a combinatorics problem with these ideas, the first step is to get clearly in mind the total set of possibilities (without regard to equivalence). Next, decide the conditions under which possibilities are to be considered equivalent (indistinguishable). This will mean membership in a common orbit under the action of some group, and so in effect this means to choose an appropriate group. Finally, compute the number of orbits relative to this group; this will be the number of distinguishable possibilities. This method derives its power from Burnside's Counting Theorem, proved below, which gives a way to compute the number of orbits relative to a group action.

In considering a group acting on a set, it is often less cumbersome to use the same symbol for a group element and the permutation assigned to it, that is, to write g for π_g; we shall do that in this section. (If this troubles you, do Problem 32.3.) *All groups and sets in this section are assumed to be finite.*

For a group G acting on a set S, we shall use $\psi(g)$ to denote the number of elements of S that are invariant under $g \in G$:

$$\psi(g) = |\{s \in S \mid g(s) = s\}|.$$

Burnside's Counting Theorem. *If a group G acts on a set S, then the number of orbits is*

$$\frac{1}{|G|} \sum_{g \in G} \psi(g).$$

Example 32.1. For the group $\{(1), (1\ \ 2), (3\ \ 4), (1\ \ 2)(3\ \ 4)\}$ acting on $\{1,2,3,4\}$,

$$\psi((1)) = 4 \qquad \psi((3\ \ 4)) = 2$$
$$\psi((1\ \ 2)) = 2 \qquad \psi((1\ \ 2)(3\ \ 4)) = 0.$$

The formula in Burnside's Theorem gives $(\frac{1}{4})(4+2+2+0) = 2$. The two orbits are $\{1,2\}$ and $\{3,4\}$.

Other examples will follow the proof of the theorem, which will depend on Theorem 31.2 and the following lemma.

Lemma 32.1. *If a group G acts on a set S, $s, t \in S$, and s and t are in the same orbit, then $|G_s| = |G_t|$.*

PROOF. Because $s \sim t$, there is an element $g \in G$ such that $g(s) = t$. If $a \in G_s$, then $(gag^{-1})(t) = (ga)(s) = g(s) = t$, and so $gag^{-1} \in G_t$. Therefore, we can define a mapping $\theta: G_s \to G_t$ by $\theta(a) = gag^{-1}$ for each $a \in G_s$. It suffices to show that θ is one-to-one and onto. One-to-one: if $\theta(a_1)=\theta(a_2)$, then $ga_1g^{-1}=ga_2g^{-1}$ so that $a_1=a_2$. Onto: if $b\in G_t$, then $(g^{-1}bg)(s)=(g^{-1}b)(t)=g^{-1}(t)=s$ so that $g^{-1}bg\in G_s$; and $\theta(g^{-1}bg)=g(g^{-1}bg)g^{-1}=b$. (In fact, θ is an isomorphism.) □

PROOF OF BURNSIDE'S THEOREM. Let n denote the number of pairs (g,s) such that $g(s)=s$ $(g\in G, s\in S)$. The number of s appearing with a given g is $\psi(g)$, so that

$$n=\sum_{g\in G}\psi(g). \tag{32.1}$$

The number of g appearing with a given s is $|G_s|$, so that

$$n=\sum_{s\in S}|G_s|. \tag{32.2}$$

If we select an orbit Orb(t), sum $|G_s|$ over all $s\in\text{Orb}(t)$, and use first Lemma 32.1 and then Theorem 31.2, we get

$$\sum_{s\in\text{Orb}(t)}|G_s|=|\text{Orb}(t)|\cdot|G_t|=|G|. \tag{32.3}$$

If we add together all sums like those in (32.3), one such sum for each orbit, we get the sum in (32.2). Therefore,

$$n=\sum_{s\in S}|G_s|=(\text{number of orbits})\cdot|G|.$$

Solve this for the number of orbits, and then use (32.1):

$$\text{number of orbits}=\frac{n}{|G|}=\frac{1}{|G|}\sum_{g\in G}\psi(g). \quad \square$$

Example 32.2. In how many distinguishable ways can the four edges of a square be painted with four different colors if there is no restriction on the number of times each color can be used, and two ways are considered indistinguishable if one can be obtained from the other by an isometry in the group of symmetries of the square? (This would be the case for a square that could be either rotated in the plane or turned over; the latter corresponds to reflection through a line in the plane.)

The appropriate set S in this case is the set of $4^4=256$ ways of painting the edges without regard to equivalence. The group is the one described at the beginning of Section 31. If ρ is a group element, then $\psi(\rho)=4^k$, where k is the number of independent choices to be made in painting the edges so as to have invariance under ρ.

$\psi(\mu_1)=4^4$	(always $\psi(\iota)=\|S\|$ if ι is the identity—four choices)
$\psi(\mu_2)=4$	(for invariance under μ_2, all edges must be of the same color—one choice)
$\psi(\mu_3)=4^2$	(pairs of opposite edges must be of the same color—two choices)
$\psi(\mu_4)=4$	(like μ_2—one choice)
$\psi(\mu_5)=4^3$	(*ab* and *dc* must be of the same color; *ad* and *bc* are independent—three choices)
$\psi(\mu_6)=4^3$	(like μ_5—three choices)
$\psi(\mu_7)=4^2$	(*ab* and *ad* must be of the same color; *cb* and *cd* must be of the same color—two choices)
$\psi(\mu_8)=4^2$	(like μ_7— two choices)

Therefore, by Burnside's Theorem, the number of distinguishable ways is

$$\left(\tfrac{1}{8}\right)(4^4+4+4^2+4+4^3+4^3+4^2+4^2) = 55.$$

Example 32.3. In how many distinguishable ways can the six faces of a cube be painted with six different colors if each face is to be a different color and two ways are considered indistinguishable if one can be obtained from the other by rotation of the cube?

The answer here is the number of orbits when the group G of all rotations of a cube acts on the set S of all cubes painted with the six different colors. The group G contains 24 elements, which can be classified as follows (the references are to Figure 32.2):

1. The identity.
2. Three 180° rotations around lines (such as *ij*) joining the centers of opposite faces.
3. Six 90° rotations around lines (such as *ij*) joining the centers of opposite faces.
4. Six 180° rotations around lines (such as *kl*) joining the mid-points of opposite edges.
5. Eight 120° rotations around lines (such as *ag*) joining opposite vertices.

The number of elements of S is $6!=720$ (the number of ways of painting the cube without regard to equivalence). If ι denotes the identity of G, then $\psi(\iota)=720$. If $\rho\neq\iota$, then $\psi(\rho)=0$, because any way of painting the cube with

Figure 32.2

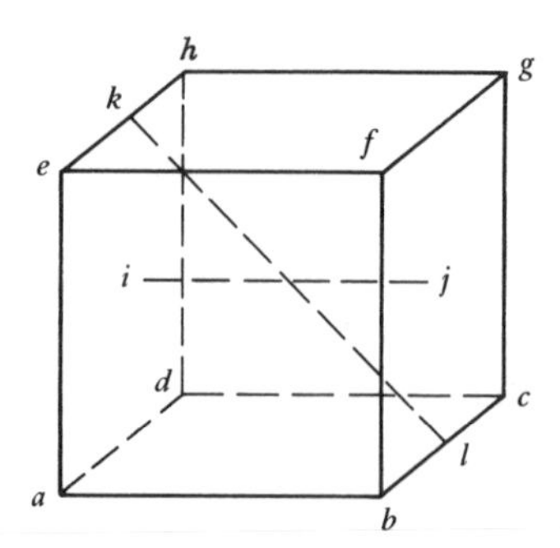

all faces different colors will be carried into a different way under the action of ρ. Therefore, by Burnside's Theorem, the number of orbits is $(\frac{1}{24})(720)=30$.

If we identify the six colors here with the integers from 1 to 6, we can interpret this answer as saying that there are 30 distinguishable ways of assigning the numbers to the faces of a die. (One of these ways is standard.)

PROBLEMS

32.1. Use Burnside's Counting Theorem to compute the number of orbits for the group $\langle(1\ \ 2\ \ 3)(4\ \ 5)\rangle$ acting on $\{1,2,3,4,5\}$. What are the orbits?

32.2. Consider the problem of painting the edges of a square so that one is red, one is white, one is blue, and one is yellow.

(a) In how many distinguishable ways can this be done if the edges of the square are distinguishable? (This is the analogue of the case of a triangle represented by the first row in Figure 32.1.)

(b) Repeat (a), except count different ways as being indistinguishable if one can be obtained from the other by rotation of the square in the plane. (Compare the second row in Figure 32.1.)

(c) Repeat (b), except permit reflections through lines as well as rotation in the plane.

32.3. Rewrite Burnside's Counting Theorem and Lemma 32.1, and their proofs, using the notation π_g in place of g wherever appropriate.

32.4. In how many distinguishable ways can the three edges of a triangle be painted with the three colors red, white, and blue if there is no restriction on the number of times each color can be used? State clearly how this problem is like, and also different from, the example at the first of this section, and also Example 32.2.

32.5. State and solve the problem that results from replacing a square by a regular pentagon in Example 32.2. (The group will have order 10. It is D_5 in the notation of Section 34.)

32.6. State and solve the problem that results from replacing a square by a regular pentagon in Problem 32.2 (with some appropriate fifth color). Compare Problem 32.5.

32.7. Repeat Problem 32.5 with a regular hexagon in place of a regular pentagon.

32.8. Repeat Problem 32.6 with a regular hexagon in place of a regular pentagon.

32.9. A bead is placed at each of the four vertices of a square, and each bead is to be painted either red (R) or blue (B). Under equivalence relative to the group of rotations of the square, there are six distinguishable patterns:

$R\ R$	$R\ R$	$R\ R$	$R\ B$	$R\ B$	$B\ B$
□	□	□	□	□	□
$R\ R$	$R\ B$	$B\ B$	$B\ R$	$B\ B$	$B\ B$

Verify that Burnside's Counting Theorem gives the correct number of distinguishable patterns.

32.10. A bead is placed at each of the six vertices of a regular hexagon, and each bead is to be painted either red or blue. How many distinguishable patterns are there under equivalence relative to the group of rotations of the hexagon? (Compare Problem 32.9.)

32.11. In how many distinguishable ways can the four faces of a regular tetrahedron be painted with four different colors if each face is to be a different color and two ways are considered indistinguishable if one can be obtained from the other by rotation of the tetrahedron? (The group of rotations in this case has order 12. In addition to the identity, there are eight 120° rotations around lines such as ae in the following figure, and three 180° rotations around lines such as fg.)

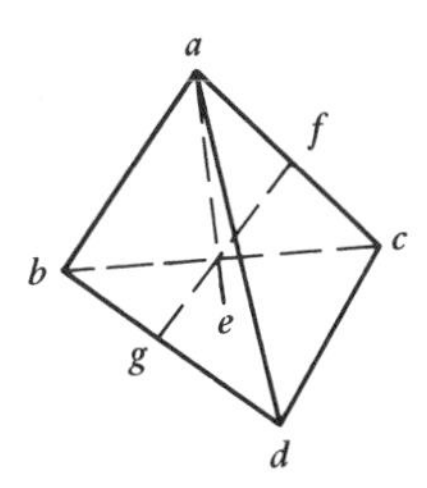

32.12. A bead is placed at each of the eight vertices of a cube, and each bead is to be painted either red or blue. Under equivalence relative to the group of rotations of the cube, how many distinguishable patterns are there?

32.13. Repeat Problem 32.12 with a regular tetrahedron in place of a cube. (The group is described in Problem 32.11.)

SECTION 33. SYLOW'S THEOREM

Sylow's Theorem. *If G is a group of order $p^k m$, where p is a prime and $p \nmid m$, then G has a subgroup of order p^k.*

This theorem should be viewed as a partial converse of Lagrange's Theorem. That theorem tells us that if H is a subgroup of G, then $|H|$ divides $|G|$. However, as we learned in Section 14, there is no guarantee of a subgroup of each possible order dividing the order of a group. Sylow's Theorem provides such a guarantee for the highest power of each prime dividing the order. In fact, the ideas we shall use in proving Sylow's Theorem can also be used to prove the existence of a subgroup of *each* prime power order—not just the highest—dividing the order of a group (Problem 33.10).

This theorem was first proved by Sylow in 1872. The proof we shall give here is due to the German mathematician Helmut Wielandt (1910–); it is a splendid application of the idea of a group acting on a set.

In the proof of the theorem, S will denote the set of all p^k-element subsets of G, and G will act on this set in a natural way. Before the proof we need some preliminary results about the set S.

The number of r-element subsets of an n-element set is the binomial coefficient

$$\binom{n}{r} = \frac{n!}{r!(n-r)!} = \frac{n(n-1)\cdots(n-r+1)}{r!}. \tag{33.1}$$

Here n can be any positive integer, and $0 \leq r \leq n$. You can construct (or review) a proof of this by doing Problem 33.3. We shall need to consider this in the special case $n = p^k m$, $r = p^k$.

Lemma 33.1. *If p is a prime and $p \nmid m$, then*

$$p \nmid \binom{p^k m}{p^k}.$$

PROOF.

$$\binom{p^k m}{p^k} = \frac{p^k m(p^k m-1)\cdots(p^k m-j)\cdots(p^k m-p^k+1)}{p^k(p^k-1)\cdots(p^k-j)\cdots(p^k-p^k+1)}.$$

Except for the factor m, which is not divisible by p, this fraction is equal to a product of fractions

$$\frac{p^k m-j}{p^k-j}$$

with $0<j\leq p^k-1$. If these fractions are reduced to lowest terms, then none has a numerator divisible by p, because for $0<j\leq p^k-1$ the highest power of p dividing p^k-j is the same as the highest power of p dividing $p^k m-j$, both being equal to the highest power of p dividing j (Problem 33.4). □

PROOF OF SYLOW'S THEOREM. Let S denote the set of all p^k-element subsets of G. By the lemma, $p\nmid|S|$. Also, G acts on S as follows: for $a\in G$ and $T\in S$,

$$\pi_a(T) = aT = \{at|t\in T\}.$$

Because $|S|$ is the sum of the numbers $|\text{Orb}(T)|$ for the different G-orbits, we must have $p\nmid|\text{Orb}(T)|$ for some $T\in S$, for otherwise we would have $p||S|$. Let $\{T_1,\ldots,T_u\}$ be an orbit with $p\nmid u$, and let $H=\{g\in G|gT_1=T_1\}$. Then H is a subgroup—in fact, in the notation of Section 31, $H=G_{T_1}$. By Theorem 31.2, $|G|=u\cdot|H|$. Since $p^k||G|$, and $p\nmid u$, we must have $p^k||H|$. Therefore $p^k\leq|H|$.

Because $hT_1=T_1$ for each $h\in H$, we can also think of H as acting on T_1. Let $t\in T_1$. Then H_t, which is $\{h\in H|ht=t\}$, contains only e, so $|H_t|=1$. By Theorem 31.2, $|H|=|\text{Orb}(t)|\cdot|H_t|$. Therefore $|H|=|\text{Orb}(t)|\leq p^k$.

From $p^k\leq H\leq p^k$ we deduce that $|H|=p^k$. □

Sylow's Theorem, as stated above, can be extended to include still more information about subgroups of prime power order. To do this, it will help to introduce the following terminology. Let p be a prime. A subgroup whose order is a power of p is called a *p-subgroup*. A p-subgroup of a finite group G whose order is the highest power of p dividing $|G|$ is called a *Sylow p-subgroup*. If H and K are subgroups of a group G, and $K=aHa^{-1}$ for some $a\in G$, then K is a *conjugate* of H (see Problem 31.10).

Sylow's Theorem (Extended Version). *Let G be a group of order p^km, where p is a prime and $p\nmid m$. Then*

(*a*) *the number n_p of Sylow p-subgroups of G satisfies* $n_p\equiv 1\ (\text{mod}\,p)$,

(*b*) $n_p|m$, *and*

(*c*) *any two Sylow p-subgroups are conjugate.*

Although we shall not prove this theorem, an application of it is given in the following example. This example gives some idea of the usefulness of Sylow's Theorem, but only more experience with the theory of finite groups can give a true appreciation of the theorem.

Example 33.1. Assume that G is a finite group of order p^km, where p is a prime, $p\nmid m$, $m>1$, and $k>0$. It is not hard to prove that if G has only one Sylow p-subgroup, then that subgroup must be normal (Problem 33.5). In particular, if G has only one Sylow p-subgroup, then G is not simple. (Simple groups, and their importance, are discussed in Section 30.) This fact can be used along with parts (a) and (b) of Sylow's Theorem (the extended version) to show that there are no simple groups of certain orders. For example, suppose that $|G|=100$. Then $n_5 \equiv 1 \pmod 5$, so that $n_5 = 1$ or 6 or 11 or $\cdots$. But also $n_5|4$. Therefore $n_5 = 1$, that is, G has only one Sylow 5-subgroup; thus G is not simple. (Problem 31.17 outlines a different proof that there is no simple group of order 100.)

PROBLEMS

33.1. What are the orders of the Sylow p-subgroups of a group of order 180?

33.2. Explain why every group of order 12 must have a subgroup of every order dividing 12 except, possibly, 6. (Problem 14.16 gives an example of a group of order 12 that has no subgroup of order 6.)

33.3. Assume that n and r are integers such that $0<r<n$.

(a) Explain why the number of ways of choosing r elements from an n-element set, taking into account the order in which the elements are chosen, is $n(n-1)\cdots(n-r+1)$.

(b) Let $\binom{n}{r}$ denote the number of ways of choosing r elements from an n-element set, without regard to order. Explain why

$$r!\binom{n}{r} = n(n-1)\cdots(n-r+1).$$

(Notice that each way of choosing r elements without regard to order corresponds to $r!$ ways of choosing r elements taking order into account.)

(c) Using part (b) and the convention $0!=1$, verify equation (33.1).

33.4. Assume that p is a prime.

(a) Prove that if $0<j\leq p^k-1$, then $p^e|j$ iff $p^e|(p^k-j)$.

(b) Prove that if $0<j\leq p^k-1$ and $p\nmid m$, then

$$p^e|j \quad \text{iff} \quad p^e|(p^km-j).$$

33.5. (a) Verify that if H is a subgroup of G, and $a\in G$, then aHa^{-1} is a subgroup of G.

(b) Prove that if H is a finite subgroup of G, and $a\in G$, then $|aHa^{-1}|=|H|$. (*Suggestion*: The mapping $h\mapsto aha^{-1}$ is one-to-one.)

(c) Explain why if H is a Sylow p-subgroup of a finite group, then so is each conjugate of H.

(d) Prove that if a finite group has only one Sylow p-subgroup for some prime p, then that subgroup must be normal.

33.6. (a) Use Sylow's Theorem (the extended version) to prove that every group of order 6 has only subgroup of order 3.

(b) Give an example of a group of order 6 that has more than one subgroup of order 2.

(c) Give an example of a group of order 6 that has only one subgroup of order 2.

33.7. Prove that there is no simple group of order 20. (See Example 33.1.)

33.8. Prove that a group of order 75 must have normal subgroups of order 3 and 25. (See Example 33.1.)

33.9. Let G be a group of order 56.

(a) Say as much as you can, based on Sylow's Theorem (the extended version), about the number of Sylow 2-subgroups and the number of Sylow 7-subgroups of G.

(b) Explain why the total number of nonidentity elements in the Sylow 7-subgroups of G is $6n_7$. (Use Lagrange's Theorem.)

(c) Using parts (a) and (b), explain why either $n_7=1$ or $n_2=1$.

(d) Using (c), explain why there is no simple group of order 56.

33.10. Prove that if p is a prime and G is a finite group such that $p^k \mid |G|$, then G has a subgroup of order p^k. [*Suggestion*: Follow the proof of Sylow's Theorem, but first prove the following generalization of Lemma 33.1: If p is a prime and p^t is the highest power of p dividing m, then p^t is also the highest power of p dividing $\binom{p^k m}{p^k}$.]

NOTES ON CHAPTER VIII

For a more complete treatment of the applications of groups in combinatorics see any of the references listed below; the survey article by N. G. deBruijn in Chapter 5 of [1] is especially recommended. More about Sylow's Theorem can be found in references [3] and [4] of Chapter IV.

1. Beckenbach, E. F., ed., *Applied Combinatorial Mathematics*, John Wiley, New York, 1964.
2. Cohen, D. I. A., *Basic Techniques of Combinatorial Theory*, John Wiley, New York, 1978.
3. Liu, C. L., *Introduction to Combinatorial Mathematics*, McGraw-Hill, New York, 1968.
4. Pólya, G., Kombinatorische Anzahlbestimmungen fur Gruppen, Graphen und Chemische Verbindungen, *Acta Mathematica*, **68** (1937) 145–254.

CHAPTER IX

Symmetry

This chapter continues the discussion of groups and symmetry that was begun in Section 8. The first section gives complete lists of the finite symmetry groups of plane figures, and the finite rotation groups of three-dimensional figures. The next section gives a complete list of the infinite discrete symmetry groups of two-dimensional figures; these are the symmetry groups of such things as border ornaments and wallpaper patterns. The third section is an introduction to the use of groups in crystallography, one of several important applications of groups to symmetry problems in science. In the last section of the chapter we shall draw on linear algebra to prove a result known as the *crystallographic restriction*; this will be done to illustrate the natural connection between groups, geometry, and linear algebra. The proofs of many of the facts presented in this chapter are too long to be included here. The goal will be to summarize some ideas and introduce some others, and to provide more experience with groups and symmetry through the examples and problems. References for proofs and other details can be found in the notes at the end of the chapter. The list of references in [10] gives sources for other applications in physics and chemistry.

SECTION 34. FINITE SYMMETRY GROUPS

The motions (isometries) leaving either a plane figure or a three-dimensional figure invariant form a group, called the group of symmetries (or symmetry group) of the figure. We saw this for plane figures in Section 8. For applications of group theory to geometry and crystallography it is important to determine just what groups can arise as symmetry groups—it turns out that most groups cannot. In this section we shall get an idea of what the possibilities are among finite groups. We begin with the two-dimensional case.

If a motion preserves orientation, it is called a *proper motion*. (A plane motion preserves orientation if an arrow on any circle pointing clockwise before the motion is applied still points clockwise after the motion is applied.) There are three basic types of plane motions: rotations, translations, and glide-reflections. We met rotations, translations, and reflections in Section 8. A *glide-reflection* is a translation in the direction of a line followed by reflection through that line; the set of footprints shown in Figure 34.1 (assumed to extend infinitely far in each direction) is invariant under a glide-reflection consisting of translation through $t/2$ units followed by reflection through the central line (axis). For convenience, the case of reflection without translation is included in the category of glide-reflections.

It is easy to verify that rotations and translations are proper motions, and that glide-reflections are improper. It is also easy to verify that if two motions are applied in succession, then the result will be proper if both are proper, proper if both are improper, and improper if one is proper and the other is improper. Problems 34.14, 34.15, and 34.16 suggest how to prove the following theorem.

Theorem 34.1. *A motion of the plane is either a rotation, a translation, or a glide-reflection. The symmetry group of a plane figure either contains only proper motions, or its proper motions form a subgroup of index two (in which case half of its motions are proper and half are improper).*

We shall restrict attention hereafter to discrete groups of motions: A group G of plane motions is *discrete* if for each point p in the plane there is a circle with center at p such that for each μ in G either $\mu(p)$ is outside that circle or $\mu(p)=p$. The symmetry group of a square (Example 8.3) is discrete. The symmetry group of a circle is not discrete—any point on the circle can be reached from points arbitrarily close to itself by rotations in the group. A group G of three-dimensional motions is *discrete* if for each

Figure 34.1

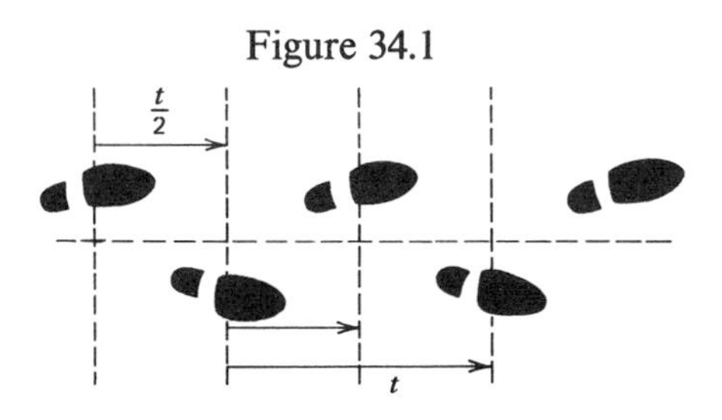

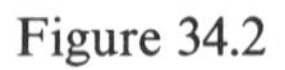

Figure 34.2

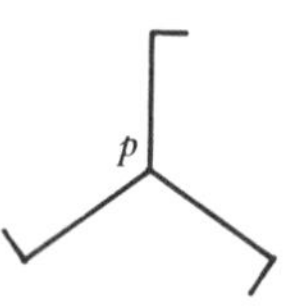

point p there is a sphere with center at p such that for each μ in G either $\mu(p)$ is outside that sphere or $\mu(p)=p$. A finite symmetry group is necessarily discrete (Problem 34.4).

Example 34.1. The group of Figure 34.2 is cyclic of order 3. It contains the identity, clockwise rotation through 120° around p, and clockwise rotation through 240° around p. Each is a proper motion. We denote this group by C_3.

Example 34.2. The group of Figure 34.3 is of order 6. It contains the group in Example 34.1 as a subgroup, and it also contains three reflections (improper motions), one through each of the three axes intersecting at p. We denote this group by D_3.

Except for the number of rotations and axes involved, the groups in the two examples just given exhaust the finite plane symmetry groups. In order to state this precisely, we introduce the following two classes of groups.

A group C_n is cyclic of order n, and consists of clockwise rotations through $k(360°/n)$, $0 \leq k < n$, around a fixed point p.

A group D_n has order $2n$, and contains the elements of C_n together with reflections through n axes that intersect at p and divide the plane into $2n$ equal angular regions. The groups D_n are called *dihedral groups.*

Example 34.3. The symmetry group of a square is D_4; the axes of symmetry are the two diagonals and the horizontal and vertical lines

Figure 34.3

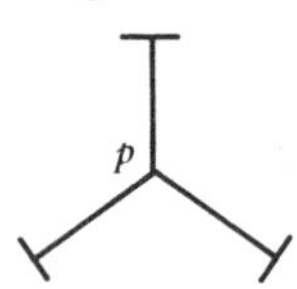

through the center of the square (Example 8.3). The symmetry group of an equilateral triangle is D_3. More generally, the symmetry group of a regular n-sided polygon is D_n. Most snowflakes have the symmetry group D_6 (Figure 34.4), although some have the symmetry group D_3. (This phenomenon was studied by Kepler [8].) The symmetry group of a rectangle (Example 8.3) is D_2, and that of a parallelogram (again Example 8.3) is C_2.

In his interesting book *Symmetry* [13], Hermann Weyl credits the discovery of the following theorem, in essence, to Leonardo da Vinci, who

Figure 34.4

wanted to determine the possible ways to attach chapels and niches to a central building without destroying the symmetry of the nucleus.

Theorem 34.2. *A finite symmetry group of a plane figure is either a cyclic group C_n or a dihedral group D_n.*

PROOF. Assume that G is a finite symmetry group of a plane figure, and assume first that G has order n and contains only rotations. Each rotation except the identity can be assumed clockwise through a positive angle of less than 360°. Let α be the one of these with smallest angle. Then G contains $\iota=\alpha^0$, α, α^2, $\alpha^3,\ldots$; in fact, every element of G will appear in this list: For suppose that $\beta\in G$ and $\beta\neq\alpha^k$ for every k. If the angles for α and β are θ_α and θ_β, respectively, then $t\theta_\alpha<\theta_\beta<(t+1)\theta_\alpha$ for some positive integer t. But then $\beta\alpha^{-t}\in G$ and $\beta\alpha^{-t}$ corresponds to a positive clockwise rotation through an angle less than θ_α, a contradiction. Therefore, $G=\{\iota=\alpha^0,\alpha,\ldots,\alpha^{n-1}\}$, with $\alpha^n=\iota$ and $\theta_\alpha=(360/n)°$.

Now assume that G contains a reflection ρ, and let H denote the set of rotations in G. Then H is a subgroup of G and by the first part of the proof we can assume that $H=\{\iota=\alpha^0,\alpha,\ldots,\alpha^{n-1}\}$ for some rotation α. Necessarily, G contains $\iota,\alpha,\ldots,\alpha^{n-1}$, $\rho,\rho\alpha,\ldots,\rho\alpha^{n-1}$. Each element $\rho\alpha^k$ $(0\leq k<n)$ is a reflection, and these elements are all distinct by the left cancellation law in G.

Let μ be a reflection in G. Then $\rho\mu$, being the product of two reflections, is a rotation (Problem 34.8). Therefore, $\rho\mu=\alpha^k$ for some k so that $\mu=\rho^{-1}\alpha^k=\rho\alpha^k$. It follows from this that $G=\{\iota,\alpha,\ldots,\alpha^{n-1},\rho,\rho\alpha,\ldots,\rho\alpha^{n-1}\}$. Problem 34.10 is designed to convince you that this makes G a dihedral group. □

We turn now to three-dimensional figures. We shall restrict our attention to rotations, and in Theorem 34.3 shall give a complete list of finite rotation groups. (As in the case of plane figures, a finite symmetry group for a three-dimensional figure either contains only rotations or the rotations form a subgroup of index 2.) In describing these groups, an axis of rotation corresponding to rotations of multiples of (360/m)° will be called an *m-fold axis*. Before Theorem 34.3 we give an example of each type of group that will occur.

Example 34.4. An *n-pyramid* $(n\geq3)$ is a right pyramid whose base is a regular n-sided polygon (for $n=3$ we also require that the edges off the base have a length different from those on the base). A 4-pyramid is shown in Figure 34.5. For completeness we define a *2-pyramid* to be a right

Figure 34.5

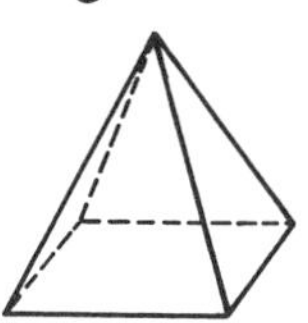

"pyramid" whose base has the shape in Figure 34.6. The rotation group of an n-pyramid is isomorphic to C_n. The line connecting the center of the base to the vertex off the base is an n-fold axis, and the elements of the group are rotations about this axis.

Figure 34.6

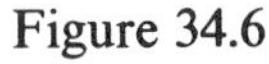

Example 34.5. An *n-prism* ($n \geq 3$) is a right cylinder whose base is a regular n-sided polygon (for $n=4$ we also require that the height be different from the length of the edges on the base). A *2-prism* is a right cylinder whose base has the shape in Figure 34.6. A 6-prism is shown in Figure 34.7. The rotation group of an n-prism is isomorphic to D_n. The line connecting the centers of the top and bottom bases is an n-fold axis. For n even, the lines connecting midpoints of opposite side-faces are 2-fold axes, and the lines connecting midpoints of opposite side-edges are also 2-fold axes; similarly for n odd. The elements of the group are the rotations about these different axes.

Figure 34.7

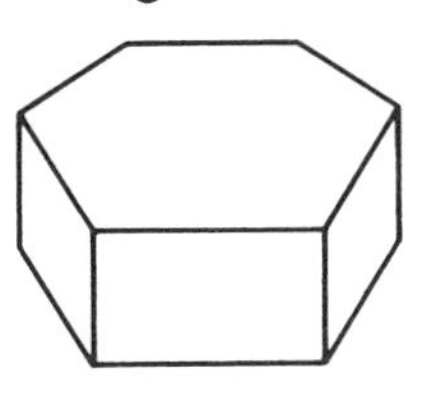

The other finite rotation groups are connected with the five regular convex polyhedra, shown in Figure 34.8. These are the only convex solids whose faces are congruent regular polygons and whose vertex angles are all equal. The rotation group of the cube was described in Example 32.3, and that of the tetrahedron in Problem 32.11.

Figure 34.8

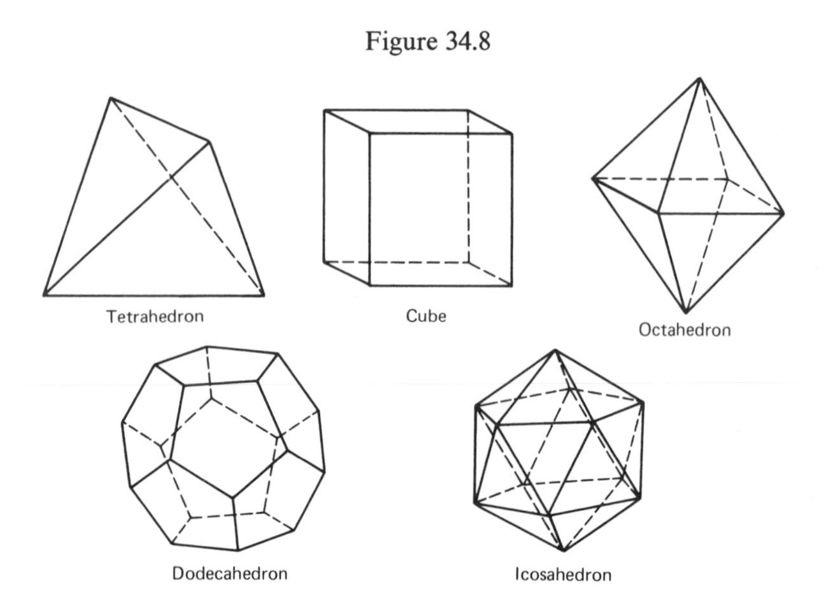

Example 34.6. The rotation group of the tetrahedron has order 12 and is called the *tetrahedral group*. There are four 3-fold axes (one through each vertex), and three 2-fold axes (joining the midpoints of nonintersecting edges).

Example 34.7. The rotation group of the cube has order 24. It is the same as the rotation group of the octahedron, and is called the *octahedral group*. For the cube, there are three 4-fold axes (joining centers of opposite faces), six 2-fold axes (joining midpoints of opposite edges), and four 3-fold axes (joining opposite vertices).

Example 34.8. The rotation group of the icosahedron has order 60. It is the same as the rotation group of the dodecahedron, and is called the *icosahedral group*. For the icosahedron, there are six 5-fold axes (joining pairs of opposite vertices), ten 3-fold axes (joining centers of opposite faces), and fifteen 2-fold axes (joining the midpoints of opposite edges).

Theorem 34.3. *A finite rotation group of a three-dimensional figure is either a cyclic group, a dihedral group, the tetrahedral group, the octahedral group, or the icosahedral group.*

Proofs of this theorem can be found in references [1], [3], [10], and [14] listed at the end of this chapter. Each of these references also gives a list and derivation of the finite symmetry groups (of three-dimensional figures) that contain improper motions, that is, reflections and reflections combined with rotations.

PROBLEMS

34.1. The groups C_2 and D_1 are isomorphic, but they are not the same as groups of motions. Show this by drawing one figure with C_2 as the symmetry group and another figure with D_1 as the symmetry group.

34.2. Determine the symmetry group of each of the following figures.

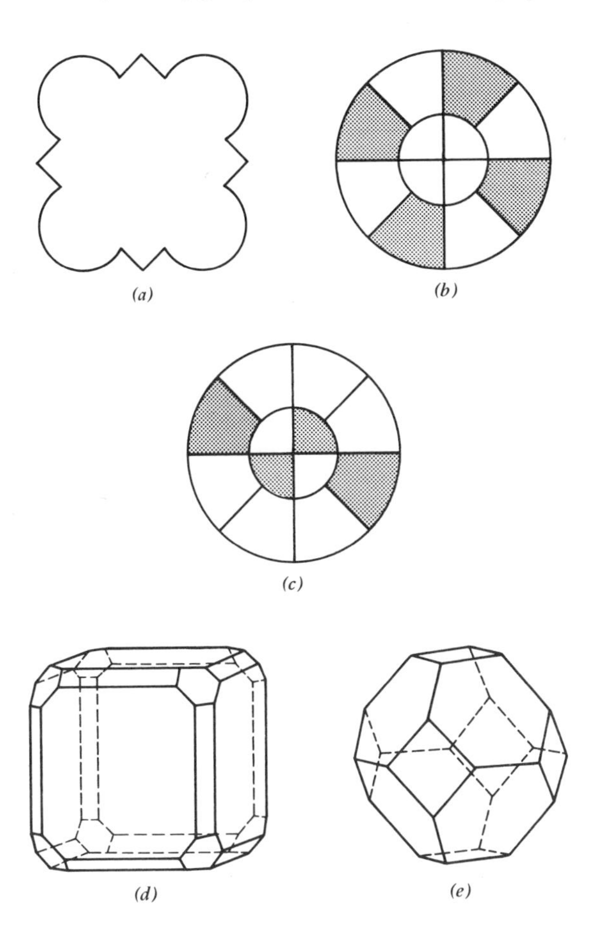

34.3. What is the symmetry group of the graph of each of the following equations?

(a) $y = x^2$ (b) $3x^2 + 4y^2 = 12$ (c) $xy = 1$

34.4. Explain why a finite symmetry group must be discrete.

34.5. True or false: Every subgroup of a discrete group is discrete.

34.6. Prove that the dihedral group D_3 is isomorphic to the symmetric group S_3.

34.7. Verify that D_n is non-Abelian for $n \geq 3$. What about D_2?

34.8. Explain why the product $\rho\mu$ in the proof of Theorem 34.2 must be a rotation. (Notice that $\rho\mu$ must be a proper motion of finite order.)

34.9. Assume α and ρ are as in the proof of Theorem 34.2, with ρ denoting reflection through L. Use the following figure to give a geometrical proof that $\rho\alpha^k$ is the same as reflection through the line bisecting L and $\alpha^{-k}(L)$. [Compute $\rho\alpha^k(p)$.]

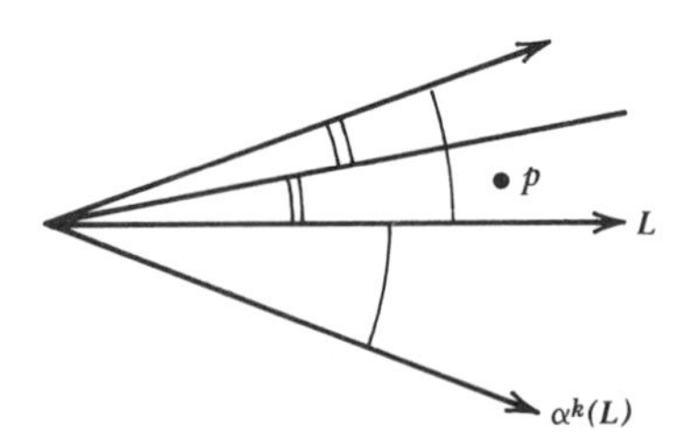

34.10. Use the notation (α and ρ) from the proof of Theorem 34.2 and compute the Cayley table for $D_6=\{\iota,\alpha,\ldots,\alpha^5,\ \rho,\ \rho\alpha,\ldots,\rho\alpha^5\}$, the symmetry group of a regular hexagon.

34.11. With the vertices of a tetrahedron labeled as in Problem 32.11, write the permutation of $\{a,b,c,d\}$ corresponding to each of the 12 rotations in the tetrahedral group.

34.12. (Dodecahedral dice. This assumes knowledge of Section 32.) In how many distinguishable ways can the 12 faces of a dodecahedron be numbered 1 through 12 if two ways are considered indistinguishable if one can be obtained from the other by rotation of the dodecahedron? (Compare Example 32.3.)

34.13. (This assumes knowledge of Section 32.) A bead is placed at each of the 12 vertices of an icosahedron, and each bead is to be painted either red or blue. Under equivalence relative to the group of rotations of the icosahedron, how many distinguishable patterns are there?

34.14. Let μ be a proper motion of the plane, and let p be a point on a directed line L in the plane. (If you stay above the plane and look in the positive direction along L, and a point r is on your right, then $\mu(r)$ will also be on your right if you look in the positive direction along $\mu(L)$; $\mu(r)$ would be on your left if μ were improper.)

(a) Explain why μ is uniquely determined by $\mu(p)$ and $\mu(L)$.

(b) Explain why μ is a rotation if $p=\mu(p)$, and a translation if L and $\mu(L)$ have the same direction.

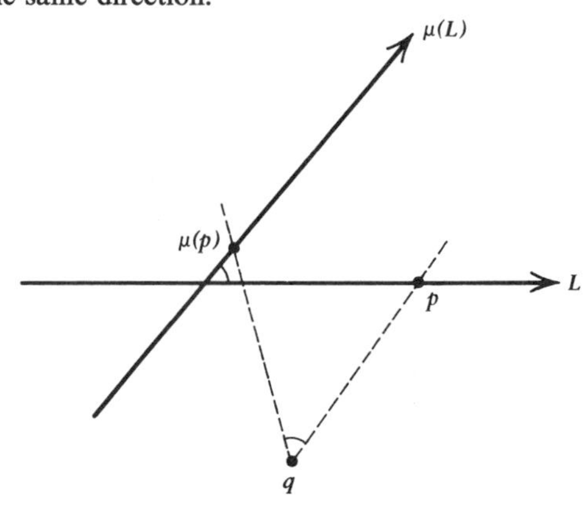

(c) Assume that $p \neq \mu(p)$, and that L and $\mu(L)$ do not have the same direction. Let q be a point equidistant from p and $\mu(p)$ such that angle $pq\mu(p)$ equals the angle between L and $\mu(L)$, as in the figure. Explain why μ is a rotation about q.

34.15. Let μ be an improper motion of the plane, and let p be a point on a directed line L in the plane. (Compare Problem 34.14.)

(a) Explain why μ is uniquely determined by $\mu(p)$ and $\mu(L)$.

(b) Let q be the midpoint of $p\mu(p)$, and let M be the line through q parallel to the bisector of the angle between the positive directions of L and $\mu(L)$, as in the following figure. Explain why μ is a glide-reflection with axis M.

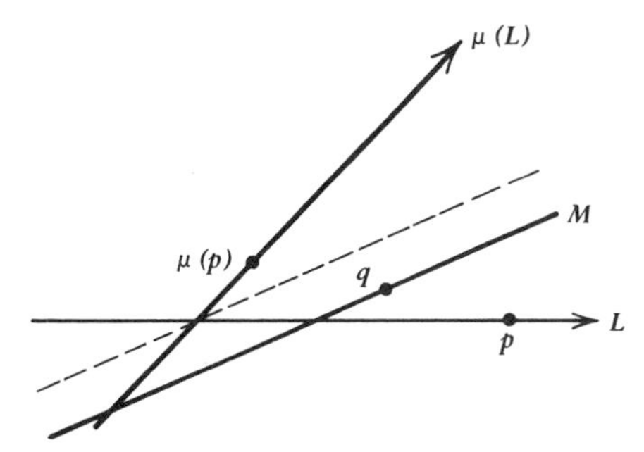

34.16. Prove Theorem 34.1, using Problems 34.14 and 34.15, and the remarks preceding the theorem.

SECTION 35. INFINITE TWO-DIMENSIONAL SYMMETRY GROUPS

If we imagine the designs on decorative borders or wallpaper to be repeated infinitely often, we obtain figures whose symmetry groups are discrete and infinite. These groups contain translations, and are conveniently divided into two classes. The groups in one class each leave a line invariant, and the groups in the other class do not. The groups in the first class are called *frieze groups*. There are seven frieze groups in all, and examples of the corresponding patterns can be seen in Figures 35.1 through 35.7. These are the symmetry groups for repeating patterns that appear on vases, bracelets, and decorative borders, and they will be described along with the examples of the patterns. Figure 35.8 gives one example of each pattern, all taken from Greek vases.

There are 17 groups that do not leave a line invariant. Figure 35.9 gives one pattern corresponding to each group. Patterns of this type will be familiar from designs on wall paper, decorative ceilings, and tile and brick arrangements; the patterns in Figure 1 of the Introduction are of this type, for example. The 17 corresponding groups will be discussed in Section 36.

Each type of pattern in this section can be found in the artwork of the past. The book [7] by Owen Jones contains an especially rich collection of such patterns. Another source is the drawings of M. C. Escher, which have been analyzed in terms of their symmetry in [9].

Descriptions of each of the seven frieze groups follow. The accompanying figures are assumed to extend infinitely far to the left and right.

TYPE I. The group in Figure 35.1 consists of translations only. If τ denotes translations through the smallest possible distance (to the right, say, in Figure 35.1), then the group is infinite cyclic with generator τ.

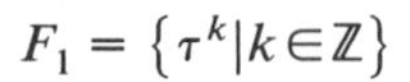

$$F_{\mathrm{I}} = \{\tau^k | k \in \mathbb{Z}\}$$

Figure 35.1

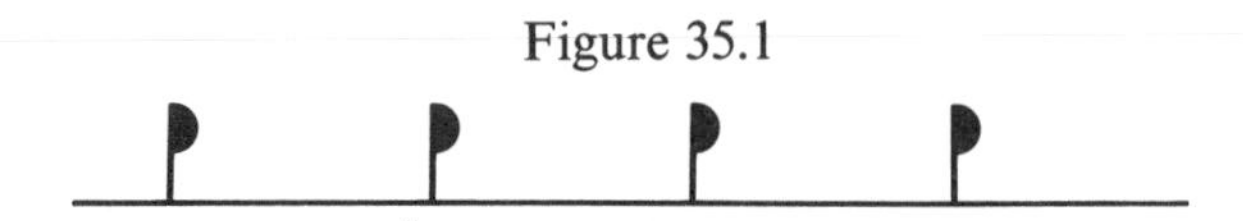

TYPE II. The group in Figure 35.2 is also infinite cyclic. It is generated by a glide-reflection; if this glide-reflection is denoted by γ, then the even powers of γ are translations.

$$F_{\mathrm{II}} = \{\gamma^k | k \in \mathbb{Z}\}$$

Figure 35.2

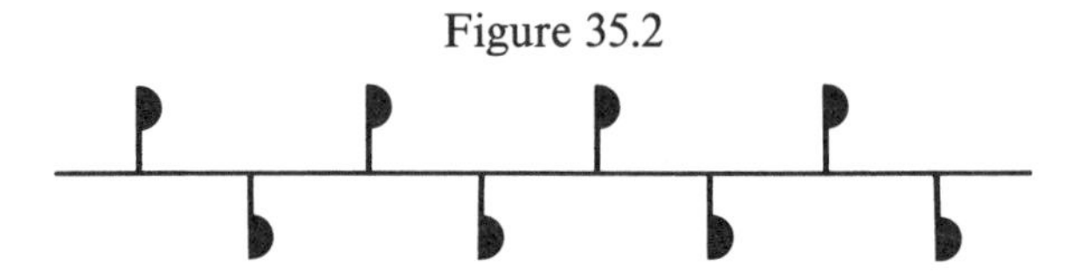

TYPE III. The group in Figure 35.3 is generated by a translation (say τ) and a reflection (say ρ) through a vertical line (such as the dotted line in the figure). This group is an *infinite dihedral group*. (Verify that $\tau\rho = \rho\tau^{-1}$.)

$$F_{\mathrm{III}} = \{\tau^k \rho^m | k \in \mathbb{Z}, m = 0, 1\}$$

Figure 35.3

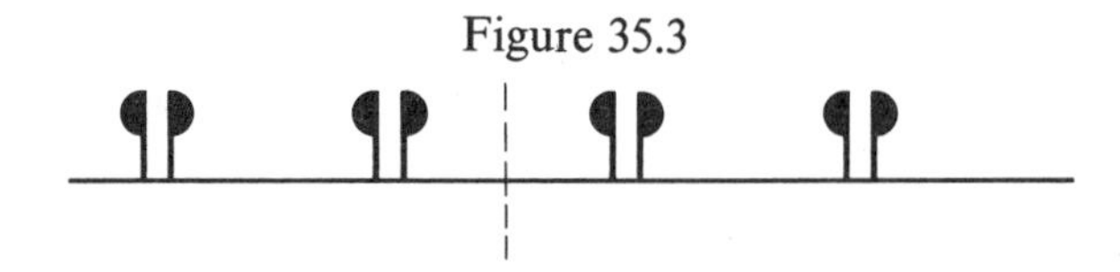

TYPE IV. The group in Figure 35.4 is generated by a translation (say τ) and a rotation (say α) through 180° around a point such as p in the figure. This is also an infinite dihedral group.

$$F_{\mathrm{IV}} = \{\tau^k \alpha^m | k \in \mathbb{Z}, m = 0, 1\}$$

Figure 35.4

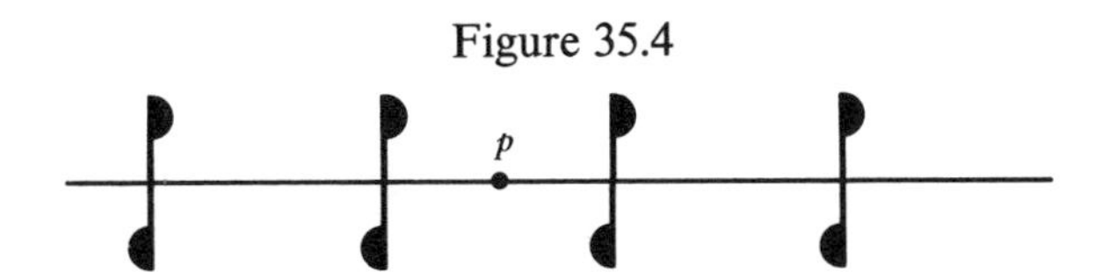

Type V. The group in Figure 35.5 is generated by a glide-reflection (say γ) and a rotation (say α) through 180° around a point such as p in the figure; F_V is another infinite dihedral group.

$$F_V = \{\gamma^k \alpha^m | k \in \mathbb{Z}, m = 0,1\}$$

Figure 35.5

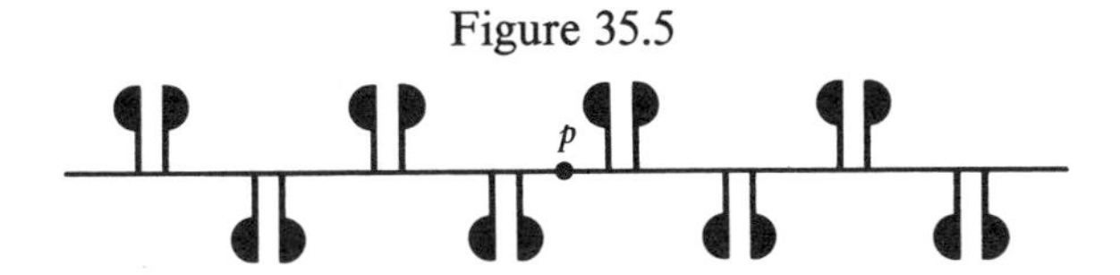

Type VI. The group in Figure 35.6 is generated by a translation (say τ) and a reflection (say β) through an axis of symmetry (which is horizontal in the figure). This group is Abelian. (Verify that $\tau\beta = \beta\tau$.)

$$F_{VI} = \{\tau^k \beta^m | k \in \mathbb{Z}, m = 0,1\}$$

Figure 35.6

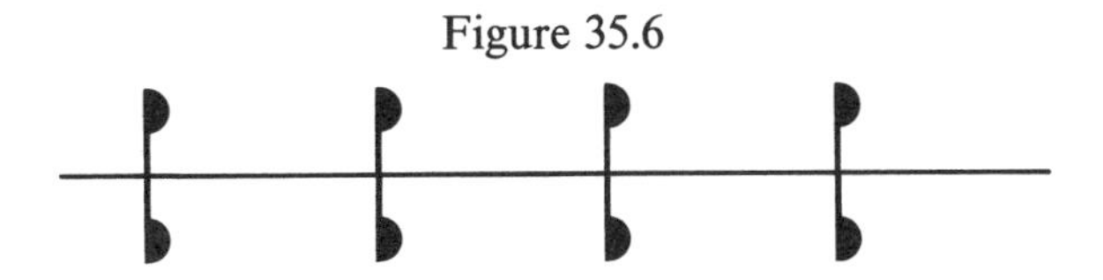

Type VII. The group in Figure 35.7 generated by a translation (say τ) and two reflections (say ρ, as in type III; and β, as in type VI). The elements τ and ρ multiply as in F_{III}; τ and β multiply as in F_{VI}; and $\rho\beta = \beta\rho$.

$$F_{VII} = \{\tau^k \beta^m \rho^n | k \in \mathbb{Z}, m = 0,1, n = 0,1\}$$

Figure 35.7

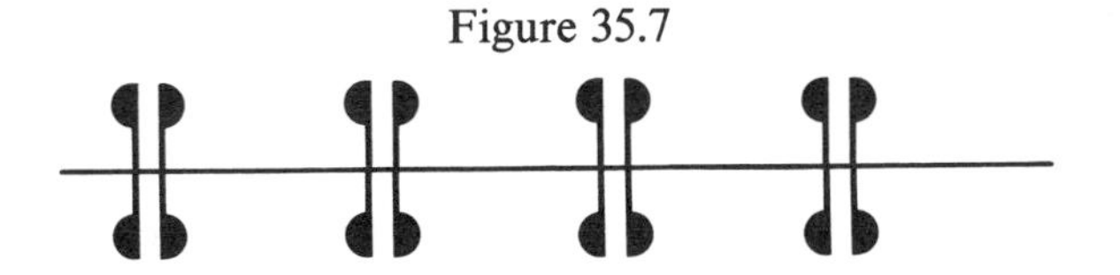

Figure 35.8

Figure 35.9

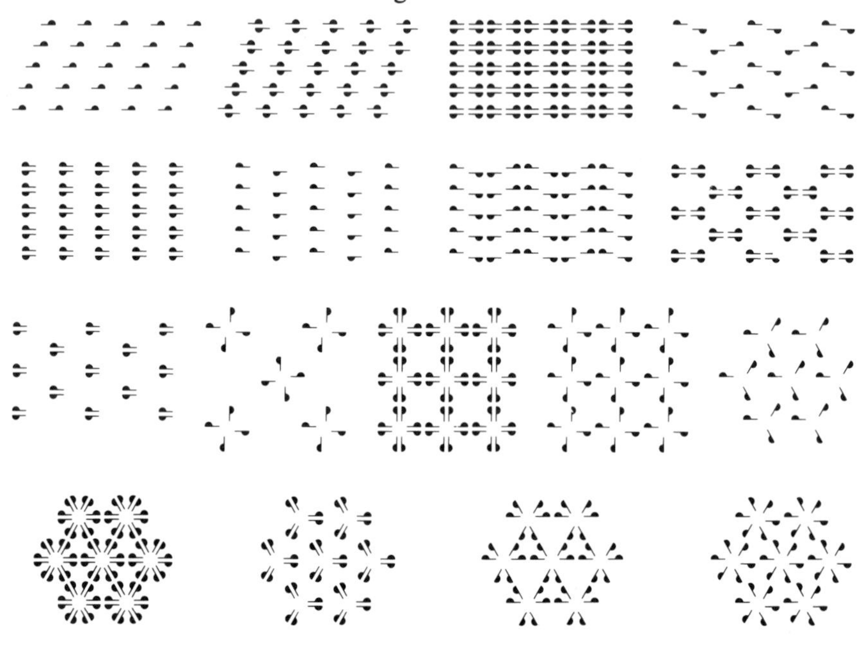

PROBLEMS

35.1. Match each part of Figure 35.8 with the appropriate group (F_{I} through F_{VII}).

35.2. Each of the following figures has one of F_{I} through F_{VII} as symmetry group, and no two parts correspond to the same group. Match each part with its group.

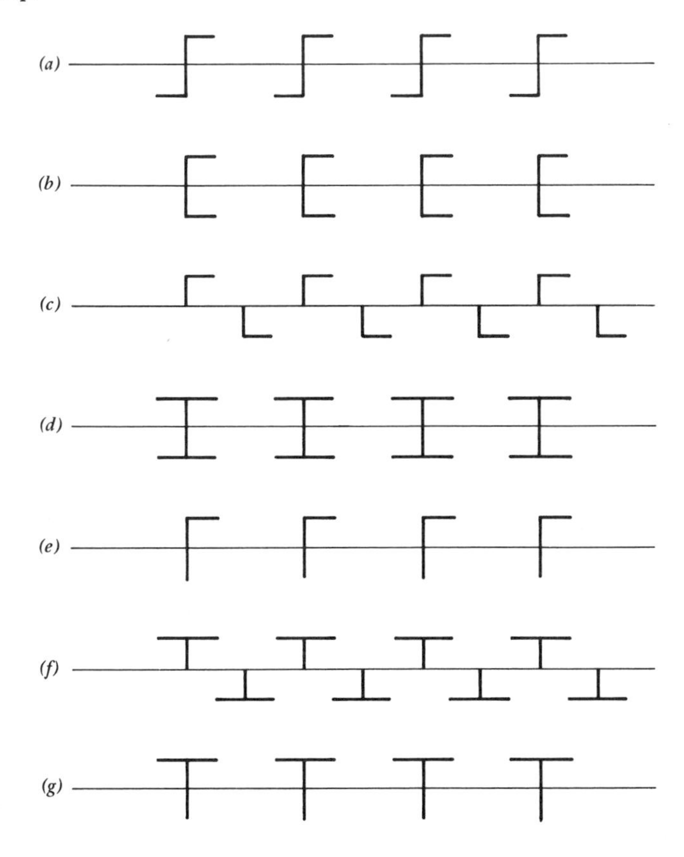

35.3. Each of the following figures has one of F_{I} through F_{VII} as its symmetry group. Find the appropriate group for each part.

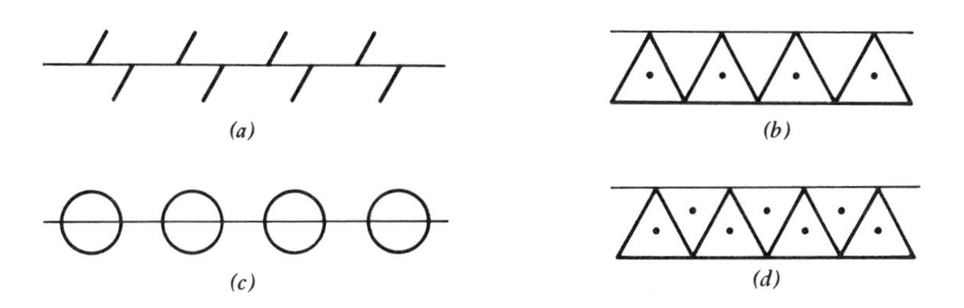

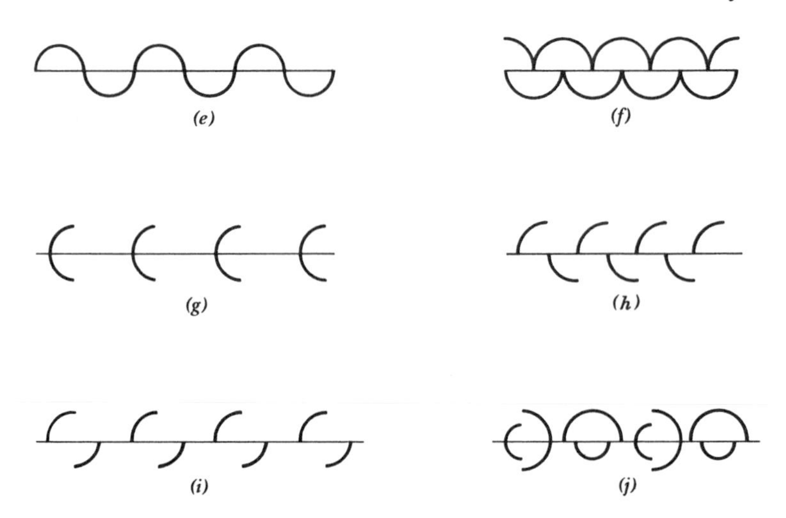

35.4. Draw seven figures, different from those in the book, illustrating the seven types of symmetry corresponding to the groups F_{I} through F_{VII}.

SECTION 36. ON CRYSTALLOGRAPHIC GROUPS

One of the most interesting applications of groups outside of mathematics is in crystallography. At the heart of this application are 32 finite (three-dimensional) symmetry groups, known as the *crystallographic point groups*, and 230 infinite symmetry groups, known as *crystallographic space groups*, which can be constructed from translation groups and the 32 crystallographic point groups. A detailed description of these groups would take us too far afield, but we can get an idea of what they involve. We begin with the two-dimensional analogue.

Consider the following problem: Fill the plane with congruent polygons having no overlap except at their edges. (In this section "polygon" will mean "polygon and its interior.") Figure 36.1 suggests solutions with equilateral triangles, squares, and regular hexagons. It also shows why there is no solution with regular pentagons. In fact, 3, 4, and 6 are the only values of n for which there is a solution with regular n-gons (Problem 36.6). Figure 36.2 offers two solutions with congruent polygons that are not regular.

Now refine the problem by demanding that the congruent polygons needed to fill the plane be those obtained by applying the motions of some group to just a single polygon; this polygon is called a *fundamental region* for the corresponding group. In the case of squares [Figure 36.1(b)], translations in perpendicular directions would suffice—the plane can be filled by starting with one square and moving it repeatedly to the left and right and up and down. For regular hexagons [Figure 36.1(c)], translations

Figure 36.1

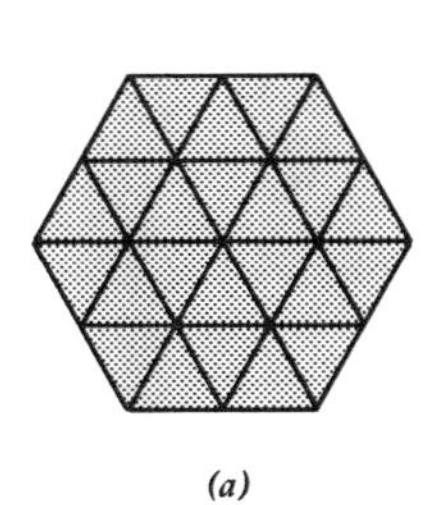

(a)

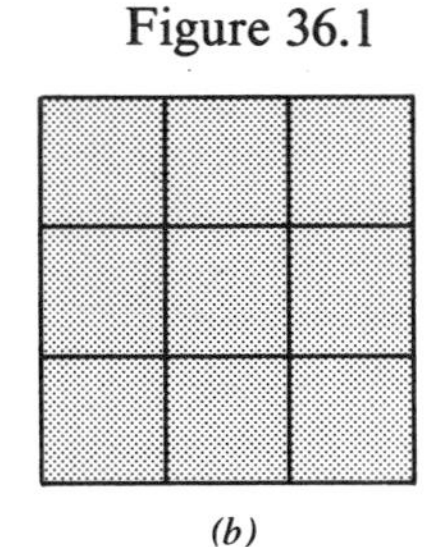

(b)

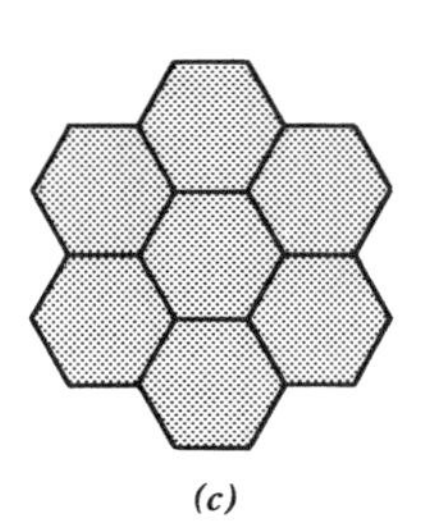

(c)

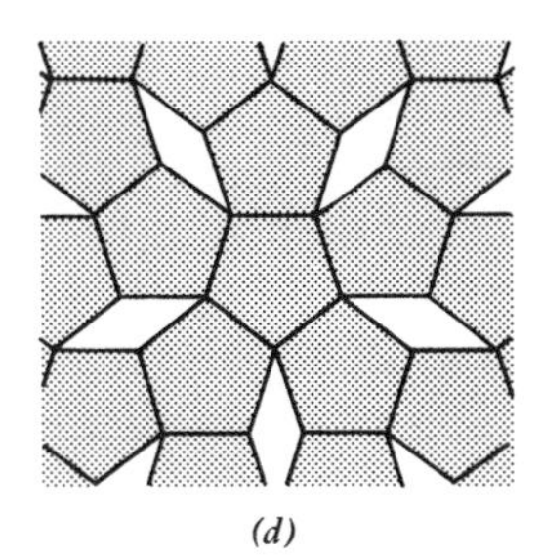

(d)

in directions meeting at angles of 60° would suffice. For equilateral triangles [Figure 36.1(a)], we need the same translations as for hexagons, and we also need rotations or reflections—rotations through multiples of 60° about a vertex of a triangle will fill out a hexagon, whose translates will fill out the plane. In Figure 36.2 the parallelograms offer a solution, but the other example does not. (We shall return to this last point at the end of the section.)

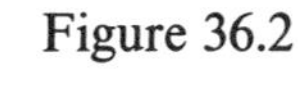

Figure 36.2

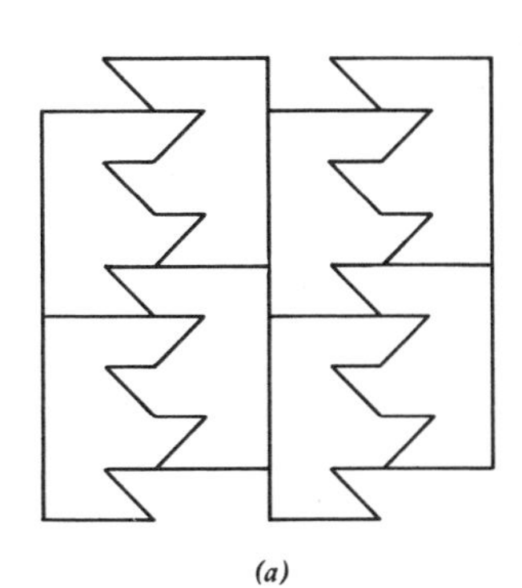

(a)

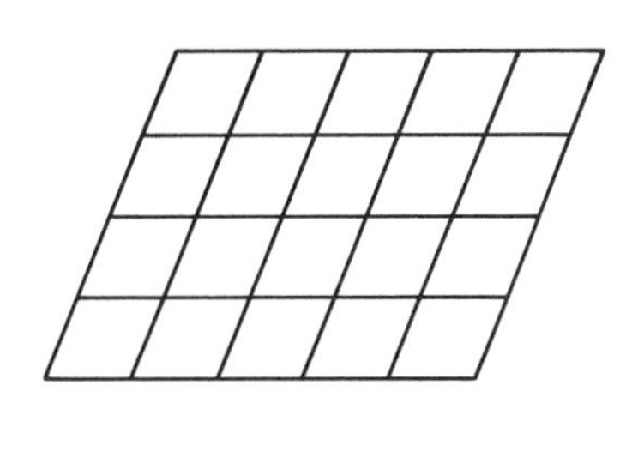

(b)

Figure 36.3

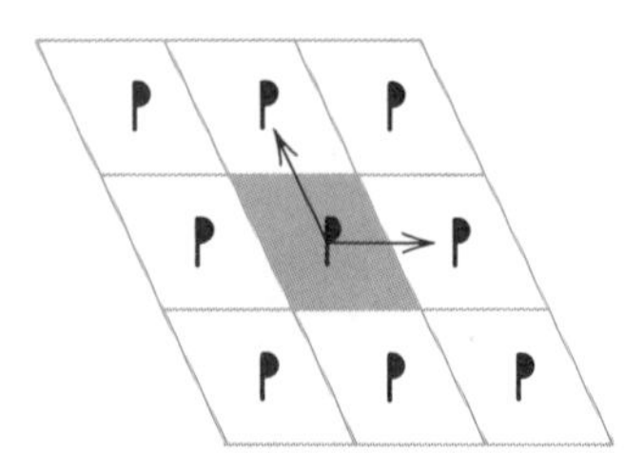

p1 generated by two translations

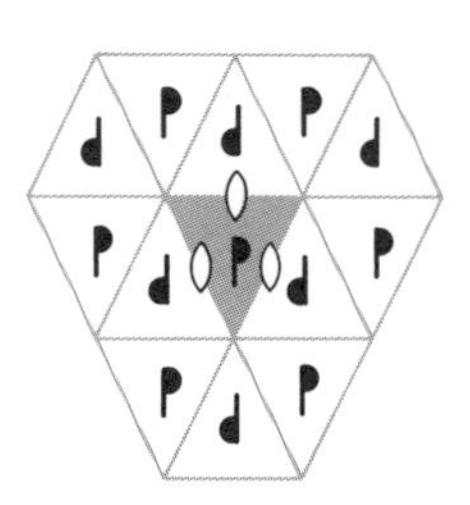

p2 generated by three half–turns

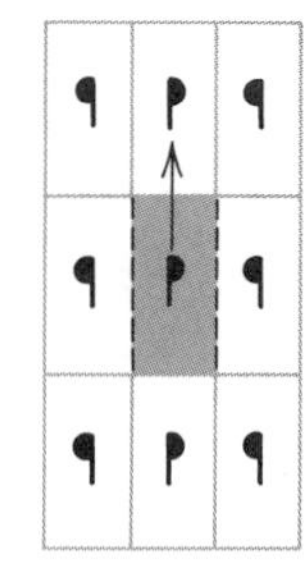

pm generated by two reflections and a translation

pg generated by two parallel glide–reflections

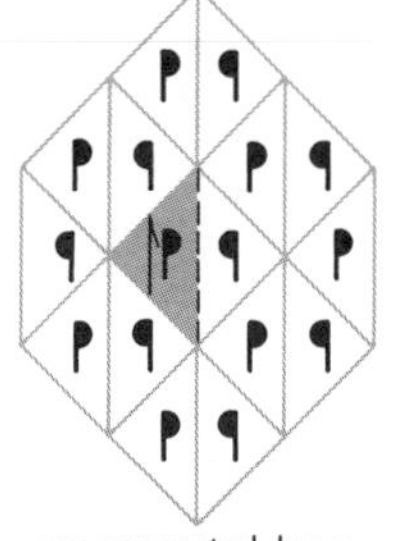

cm generated by a reflection and glide–reflection

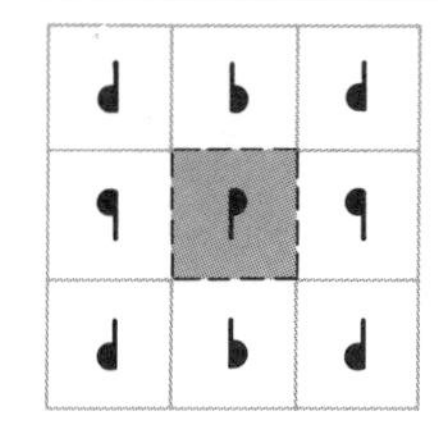

pmm generated by four reflections

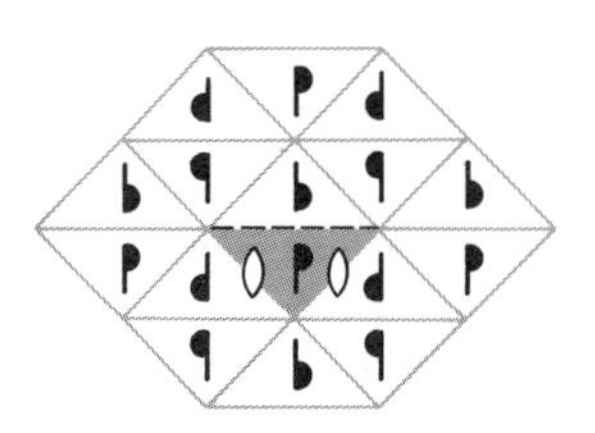

pmg generated by a reflection and two half–turns

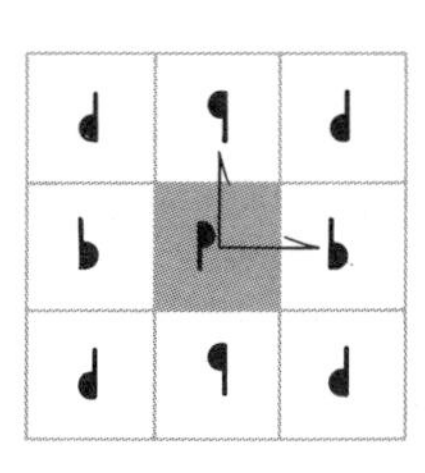

pgg generated by two perpendicular glide reflections

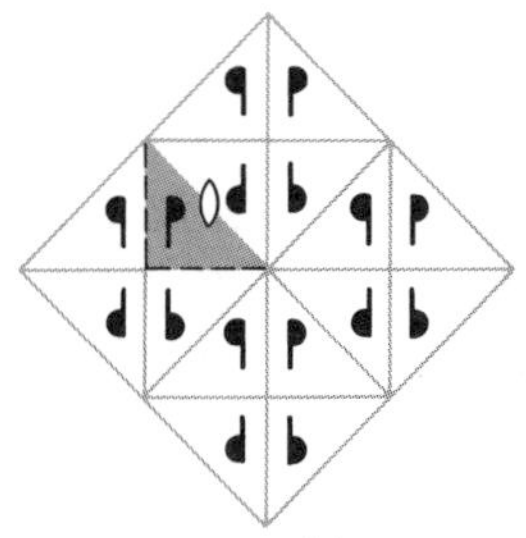

cmm generated by two perpendicular reflections and a half–turn

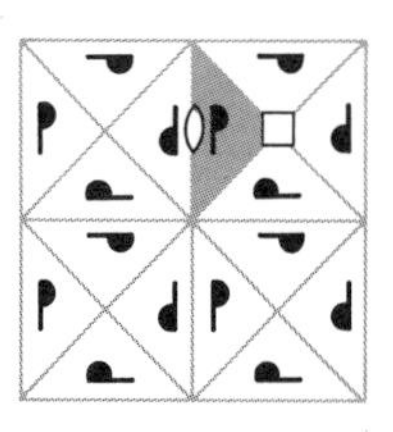

p4 generated by a half–turn and a quarter–turn

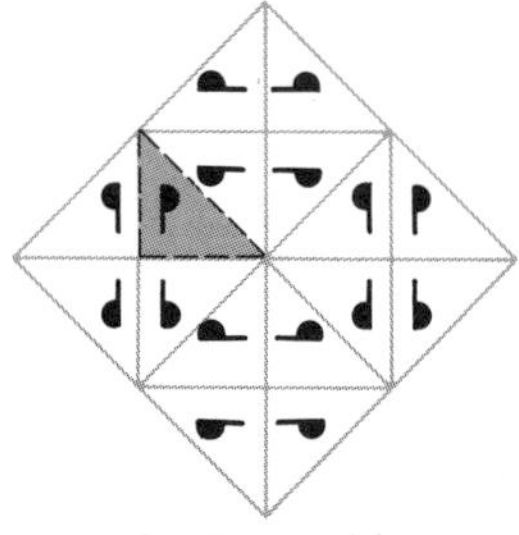

p4m generated by reflections in the sides of a (45°, 45°, 90°) triangle

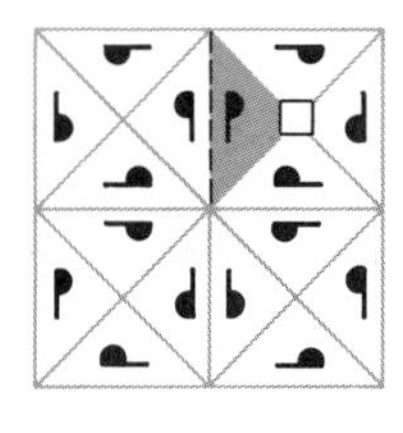

p4g generated by a reflection and a quarter–turn

Figure 36.3 (*Continued.*)

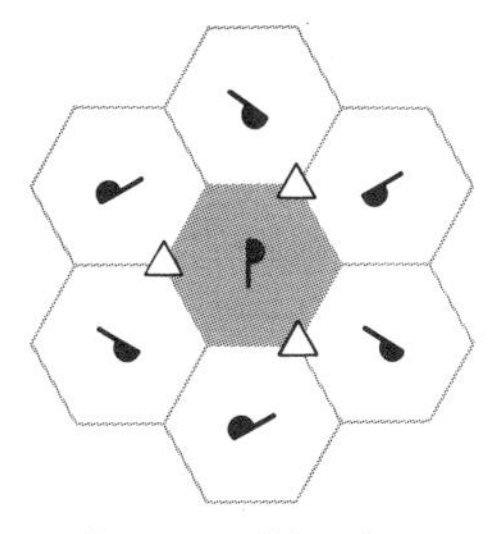

p3 generated by three rotations through 120°

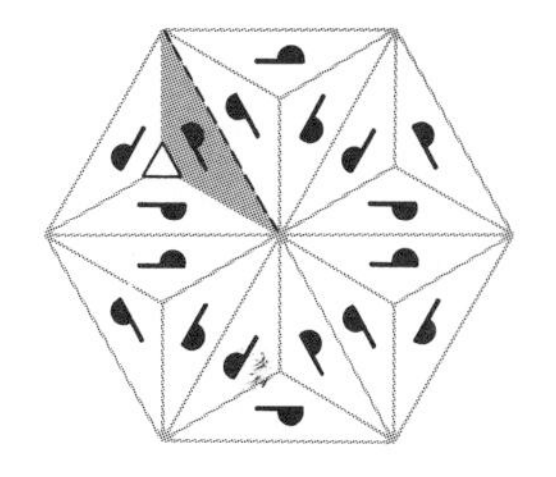

p31m generated by a reflection and a rotation through 120°

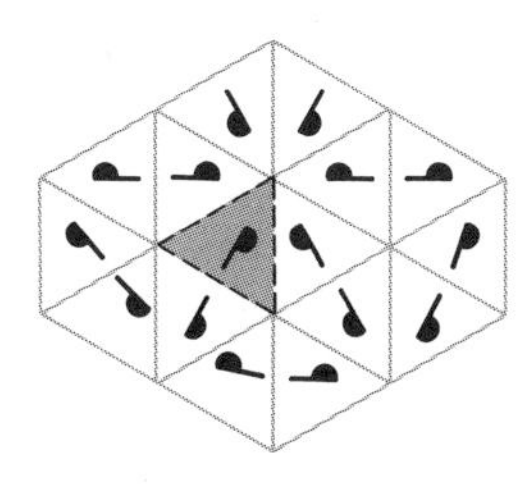

p3m1 generated by three reflections in the sides of an equilateral triangle

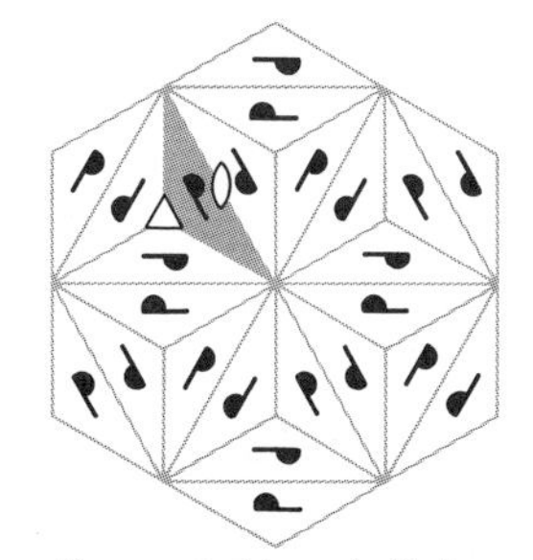

p6 generated by a half–turn and a rotation of 120°

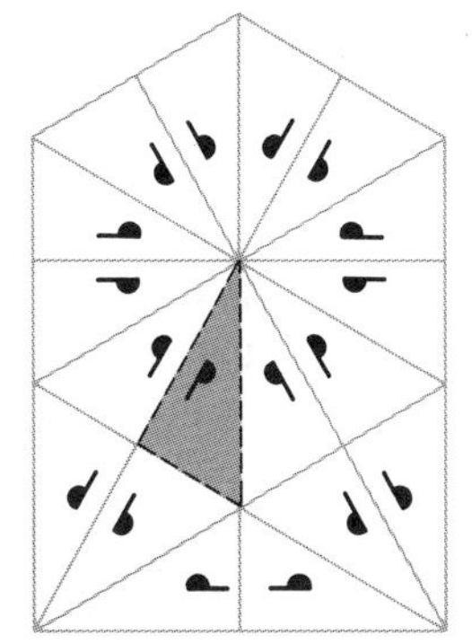

p6m generated by reflections in the sides of a (30°, 60°, 90°) triangle

The group associated with any solution of this problem will be the symmetry group of the figure formed by the edges of all the polygons making up the solution. The fundamental fact connecting group theory and two-dimensional crystallography is this: there are just 17 groups that arise in this way. These are the *two-dimensional crystallographic groups*, and it turns out that they are precisely the discrete symmetry groups (for plane figures) that leave neither a point nor a line invariant. Thus they are the groups of the patterns in Figure 35.9. (Each of the finite groups in Section 34 leaves a point invariant, and each of the frieze groups in Section 35 leaves a line invariant.)

Figure 36.3 contains diagrams corresponding of each of the 17 groups. These diagrams are based on figures in [2] and give the following information.

1. The shaded area is a fundamental region.
2. The generators of the group are given by the following scheme:

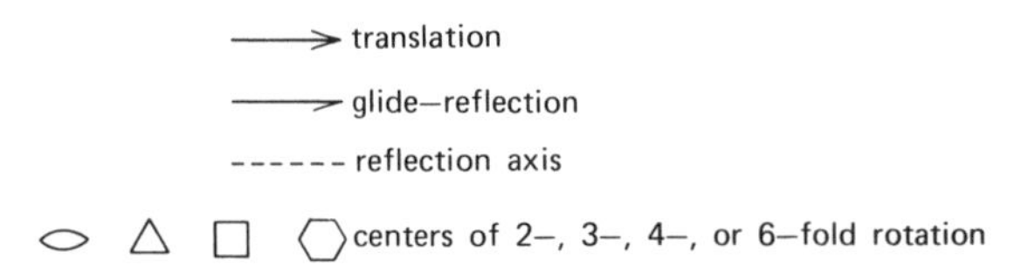

By starting with a fundamental region and subjecting it repeatedly to the motions suggested by the generators, the entire plane will be filled. Each element of the group will correspond to a different congruent copy of the fundamental region. The notation **p1**, **p2**,... for the groups is based on [5], and belongs to one of several different systems used to describe such groups. The notation is given here for convenience; it will not be explained.

The groups that are of interest in crystallography arise from the three-dimensional version of the problem just considered. These groups can be described as follows. First, a *lattice group* is a nontrivial discrete motion group whose elements are translations. If, for example, a rectangular coordinate system is chosen for three-dimensional space, then the three translations of unit length in the positive directions parallel to the coordinate axes generate a lattice group. If G is a lattice group, then any G-orbit is called a *lattice*.† (A G-orbit consists of all the points that can be reached by applying motions of G to a single point of the lattice. See Section 31.) For the example mentioned, the lattice containing the origin would consist of all points with integral coordinates.

†This use of the word *lattice* is different from that in Chapter XVII. Both uses of the word are well established. Confusion is unlikely.

Figure 36.4

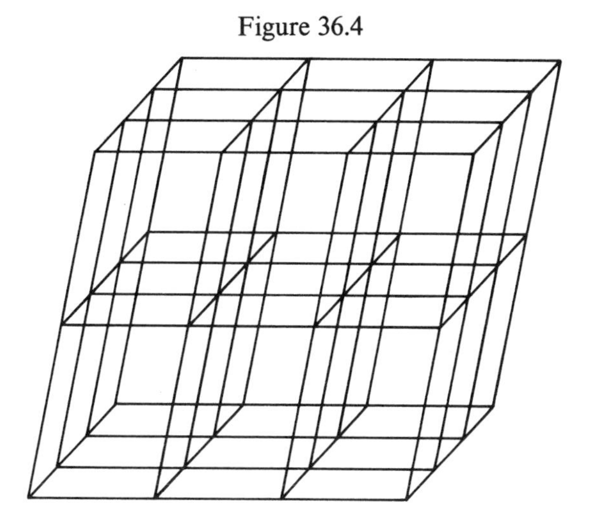

The lattice and associated group are called one-, two-, or three-dimensional depending on whether the points of the lattice are collinear, coplanar but not collinear, or not coplanar, respectively. The points of intersection of the lines forming the parallelograms in Figure 36.2 form part of a two-dimensional lattice. The points of intersection of the lines in Figure 36.4 form part of a three-dimensional lattice.

A discrete group of motions is a *crystallographic point group* if it has an invariant point p and leaves invariant some three-dimensional lattice containing p (the individual lattice points other than p will in general not be invariant, of course). A discrete group of motions is a *crystallographic space group* if its translations form a three-dimensional lattice group.

As stated at the outset, there are 32 crystallographic point groups, each of which is finite. Eleven of these groups contain only rotations and are among the groups in Theorem 34.3 ($C_1, C_2, C_3, C_4, C_6, D_2, D_3, D_4, D_6$, the tetrahedral group, and the octahedral group). Each of the other 21 point groups contains improper motions (reflections through a plane or a point) and contains one of the 11 listed groups of proper motions as a subgroup of index 2.

If G denotes a crystallographic space group, then the translations in G form a normal subgroup T that is a lattice group, and the quotient group G/T is one of the 32 crystallographic point groups. In the 1890s, E. S. Fedorov (Russian), A. Schönflies (German), and W. Barlow (British) showed, independently, that there are only 230 crystallographic space groups. Of these, 65 contain only proper motions and the remaining 165 contain improper motions. These groups are the basis for the classification of crystals by symmetry type and have been the subject of extensive study since their importance was first realized in the middle of the last century. The Introduction contains some brief remarks about this use of groups; for more details see [6] and the references given there.

The proof that there are only 230 crystallographic space groups is, as could be guessed, not particularly easy. One of the restricting factors is the following theorem, which will be proved in Section 37.

The Crystallographic Restriction. *Any nontrivial rotation in a two- or three-dimensional crystallographic point group is of order* 2, 3, 4, *or* 6.

At the International Congress of Mathematicians in 1900, the German mathematician David Hilbert (1862–1943) presented a collection of 23 problems, which have attracted the interest of many mathematicians throughout this century. The eighteenth problem on Hilbert's list was divided into three parts; the second part was this: "Whether polyhedra also exist which do not appear as fundamental regions of groups of motions, by means of which nevertheless by a suitable juxtaposition of congruent copies a complete filling up of all [Euclidean] space is possible." A complicated three-dimensional example was given by K. Reinhardt in 1928. Figure 36.2(a) is a two-dimensional example, which was constructed by H. Heesch in 1935. Further references are given in the paper by John Milnor on Hilbert's Eighteenth Problem in [11].

PROBLEMS

36.1. Determine the symmetry group of each of the designs in Figure 35.9. (Use the notation from Figure 36.3.)

36.2. Determine the symmetry group of each of the three designs in Figure 1 (of the Introduction).

36.3. Which of the 17 two-dimensional crystallographic groups contain rotations of order 4? 6? Which contain reflections? Which contain glide-reflections that are not reflections?

36.4. Draw 17 figures, different from those in the book, illustrating the 17 types of symmetry shown in Figure 36.3. (Be sure the figures do not have too much symmetry in each case.)

36.5. Verify that the rotations in the 32 crystallographic point groups are consistent with the crystallographic restriction.

36.6. Prove that the only values of n for which the plane can be filled with congruent regular n-gons having no overlap except at their edges are $n=3$, 4, and 6. [Each of the interior angles of a regular n-gon has measure $(n-2)\pi/n$. If r such polygons have a vertex in common, then $[r(n-2)\pi/n]=2\pi$. For which integral values of n and r is this possible?]

SECTION 37. THE EUCLIDEAN GROUP

The distance-preserving mappings of three-dimensional space form a group, called the *Euclidean group*, which is implicit in all of the mathematics and science based on Euclidean geometry. In agreeing not to distinguish between two congruent figures, we are in essence agreeing not to distinguish between the figures if there is an element of the Euclidean group that maps one of the figures onto the other. All of the symmetry groups of two-or three-dimensional figures are subgroups of the Euclidean group. In this section this group will be studied in the context of coordinate geometry and matrices. It will be shown that the Euclidean group is built up from several of its important subgroups, and the crystallographic restriction will be proved. The section uses ideas and notation from linear algebra that are reviewed in Appendix D.

Let $E(3)$ denote the Euclidean group, and let $\mathbb{R}^3$ denote the vector space of all 3-tuples of real numbers. We shall begin by studying those elements of $E(3)$ that are also linear transformations. It will be convenient to identify the points of three-dimensional Euclidean space with the elements of $\mathbb{R}^3$ in the usual way. To do this, first choose a unit of length and three mutually perpendicular coordinate axes. Then associate the geometric unit vectors along the coordinate axes with the unit vectors $e_1=(1,0,0)$, $e_2=(0,1,0)$, and $e_3=(0,0,1)$ in $\mathbb{R}^3$. Relative to the basis $\{e_1,e_2,e_3\}$, each linear transformation of $\mathbb{R}^3$ is represented by a unique matrix in $M(3,\mathbb{R})$. The elements of $E(3)$ are invertible, and so if a linear transformation is in $E(3)$, then the corresponding matrix must be in $GL(3,\mathbb{R})$. Therefore, relative to a fixed coordinate system, the set of linear transformations in $E(3)$ is $E(3)\cap GL(3,\mathbb{R})$. This is an intersection of groups so that it is also a group. It is called the (*real*) *orthogonal group* and is denoted by $O(3)$. To say more about the elements of $O(3)$, we shall look at them in terms of coordinates and matrices.

The *inner product* of vectors $v=(x_1,x_2,x_3)$ and $w=(y_1,y_2,y_3)$ is defined by

$$\langle v,w\rangle = \sum_{i=1}^{3} x_i y_i.$$

In particular, then,

$$\langle e_i,e_j\rangle = \delta_{ij} \tag{37.1}$$

where $\delta_{ii}=1$ and $\delta_{ij}=0$ for $i\neq j$.

The *length* of a vector $v=(x,y,z)\in\mathbb{R}^3$ is given by

$$\|v\| = \langle v,v\rangle^{\frac{1}{2}} = (x^2+y^2+z^2)^{\frac{1}{2}}. \tag{37.2}$$

The *distance* between vectors v and w is given by

$$d(v,w) = \|v - w\|. \tag{37.3}$$

Therefore, a mapping $\alpha: \mathbb{R}^3 \to \mathbb{R}^3$ is in $E(3)$ iff

$$\|\alpha(v) - \alpha(w)\| = \|v - w\| \tag{37.4}$$

for all $v, w \in \mathbb{R}^3$. If the mapping α is linear, then $\alpha(v) - \alpha(w) = \alpha(v - w)$. Therefore, a linear mapping is distance-preserving iff it is length-preserving (Problem 37.1). It follows that $\alpha \in O(3)$ iff α is linear and

$$\|\alpha(v)\| = \|v\| \tag{37.5}$$

for all $v \in \mathbb{R}^3$. It can also be proved (Problem 37.3) that if $\alpha \in O(3)$, then α preserves inner products; that is,

$$\langle \alpha(v), \alpha(w) \rangle = \langle v, w \rangle \tag{37.6}$$

for all $v, w \in \mathbb{R}^3$. The matrix form of α is $A = [a_{ij}]$, where

$$\alpha(e_i) = \sum_{j=1}^{3} a_{ji} e_j \qquad (1 \le i \le 3). \tag{37.7}$$

If we compute $\langle \alpha(e_i), \alpha(e_j) \rangle$, and make use of (37.1), (37.6), and (37.7), we obtain

$$\sum_{k=1}^{3} a_{ki} a_{kj} = \delta_{ij} \tag{37.8}$$

for $1 \le i \le 3$, $1 \le j \le 3$ (Problem 37.4). From these remarks and calculations it follows that

$$\alpha \in O(3) \qquad \text{iff} \qquad A'A = I_3, \tag{37.9}$$

where A is the matrix of α relative to $\{e_1, e_2, e_3\}$.

A matrix A such that $A'A = I_3$ is called *orthogonal*. Any such matrix is invertible, with $A^{-1} = A'$. If $\det A$ is used to denote the determinant of a matrix A, then $\det A = \pm 1$ if A is orthogonal, because if A is orthogonal then $\det(A'A) = (\det A')(\det A) = (\det A)^2$. We recall that if A_1 and A_2 are matrices representing the same linear transformation α relative to different bases, then $\det A_1 = \det A_2$ (Problem 37.5); therefore, $\det \alpha$ can be defined to be $\det A$, where A is the matrix representing α relative to any basis.

The mapping $\alpha \mapsto \det \alpha$ defines a homomorphism from $O(3)$ onto $\{1, -1\}$, because $\det(\alpha\beta) = (\det \alpha)(\det \beta)$. The kernel of this homomorphism is a normal subgroup of $O(3)$, called the *special orthogonal group* and denoted by $SO(3)$. Notice that $[O(3):SO(3)] = 2$ because $O(3)/SO(3) \approx \{1, -1\}$. We shall now prove that $SO(3)$ consists of the rotations of Euclidean space about axes passing through the origin.

Lemma 37.1. *If $\alpha \in SO(3)$, then there is a unit vector u_3 such that $\alpha(u_3) = u_3$ and α is a rotation with u_3 as the axis.*

PROOF. Let A be the matrix of α relative to the basis $\{e_1,e_2,e_3\}$. Then $\det(A-I_3)=\det(A-I_3)'=\det(A'-I_3')=\det(A^{-1}-I_3)=(\det A^{-1})(\det(I_3-A))=(\det A^{-1})(\det(-I_3))(\det(A-I_3))=-\det(A-I_3)$. Therefore, $\det(A-I_3)=0$ so that $\lambda=1$ is a characteristic value of A. A characteristic vector u corresponding to $\lambda=1$ will satisfy $\alpha(u)=u$, as required.

Let u_3 be a unit vector in the direction of u, and choose u_1 and u_2 so that $\{u_1,u_2,u_3\}$ is an orthonormal basis for $\mathbb{R}^3$. Then

$$\langle\alpha(u_i),\alpha(u_j)\rangle = \langle u_i,u_j\rangle = \delta_{ij}. \tag{37.10}$$

The relations given by (37.10) can be used to show that the matrix of α relative to the basis $\{u_1,u_2,u_3\}$ is

$$\begin{bmatrix} \cos\theta & -\sin\theta & 0 \\ \sin\theta & \cos\theta & 0 \\ 0 & 0 & 1 \end{bmatrix} \tag{37.11}$$

for $0\leq\theta<2\pi$(Problem 37.6). This shows that α is a clockwise rotation through angle θ about the axis u_3. □

Another important subgroup of $E(3)$ is $T(3)$, the group of *translations*. It can be proved that

$$T(3) \lhd E(3), \qquad \frac{E(3)}{T(3)} \approx O(3), \qquad \text{and} \qquad |T(3)\cap O(3)| = 1.$$

In the language of Section 30, $E(3)$ is an extension of $T(3)$ by $O(3)$. (Proofs of these facts can be found in reference [10].) Instead of pursuing this, we return now to the following fact, which was stated without proof in Section 36.

The Crystallographic Restriction. *Any nontrivial rotation in a two- or three-dimensional crystallographic point group is of order* 2, 3, 4, *or* 6.

PROOF. The proof will be written for three-dimensional groups; the two-dimensional case follows from this (Problem 37.7).

Let α be a rotation in a three-dimensional crystallographic point group, and assume that α leaves the point p fixed and the lattice L invariant. Let $\{v_1,v_2,v_3\}$ be a basis for L; then $\{v_1,v_2,v_3\}$ will also be a basis for $\mathbb{R}^3$. The matrix for α relative to this basis is $A=[a_{ij}]$, defined by

$$\alpha(v_i) = \sum_{j=1}^{n} a_{ji}v_j; \tag{37.12}$$

and each a_{ji} must be an integer by the invariance of L. As shown in the proof of Lemma 37.1, there is a basis for $\mathbb{R}^3$ relative to which the matrix of α has the form (37.11). The trace of the matrix of a linear transformation is invariant under a change of basis; therefore, since the trace $a_{11}+a_{22}+a_{33}$

is an integer, $2\cos\theta+1$ must be an integer, where θ is the angle of rotation for α(from Lemma 37.1). The only values of θ $(0\leq\theta<2\pi)$ for which $2\cos\theta+1$ is an integer are $\theta=0$, $\pi/2$, $3\pi/2$, $\pi/3$, $2\pi/3$, π, $4\pi/3$, and $5\pi/3$. The orders of the rotations through these angles are those in the crystallographic restriction. □

PROBLEMS

37.1. Prove that a linear mapping is distance-preserving iff it is length-preserving. [See the discussion following Equation (37.4).]

37.2. Verify that the inner product used in this section has these properties (for all $u,v,w\in\mathbb{R}^3$ and all $a\in\mathbb{R}$).

(a) $\langle u+v,w\rangle=\langle u,w\rangle+\langle v,w\rangle$

(b) $\langle u,v+w\rangle=\langle u,v\rangle+\langle u,w\rangle$

(c) $\langle au,v\rangle=\langle u,av\rangle=a\langle u,v\rangle$

(d) $\langle u,v\rangle=\langle v,u\rangle$

37.3. Prove that if $\alpha\in O(3)$, then α preserves inner products. [See Equation (37.6). *Suggestion*: Use $\langle\alpha(v+w),\,\alpha(v+w)\rangle=\langle v+w,v+w\rangle$ and Problem 37.2.]

37.4. Verify Equation (37.8).

37.5. Prove that if A_1 and A_2 are matrices representing the same linear transformation relative to different bases, then $\det A_1=\det A_2$. (Recall that $A_2=BA_2B^{-1}$ for an appropriate B, and $\det(XY)=\det(X)\det(Y)$.)

37.6. Prove the statement containing Equation (37.11).

37.7. Explain why the crystallographic restriction for three-dimensional groups implies the crystallographic restriction for two-dimensional groups. (*Suggestion*: Consider Figure 34.7.)

NOTES ON CHAPTER IX

1. Coxeter, H. S. M., *Introduction to Geometry*, 2nd ed., John Wiley, New York, 1969.
2. Coxeter, H. S. M., and W. O. J. Moser, *Generators and Relations for Discrete Groups*, 3rd ed., Springer-Verlag, Berlin, 1972.
3. Fejes Tóth, L., *Regular Figures*, Macmillan, New York, 1964.
4. Griesbach, C. B., *Historic Ornament: A Pictorial Archive*, Dover, New York, 1975.
5. Henry, N. F. M. and K. Lonsdale, eds., *International Tables for X-Ray Crystallography*, Kynock Press, Birmingham, England, 1965.
6. Hurlbut, C. S., Jr., and C. Klein, after J. D. Dana, *Manual of Mineralogy*, 19th ed., John Wiley, New York, 1977.
7. Jones, O., *The Grammar of Ornament,* Bernard Quaritch Ltd., London, 1910, 1928.
8. Kepler, J., *The Six-Cornered Snowflake*, English trans., Oxford University Press, London, 1966.

9. MacGillavry, C. H., *Fantasy & Symmetry: The Periodic Drawings of M. C. Escher*, Harry N. Abrams, New York, 1976.
10. Miller, W., Jr., *Symmetry Groups and Their Applications*, Academic Press, New York, 1972.
11. Milnor, J., Hilbert's Problem 18: On crystallographic groups, fundamental domains, and on sphere packing, *Proceedings of Symposia in Pure Mathematics, Vol. XXVIII*, American Mathematical Society, 1976.
12. Shubnikov, A. V., and V. A. Koptsik, *Symmetry in Art and Science*, Plenum Press, New York, 1974. (Translated from the 1972 Russian edition.)
13. Weyl, H., *Symmetry*, Princeton University Press, Princeton, 1952.
14. Yale, P. B., *Geometry and Symmetry*, Holden-Day, San Francisco, 1968.

CHAPTER X

Factorization of Integers

This chapter contains proofs of the basic facts about divisibility and factorization of the integers, culminating with a proof that every integer greater than 1 can be written uniquely (except for the order of the factors) as a product of primes. These facts are proved here not only because they are often needed in algebra, but also because they will serve as a model for similar facts about polynomials that will be proved in Chapter XI. It will be shown in Sections 43 and 47, moreover, that questions about divisibility and factorization are important in even broader contexts than that of the integers and polynomials. The only prerequisite for this chapter is Section 10.

SECTION 38. GREATEST COMMON DIVISORS. THE EUCLIDEAN ALGORITHM

Theorem 38.1. *If a and b are integers, not both zero, then there is a unique positive integer d such that*

(a) $d|a$ and $d|b$, and

(b) if c is an integer such that $c|a$ and $c|b$, then $c|d$.

Property (a) states that d is a *common divisor* of a and b; property (b) ensures that d is the greatest such divisor. Therefore, the integer d in the theorem is called the *greatest common divisor* of a and b. It is denoted (a,b). (The context will usually make it clear whether this or some other interpretation of the ordered pair notation is intended.) Examples are $(4,-6)=2$, $(-7,0)=7$, and $(25,33)=1$.

The following proof of Theorem 38.1 shows how to compute (a,b) by a systematic procedure known as the *Euclidean Algorithm*. Another proof, which shows the existence of (a,b), but not how to compute it, is outlined in Problem 38.13.

PROOF OF THEOREM 38.1. First consider the case $b>0$. By the Division Algorithm (Section 10), there are unique integers q_1 and r_1 such that

$$a = bq_1 + r_1, \qquad 0 \le r_1 < b.$$

If $r_1 = 0$, then $b|a$, and b will satisfy the conditions for d in parts (a) and (b). If $r_1 \neq 0$, we can apply the Division Algorithm again, getting integers q_2 and r_2 such that

$$b = r_1q_2 + r_2, \qquad 0 \le r_2 < r_1.$$

Repeated application of the Division Algorithm in this way produces a sequence of pairs of integers $q_1, r_1;\ q_2, r_2;\ q_3, r_3; \ldots$ such that

$$\begin{array}{ll} a = bq_1 + r_1, & 0 \le r_1 < b \\ b = r_1q_2 + r_2, & 0 \le r_2 < r_1 \\ r_1 = r_2q_3 + r_3, & 0 \le r_3 < r_2 \\ \cdots & \end{array} \tag{38.1}$$

Because each remainder is nonnegative, and $r_1 > r_2 > r_3 > \ldots$, we must eventually reach a remainder that is zero. If r_{k+1} denotes the first zero remainder, then the process terminates with

$$\begin{array}{ll} r_{k-2} = r_{k-1}q_k + r_k, & 0 < r_k < r_{k-1} \\ r_{k-1} = r_kq_{k+1}. & \end{array}$$

We shall show that r_k, *the last nonzero remainder*, satisfies requirements (a) and (b) for d in the theorem.

Notice first that $r_k | r_{k-1}$, because $r_{k-1} = r_kq_{k+1}$. But then $r_k | r_{k-2}$, because $r_{k-2} = r_{k-1}q_k + r_k$ and $r_k | r_{k-1}$ and $r_k | r_k$. Continuing in this way, we can work through the equations in (38.1), from the end, and obtain $r_k | r_{k-1}, r_k | r_{k-2}, r_k | r_{k-3}, \ldots$, until we arrive at $r_k | r_1$, $r_k | b$, and finally $r_k | a$. Thus r_k is a common divisor of a and b.

Now, moving to condition (b), assume that $c|a$ and $c|b$. Then $c|r_1$ because $r_1 = a - bq_1$. But if $c|b$ and $c|r_1$, then $c|r_2$, because $r_2 = b - r_1q_2$. Continuing in this way, we can work through the equations in (38.1), from the beginning, and obtain $c|r_1$, $c|r_2$, $c|r_3, \ldots$, and finally $c|r_k$. This verifies condition (b).

If $b<0$, we simply go through the same process starting with a and $-b$, rather than a and b; and then observe that since b and $-b$ have the same divisors, a greatest common divisor of a and $-b$ will also be a greatest common divisor of a and b.

If $b=0$, then $|a|$ satisfies the requirements (a) and (b) for d in the theorem.

To prove the uniqueness of (a,b), assume that d_1 and d_2 are integers each satisfying both of the requirements (a) and (b) for d. Then $d_1|d_2$ and $d_2|d_1$. Therefore, since both d_1 and d_2 are positive, $d_1=d_2$. □

Example 38.1. Here is the Euclidean Algorithm applied to compute (1001, 357).

$$\begin{aligned} 1001 &= 357\cdot 2+287 \\ 357 &= 287\cdot 1+70 \\ 287 &= 70\cdot 4\ +7 \\ 70 &= 7\cdot 10 \end{aligned}$$

Therefore $(1001, 357)=7$.

If a and b are integers, then any integer that is equal to $am+bn$ for some integers m and n is said to be a *linear combination* of a and b. The equations that arise in the Division Algorithm can be used to express (a,b) as a linear combination of a and b. Before proving this, let us illustrate the idea with an example.

Example 38.2. Using the equations in Example 38.1, starting with the first, we can express each of the successive remainders, 287, 70, and 7, as a linear combination of 1001 and 357.

$$\begin{aligned} 287 &= 1001-357\cdot 2 \\ 70 &= 357-287\cdot 1 \\ &= 357-(1001-357\cdot 2)\cdot 1 \\ &= 357\cdot 3+1001(-1) \\ 7 &= 287-70\cdot 4 \\ &= (1001-357\cdot 2)-[357\cdot 3+1001(-1)]4 \\ &= 1001\cdot 5+357(-14) \end{aligned}$$

Thus $(1001, 357)=1001m+357n$ for $m=5$ and $n=-14$.

Theorem 38.2. *The greatest common divisor of integers a and b, not both zero, can be expressed as a linear combination of a and b*:

$$(a,b) = am+bn \text{ for some integers } m \text{ and } n.$$

Proof. We shall work through the equations in (38.1), starting with the first, and show in turn that each of the remainders $r_1, r_2, \ldots$ can be expressed as a linear combination of a and b. Since $r_k=(a,b)$, this will prove the theorem.

The first equation in (38.1) yields

$$r_1 = a - bq_1,$$

a linear combination of a and b. Using this result in the second equation of (38.1), we can write

$$\begin{aligned} r_2 &= b - r_1q_2 \\ &= b-(a-bq_1)q_2 \\ &= a(-q_2)+b(1+q_1q_2), \end{aligned}$$

again a linear combination of a and b.

In general, if each of the remainders $r_1, r_2, \ldots, r_j$ has been expressed as a linear combination of a and b, then, in particular,

$$r_{j-1} = am_{j-1} + bn_{j-1}$$

and

$$r_j = am_j + bn_j$$

for some integers m_{j-1}, n_{j-1}, m_j, and n_j. Then the equation $r_{j-1}=r_jq_{j+1}+r_{j+1}$ yields

$$\begin{aligned} r_{j+1} &= r_{j-1} - r_jq_{j+1} \\ &= (am_{j-1}+bn_{j-1})-(am_j+bn_j)q_{j+1} \\ &= a(m_{j-1}-m_jq_{j+1})+b(n_{j-1}-n_jq_{j+1}), \end{aligned}$$

a linear combination of a and b. This ensures that each successive remainder, including r_k, can be expressed as a linear combination of a and b, which thereby proves the theorem. □

An application of Theorem 38.2 will come at the beginning of the next section. For the moment, we settle for this observation: Although the Euclidean algorithm can be used to calculate integers m and n such that $(a,b)=am+bn$, it is the mere existence of such m and n, not their actual calculation, that is most often important. In trying to prove statements involving greatest common divisors, it is frequently helpful to begin just by trying to make use of Theorem 38.2. Lemma 39.1 illustrates this point.

PROBLEMS

38.1. Use the Euclidean algorithm to compute each of the following greatest common divisors.

(a) $(1001,33)$ (b) $(56,126)$ (c) $(1386,90)$
(d) $(27,0)$ (e) $(234,-2341)$ (f) $(-2860,-2310)$

38.2. (a) to (f). Express each greatest common divisor in Problem 38.1 as a linear combination of the two given integers.

38.3. Verify that $(2,3)$ can be expressed in at least two different ways as a linear combination of 2 and 3. (This shows that the integers m and n in Theorem 38.2 are not uniquely determined by a and b. Problem 38.4 gives a much stronger statement.)

38.4. Prove that if a and b are integers, not both zero, then there are infinitely many pairs of integers m,n such that $(a,b)=am+bn$.

38.5. Prove that if a and b are integers, then $(a,b)=1$ iff there are integers m and n such that $am+bn=1$. (Integers having greatest common divisor 1 are said to be *relatively prime*.)

38.6. Prove that if c is a positive integer then $(ac,bc)=(a,b)c$.

38.7. Prove that if d is a positive integer, $d|a$, and $d|b$, then $(a,b)=d$ iff $(a/d,b/d)=1$.

38.8. Prove that if p is a prime and a is an integer, and $p\nmid a$, then $(a,p)=1$.

38.9. Prove that if $(a,c)=1$ and $(b,c)=1$, then $(ab,c)=1$.

38.10. Prove that if $c|ab$ and $(a,c)=d$, then $c|bd$.

38.11. (a) Prove that if a and b are integers, with $n>0$ and $d=(a,n)$, then $\langle[a]\rangle=\langle[d]\rangle$ in $\mathbb{Z}_n$.

(b) Prove that $\langle[a]\rangle=\mathbb{Z}_n$ iff $(a,n)=1$.

(c) Prove that in $\mathbb{Z}_n$, $\langle[a]\rangle=\langle[b]\rangle$ iff $(a,n)=(b,n)$.

38.12. For each positive integer n, let $\phi(n)$ denote the number of positive integers that are less than n and relatively prime to n (see Problem 38.5). For example, $\phi(3)=2$ and $\phi(4)=2$.

(a) Compute $\phi(k)$ for $1\leq k\leq 10$.

(b) If p is a prime, what is $\phi(p)$?

(c) Prove that $\mathbb{Z}_n$ has $\phi(n)$ different generators for $n>1$. (See Problem 38.11. Also compare Problem 27.7.)

38.13. Give an alternate proof of Theorem 38.1 by justifying each of the following steps.

(a) Let $S=\{ax+by|x,y\in\mathbb{Z}\}$. The set S contains at least one positive integer.

(b) There is a least positive integer in S; let it be $d=am+bn(m,n\in\mathbb{Z})$. (This will be the d of Theorem 38.1.)

(c) Because $a\in S$ and $b\in S$, to prove Theorem 38.1(a) it suffices to prove that d divides each positive integer in S.

(d) Let $k=ua+vb$ denote a positive integer in S. There are integers q and r such that

$$k=dq+r,\ 0\leq r<d.$$

Therefore,

$$(au+bv)=(am+bn)q+r,$$

and

$$r=(u-mq)a+(v-nq)b,$$

which implies $r\in S$.

(e) Therefore $r=0$, and d does divide each positive integer in S. (As stated, this shows that $d|a$ and $d|b$.)
(f) If $c|a$ and $c|b$, then $c|d$.

SECTION 39. THE FUNDAMENTAL THEOREM OF ARITHMETIC

Fundamental Theorem of Arithmetic. *Each integer greater than 1 can be written as a product of primes, and, except for the order in which these primes are written, this can be done in only one way.*

Thus $15=3\cdot5=5\cdot3$, $16=2^4$, and $17=17$, the last example showing that "product" is taken to include the possibility that there is only one factor present. We shall prove two lemmas before proving the theorem.

Lemma 39.1. *If a, b, and c are integers, with $a|bc$ and $(a,b)=1$, then $a|c$.*

PROOF. Since $(a,b)=1$, by Theorem 38.2 there are integers m and n such that $1=am+bn$. On multiplying both sides of this equation by c, we get $c=amc+bnc$. Certainly $a|amc$. And we are assuming that $a|bc$, so that $a|bnc$. Therefore, a divides $amc+bnc$, which is equal to c. □

Lemma 39.2. *If p is a prime, $a_1,a_2,\ldots,a_n$ are integers, and $p|a_1a_2\cdots a_n$, then $p|a_i$ for some $i(1\leq i\leq n)$.*

PROOF. Induct on n. The case $n=1$ is obvious. Assume that $n>1$. If $p|a_1a_2\cdots a_{n-1}$, then $p|a_i$ for some i $(1\leq i\leq n-1)$ by the induction hypothesis. If $p\nmid a_1a_2\cdots a_{n-1}$, then $(p,a_1a_2\cdots a_{n-1})=1$ because p is a prime. In this case, by Lemma 39.1 (with $a=p$, $b=a_1a_2\cdots a_{n-1}$, $c=a_n$), $p|a_n$. □

PROOF OF THE THEOREM. Let S denote the set of those integers greater than 1 that *cannot* be written as a product of primes. To prove the first part of the theorem we must show that S is empty. Assume otherwise. Then, by the Least Integer Principle (Section 10), S contains a least element, which we denote by n. The set S contains no primes, and so n can be factored as $n=n_1n_2$, where both n_1 and n_2 are integers and $1<n_1<n$ and $1<n_2<n$. Because $n_1<n$ and $n_2<n$, and n is the least element of S, it follows that

$n_1 \notin S$ and $n_2 \notin S$. Thus n_1 and n_2 both can be written as products of primes, and so the same is true of $n = n_1 n_2$. This contradicts the fact that $n \in S$, and so we conclude that S must be empty, as required.

To prove the last part of the theorem, assume m to be an integer greater than 1, and assume $m = p_1 p_2 \cdots p_s = q_1 q_2 \cdots q_t$, where the p_i and q_j are primes. Then $p_1 | q_1 q_2 \cdots q_t$ because $p_1 | p_1 p_2 \cdots p_s$. Thus $p | q_j$ for some j $(1 \le j \le t)$, by Lemma 39.2. But then $p_1 = q_j$ because both p_1 and q_j are primes. It follows that

$$p_2 p_3 \cdots p_s = q_1 q_2 \cdots q_{j-1} q_{j+1} \cdots q_t,$$

where we have canceled p_1 on the left and q_j on the right. Repeating the argument just used, we see that p_2 must equal one of the remaining prime factors on the right. Continuing in this way, we can pair each prime on the left with a prime on the right. The primes on one side cannot all be canceled before those on the other side because that would imply that 1 is equal to a product of primes, an impossibility. Thus $s = t$ and the lists $p_1, p_2, \ldots, p_s$ and $q_1, q_2, \ldots, q_t$ must be the same except possibly for their arrangement. □

By arranging the prime factors in increasing order, we see that each integer $n > 1$ can be written in the form

$$n = p_1^{e_1} p_2^{e_2} \cdots p_k^{e_k} \qquad (p_1 < p_2 < \cdots < p_k), \tag{39.1}$$

where the primes $p_1, p_2, \ldots, p_k$ and the positive integers $e_1, e_2, \ldots, e_k$ are uniquely determined by n. We shall call this the *standard form* for n.

PROBLEMS

39.1. Prove that if $(a,b) = 1$, $a|m$, and $b|m$, then $ab|m$. (*Suggestion*: If $m = ak$, then $b|k$ by Lemma 39.1.)

39.2. Determine the standard form (39.1) for each of the following integers.
(a) 105 (b) 684 (c) 1375 (d) 139

39.3. Let

$$m = p_1^{s_1} p_2^{s_2} \cdots p_k^{s_k}$$

and

$$n = p_1^{t_1} p_2^{t_2} \cdots p_k^{t_k},$$

where $p_1, p_2, \ldots, p_k$ are distinct prime numbers, $s_i \ge 0$ for $1 \le i \le k$, and $t_i \ge 0$ for $1 \le i \le k$. Prove that $m|n$ iff $s_i \le t_i$ for $1 \le i \le k$.

39.4. Determine all positive integral divisors of each of the following integers.
(a) 16 (b) 27 (c) $2^3 3^2$ (d) $2^r 3^s$

39.5. Determine the number of positive integral divisors of an integer n having the standard form $p_1^{e_1} p_2^{e_2} \cdots p_k^{e_k}$. (Compare Problem 39.4.)

39.6. An integer is *square-free* if it is not divisible by the square of any integer greater than 1. An integer n is a *perfect square* if $n=k^2$ for some integer k.
(a) Prove that n is square-free iff in the standard form (39.1) each $e_i=1$.
(b) Prove that n is a perfect square iff in the standard form (39.1) each e_i is even.
(c) Prove that every integer greater than 1 is the product of a square-free integer and a perfect square.

39.7. Prove that if n is an integer, then $\sqrt{n}$ is rational iff n is a perfect square (see Problem 39.6). (*Suggestion*: Apply the Fundamental Theorem of Arithmetic to $a^2=nb^2$. Compare Theorem 25.1.)

39.8. Prove that $\sqrt[3]{2}$ is irrational. (*Suggestion*: Apply the Fundamental Theorem of Arithmetic to $a^3=2b^3$.)

39.9 State and prove a theorem characterizing those integers n for which $\sqrt[3]{n}$ is rational. (Compare Problems 39.7 and 39.8.)

39.10. Prove that if a and b are integers, not both zero, then there is a unique positive integer m such that
(a) $a|m$ and $b|m$, and
(b) if c is an integer such that $a|c$ and $b|c$, then $m|c$. The integer m is called the *least common multiple* of a and b. It is denoted $[a,b]$.

39.11. Let m and n be as in Problem 39.3, and let

$$u_i = \text{the minimum of } s_i \text{ and } t_i \qquad \text{for each } i$$

and

$$v_i = \text{the maximum of } s_i \text{ and } t_i \qquad \text{for each } i.$$

(a) Prove that $(m,n)=p_1^{u_1}p_2^{u_2}\cdots p_k^{u_k}$.
(b) Prove that $[m,n]=p_1^{v_1}p_2^{v_2}\cdots p_k^{v_k}$. (Here $[m,n]$ denotes the least common multiple, defined in Problem 39.10.)

39.12. Compute each of the following least common multiples. (See Problems 39.10 and 39.11.)

(a) $[6,9]$ (b) $[10,105]$ (c) $[0,20]$
(d) $[-39,54]$ (e) $[56,126]$ (f) $[-2387,143]$

39.13. Prove that if a and b are positive integers, then

$$(a,b)[a,b] = ab.$$

(See Problem 39.11.)

39.14. Prove that if $G=\langle a\rangle$ is a cyclic group of order n, and $0\leq k\leq n$, then $o(a^k)=n/(n,k)$.

39.15. (a) Prove that if $(m,n)=1$, then the rings $\mathbb{Z}_{mn}$ and $\mathbb{Z}_m\times\mathbb{Z}_n$ are isomorphic. (Example 21.3 is the special case $m=2$, $n=3$.)
(b) Prove that if $(m,n)\neq 1$, then $\mathbb{Z}_{mn}\not\approx\mathbb{Z}_m\times\mathbb{Z}_n$.

CHAPTER XI

Polynomials

This chapter presents the facts about polynomials that are necessary for studying fields (Chapter XIII) and polynomial equations (Chapter XIV). Polynomials with real numbers as coefficients will be familiar from elementary algebra and calculus; now we must allow for the possibility that the coefficients are from some ring other than $\mathbb{R}$. A similarity between the theorems in this chapter and those in Chapter X will be obvious; more will be said about this similarity in the last section of this chapter.

SECTION 40. DEFINITION AND ELEMENTARY PROPERTIES

If R is a commutative ring, $a_0, a_1, \ldots, a_n \in R$, and $x \in R$, then

$$a_0 + a_1x + a_2x^2 + \cdots + a_nx^n \tag{40.1}$$

is an element of R. The expression in (40.1) is a polynomial in x: it is a sum of terms, each of which is some element of R times a nonnegative integral power of x. We become acquainted with such expressions, and how to add and multiply them, in elementary algebra. In what follows we shall need to work with these expressions when the symbol x denotes, possibly, something other than an element of R. Whether x denotes an element of R or not, we shall define operations on the set of all such expressions in a way that will yield a commutative ring.

Our first problem is that if x is not an element of R, then terms such as a_1x and a_nx^n, as well as "sums" of such terms, may not have a predetermined meaning. One way around this is to consider not (40.1), but rather the sequence $(a_0, a_1, \ldots, a_n, 0, \ldots)$ of elements of R arising from (40.1), and to define appropriate ring operations on the set of all these sequences. This procedure is outlined in the appendix to this section. It has the advantage that it avoids questions about the precise meaning of expressions such as that in (40.1), but it is not the way polynomials are handled in practice. For most purposes the discussion that follows will be more satisfactory.

Definition. Let R be a commutative ring. A *polynomial in indeterminate* x *over* R is an expression of the form (40.1), where the *coefficients* $a_0, a_1, \ldots, a_n$ are elements of R. If $a_n \neq 0$, then the integer n is the *degree* of the polynomial, and a_n is its *leading* coefficient. A polynomial over a field is said to be *monic* if its leading coefficient is the unity of the field. Two polynomials in x are *equal* iff coefficients on like powers of x are equal. The set of all polynomials in x over R will be denoted $R[x]$.

Polynomials are added by adding coefficients on like powers of x. They are multiplied by assuming that the laws of a commutative ring apply to all symbols present (the elements of R, the powers of x, the $+$ sign, and the juxtaposition of the coefficients with powers of x). Before stating the formal definition that follows from this assumption, let us look at an example.

Example 40.1. In $\mathbb{Z}[x]$,

$$\begin{aligned}(2x+5x^2)+(1-3x^2-x^3) &= (0+2x+5x^2+0x^3)+(1+0x+(-3)x^2+(-1)x^3)\\ &= (0+1)+(2+0)x+(5-3)x^2+(0-1)x^3\\ &= 1+2x+2x^2-x^3\end{aligned}$$

and

$$\begin{aligned}(2x+5x^2)(1-3x^2-x^3) &= 2x(1-3x^2-x^3)+5x^2(1-3x^2-x^3)\\ &= (2x-6x^3-2x^4)+(5x^2-15x^4-5x^5)\\ &= 2x+5x^2-6x^3-17x^4-5x^5.\end{aligned}$$

Definition. Let

$$p(x) = a_0 + a_1x + \cdots + a_mx^m$$

and

$$q(x) = b_0 + b_1x + \cdots + b_nx^n$$

be polynomials over a commutative ring R. Then

$$p(x)+q(x)=(a_0+b_0)+(a_1+b_1)x+\cdots+(a_n+b_n)x^n + a_{n+1}x^{n+1}+\cdots+a_mx^m, \qquad \text{for } m \geq n, \tag{40.2}$$

with a similar formula if $m<n$. And

$$p(x)q(x) = a_0b_0+(a_0b_1+a_1b_0)x+\cdots+a_mb_nx^{m+n}, \tag{40.3}$$

the coefficient of x^k being

$$a_0b_k+a_1b_{k-1}+a_2b_{k-2}+\cdots+a_kb_0.$$

Example 40.2. In $\mathbb{Z}_4[x]$,

$$\begin{aligned}([2]+[2]x)+([2]+[3]x-[1]x^2) &= ([2]\oplus[2])+([2]\oplus[3])x+(-[1])x^2\\ &=[0]+[1]x+[-1]x^2\\ &=[1]x+[3]x^2\end{aligned}$$

and

$$\begin{aligned}([2]+[2]x)([2]+[3]x-[1]x^2) &= ([2]\odot[2])+([2]\odot[3]\oplus[2]\odot[2])x\\ &\quad+([2]\odot[-1]\oplus[2]\odot[3])x^2\\ &\quad+([2]\odot[-1])x^3\\ &=[0]+[2]x+[0]x^2+[-2]x^3\\ &=[2]x+[2]x^3.\end{aligned}$$

Notice that $-[1]x^2=[-1]x^2=[3]x^2$. More generally, $-ax^n=(-a)x^n$ in any polynomial ring.

In working with polynomials over any ring $\mathbb{Z}_n[x]$, we shall frequently omit brackets from the coefficients, write $+$ instead of $\oplus$, use juxtaposition instead of $\odot$, and rely on the context to remind us that all calculations are to be done modulo n. With this abbreviated notation the results of the calculations above would be written

$$(2+2x)+(2+3x-x^2) = x+3x^2$$

and

$$(2+2x)(2+3x-x^2) = 2x+2x^3.$$

Theorem 40.1. *If R is a commutative ring, then $R[x]$ is a commutative ring with respect to the operations defined by* (40.2) *and* (40.3). *If R is an integral domain, then $R[x]$ is an integral domain.*

PROOF. The zero of $R[x]$ is the polynomial having all coefficients equal to the zero of R. (This polynomial has no degree.) The details of proving that $R[x]$ is a commutative ring are left as an exercise (Problem 40.6).

Assume that R is an integral domain, with unity e. Then it is easy to verify that the polynomial of degree zero with coefficient e is a unity for $R[x]$. Also, if $p(x)$ and $q(x)$ are nonzero elements of $R[x]$, with leading coefficients $a_m \neq 0$ and $b_n \neq 0$, then (40.3) shows that $p(x)q(x)$ has leading coefficient $a_m b_n \neq 0$, and thus $p(x)q(x)$ is also nonzero. Therefore $R[x]$ has no zero divisors. This proves that $R[x]$ is an integral domain. □

Notice, in particular, that $F[x]$ is an integral domain if F is a field. However, $F[x]$ is not a field, no matter what F is (Problem 40.7). The ring $R[x]$ is called the *ring of polynomials in x over R*.

APPENDIX TO SECTION 40

Here is an alternative way to define the ring of polynomials in an indeterminate over a commutative ring R.

If R is a commutative ring, then a *sequence* of elements of R is a mapping from the set of nonnegative integers to R. Such sequences can be denoted by $(a_0, a_1, \ldots, a_n, \ldots)$, where a_n, the nth *term* of the sequence, is the ring element corresponding to the nonnegative integer n.

Definition. Let R be a commutative ring. A *polynomial over R* is a sequence

$$(a_0, a_1, \ldots, a_n, \ldots) \tag{40.1A}$$

of elements of R such that only finitely many of its terms are different from the zero element of R. If $a_n \neq 0$, but all terms with subscripts greater than n are zero, then the polynomial is said to be of *degree n*; the terms $a_0, a_1, \ldots, a_n$ are called the *coefficients* of the polynomial, with a_n the *leading* coefficient. Denote the set of all polynomials over R by $R[X]$.

By the definition of equality of mappings (Section 1), two polynomials over R are equal iff they have the same degree and their corresponding terms are equal. (This uses the fact that polynomials are sequences, that is, mappings from the set of nonnegative integers to R.)

Definition. Let

$$p = (a_0, a_1, \ldots, a_n, \ldots)$$

and

$$q = (b_0, b_1, \ldots, b_n, \ldots)$$

be polynomials over the commutative ring R. Then

$$p+q=(a_0+b_0,a_1+b_1,\dots,a_n+b_n,\dots) \tag{40.2A}$$

and

$$pq=(a_0b_0,a_0b_1+a_1b_0,\dots), \tag{40.3A}$$

where the kth term of pq is

$$a_0b_k+a_1b_{k-1}+\cdots+a_kb_0.$$

Because the sequences p and q each have only finitely many nonzero terms, the same is true for both $p+q$ and pq. Therefore, both $p+q$ and pq are members of $R[X]$.

Theorem 40.1A. *If R is a commutative ring, then $R[X]$ is a commutative ring with respect to the operations defined by* (40.2*A*) *and* (40.3*A*). *If R is an integral domain, then $R[X]$ is an integral domain.*

PROOF. The zero of $R[X]$ is the sequence $(0,0,\dots,0,\dots)$. If R has a unity e, then $R[X]$ has a unity $(e,0,0,\dots,0,\dots)$. If R is an integral domain, and p and q are nonzero elements of $R[X]$ having degree m and n, then pq has degree mn, and thus is also nonzero. The other details of the proof are left as an exercise (Problem 40.14). □

The mapping that assigns the "polynomial" in (40.1) to the "polynomial" in (40.1A), with all terms after a_n assumed to be zero in (40.1A), is an isomorphism of the ring $R[x]$ onto the ring $R[X]$ (Problem 40.15).

Assume that R has unity e, and let X denote the sequence $(0,e,\dots,0,\dots)$. Then it follows from (40.3A) that $X^2=XX=(0,0,e,\dots,0,\dots)$, $X^3=(0,0,0,e,\dots,0,\dots)$, and, in general, $X^k=(0,0,\dots,0,e,0,\dots)$, the sequence (element of $R[X]$) having the $(k+1)$st term e and all other terms zero. If the product of an element $a\in R$ and a sequence $(b_0,b_1,\dots,b_n,\dots)$ is defined to be $(ab_0,ab_1,\dots,ab_n,\dots)$, then it follows that

$$a_0+a_1X+\cdots+a_nX^n=(a_0,a_1,\dots,a_n,0,\dots).$$

This can be interpreted as giving a more precise meaning to the left side than that given by calling it an "expression," which was the way we referred to $a_0+a_1x+\cdots+a_nx^n$.

PROBLEMS

40.1. There are four different polynomials of degree 2 in $\mathbb{Z}_2[x]$. List them all.

40.2. (a) In $\mathbb{Z}_n[x]$, how many different polynomials are there of degree $\le d$?

(b) In $\mathbb{Z}_n[x]$, how many different monic polynomials are there of degree d?
(c) In $\mathbb{Z}_n[x]$, how many different polynomials are there of degree d?

40.3. The following are polynomials over $\mathbb{Z}$. Express each in the form (40.1).
(a) $(1+2x)+(2-x+2x^2)$
(b) $(2x+x^3)+(x+2x^4)$
(c) $(1+2x)(2-x+2x^2)$
(d) $(2x+x^3)(x+2x^4)$
(e) $(2x+x^2)^3$

40.4. (a) to (e). Interpret the polynomials in Problem 40.3 as being over $\mathbb{Z}_3$ rather than $\mathbb{Z}$, and write each in the form (40.1). (See the remark on notation at the end of Example 40.2.)

40.5. (a) to (e). Repeat Problem 40.4 with $\mathbb{Z}_4$ in place of $\mathbb{Z}_3$.

40.6. Complete the proof of Theorem 40.1 (that is, prove that $R[x]$ is a commutative ring.)

40.7. Explain why a polynomial ring $R[x]$ cannot be a field.

40.8. (a) True or false: The degree of the sum of two polynomials is at least as large as the degree of each of the two polynomials.
(b) True or false: The degree of the product of two polynomials is the sum of the degrees of the two polynomials.

40.9. Prove that if R is a commutative ring and $R[x]$ is an integral domain, then R must be an integral domain. (Compare the last sentence of Theorem 40.1.)

40.10. Prove that if R is any commutative ring, then the characteristic of $R[x]$ is equal to the characteristic of R.

40.11. Prove that if R and S are commutative rings, and $R \approx S$, then $R[x] \approx S[x]$.

40.12. The *formal derivative* of a polynomial

$$p(x) = a_0 + a_1 x + \cdots + a_n x^n$$

over R is defined to be

$$p'(x) = a_1 + 2a_2 x + \cdots + n a_n x^{n-1}.$$

Prove that

$$[p(x)+q(x)]' = p'(x) + q'(x)$$

and

$$[p(x)q(x)]' = p(x)q'(x) + p'(x)q(x).$$

(An application of formal derivatives will be given in Problem 50.6.)

40.13. Let D be an ordered integral domain (Section 22), and let $D[x]^p$ consist of all nonzero polynomials in $D[x]$ that have leading coefficient in D^p, the set of positive elements of D.
(a) Prove that this makes $D[x]$ an ordered integral domain with $D[x]^p$ as the set of positive elements.
(b) Prove that the polynomial 1 is a least positive element of $\mathbb{Z}[x]$.
(c) Prove that $\mathbb{Z}[x]$ is not well-ordered.

40.14. Complete the proof of Theorem 40.1A.

40.15. Prove that for any commutative ring R, $R[x] \approx R[X]$. (See the paragraph following the proof of Theorem 40.1A.)

SECTION 41. THE DIVISION ALGORITHM

In the next two sections we shall concentrate on rings of polynomials over fields, proving divisibility and factorization theorems for these rings that are analogous to the divisibility and factorization theorems that were proved in Chapter X for the ring of integers. We shall use $\deg f(x)$ to denote the degree of a polynomial $f(x)$.

Division Algorithm. *If $f(x)$ and $g(x)$ are polynomials over a field F, with $g(x)\neq 0$, then there exist unique polynomials $q(x)$ and $r(x)$ over F such that*

$$f(x) = g(x)q(x) + r(x), \quad \text{with} \quad r(x) = 0 \quad \text{or} \quad \deg r(x) < \deg g(x).$$

The polynomials $q(x)$ and $r(x)$ are called, respectively, the *quotient* and *remainder* in the division of $f(x)$ by $g(x)$. The following example illustrates how they can be computed. The same idea will be used in the proof that follows the example.

Example 41.1. Let $f(x)=2x^4+x^2-x+1$ and $g(x)=2x-1$.

$$\begin{array}{r|l}
 & x^3+\tfrac{1}{2}x^2+\tfrac{3}{4}x-\tfrac{1}{8} \\
\hline
2x-1 & 2x^4 \qquad\quad + x^2 - x + 1 \\
 & 2x^4 - x^3 \\
\cline{2-2}
 & \qquad x^3 + x^2 - x + 1 \\
 & \qquad x^3 - \tfrac{1}{2}x^2 \\
\cline{2-2}
 & \qquad\quad \tfrac{3}{2}x^2 - x + 1 \\
 & \qquad\quad \tfrac{3}{2}x^2 - \tfrac{3}{4}x \\
\cline{2-2}
 & \qquad\qquad -\tfrac{1}{4}x + 1 \\
 & \qquad\qquad -\tfrac{1}{4}x + \tfrac{1}{8} \\
\cline{2-2}
 & \qquad\qquad\qquad \tfrac{7}{8}
\end{array}$$

Therefore, $q(x)=x^3+\frac{1}{2}x^2+\frac{3}{4}x-\frac{1}{8}$ and $r(x)=\frac{7}{8}$. Notice that at each step we eliminated the highest remaining power of $f(x)$ by subtracting an appropriate multiple of $g(x)$. For the proof we simply show that this can be done no matter what $f(x)$ and $g(x)$.

PROOF OF THE DIVISION ALGORITHM. Let $f(x)=a_0+a_1x+\cdots+a_mx^m$ and $g(x)=b_0+b_1x+\cdots+b_nx^n$. Since $g(x)\neq 0$, we can assume that $b_n\neq 0$

so that $\deg g(x)=n$. The theorem is trivial for $f(x)=0$, and therefore we also assume that $a_m \neq 0$ so that $\deg f(x)=m$.

We shall first prove the existence of $q(x)$ and $r(x)$, using induction on m. If $m<n$, then $f(x)=g(x)\cdot 0+f(x)$ gives the required representation; that is, we can take $q(x)=0$ and $r(x)=f(x)$. Thus assume that $m \geq n$. If $m=0$, then $f(x)=a_0$ and $g(x)=b_0$; in this case $a_0=b_0 \cdot b_0^{-1}a_0+0$; hence we can take $q(x)=b_0^{-1}a_0$ and $r(x)=0$. It remains to prove the statement for $\deg f(x)= m$, on the basis of the induction hypothesis that it is true whenever $f(x)$ is replaced by a polynomial of degree less than m. Let $f_1(x)=f(x)- a_m b_n^{-1}x^{m-n}g(x)$. Then $\deg f_1(x)<\deg f(x)$ (Problem 41.2). Therefore, by the induction hypothesis, there exist polynomials $q_1(x)$ and $r_1(x)$ such that

$$f_1(x) = g(x)q_1(x) + r_1(x), \text{ with } r_1(x) = 0 \text{ or } \deg r_1(x) < \deg g(x).$$

This implies that

$$\begin{aligned} f(x) - a_m b_n^{-1}x^{m-n}g(x) &= g(x)q_1(x) + r_1(x) \\ f(x) &= g(x)\left[a_m b_n^{-1}x^{m-n} + q_1(x)\right] + r_1(x). \end{aligned}$$

Thus we can take $q(x)=a_m b_n^{-1}x^{m-n}+q_1(x)$ and $r(x)=r_1(x)$. This proves the existence of $q(x)$ and $r(x)$.

To prove that the polynomials $q(x)$ and $r(x)$ are unique, assume that $q^*(x)$ and $r^*(x)$ are also polynomials over F, and that

$$f(x) = g(x)q^*(x) + r^*(x), \text{ with } r^*(x) = 0 \text{ or } \deg r^*(x) < \deg g(x).$$

Then

$$g(x)q(x) + r(x) = g(x)q^*(x) + r^*(x)$$

and

$$g(x)\left[q(x) - q^*(x)\right] = r^*(x) - r(x).$$

The right side of this equation is zero or of degree less than $\deg g(x)$. Since the left side is zero or of degree at least $g(x)$, this forces $q(x)=q^*(x)$. Then we must also have $r^*(x)=r(x)$. $\square$

Example 41.2. Let $f(x)=[2]x^4+[1]x^2+[-1]x+[1] \in \mathbb{Z}_5[x]$ and $g(x)=[2]x+[-1] \in \mathbb{Z}_5[x]$. (Compare Example 41.1.) If we use the abbreviated notation a for $[a]$, $+$ for $\oplus$, and juxtaposition for $\odot$, then, for instance, since we are in $\mathbb{Z}_5[x]$, $2^{-1}=3$, $3^{-1}=2$, and $4^{-1}=4$. The division of $f(x)$ by $g(x)$

follows.

$$
\begin{array}{r|rrrrr}
 & & x^3 & +3x^2 & +2x & +3 \\
\hline
2x-1 & 2x^4 & & +x^2 & -x & +1 \\
 & 2x^4 & -x^3 & & & \\
\cline{2-6}
 & & x^3 & +x^2 & -x & +1 \\
 & & x^3 & -3x^2 & & \\
\cline{3-6}
 & & & 4x^2 & -x & +1 \\
 & & & 4x^2 & -2x & \\
\cline{4-6}
 & & & & x & +1 \\
 & & & & x & -3 \\
\cline{5-6}
 & & & & & 4
\end{array}
$$

Therefore $q(x)=[1]x^3+[3]x^2+[2]x+[3]$, and $r(x)=[4]$.

The indeterminate x in a polynomial $f(x)=a_0+a_1x+\cdots+a_nx^n \in F[x]$ need not denote an element of F. It *may* denote an element of F, however, and in that case the polynomial determines an element of f: if $x=c\in F$, then $f(x)=f(c)=a_0+a_1c+\cdots+a_nc^n\in F$.

Example 41.3. (a) If $f(x)=x^3-2x^2+2\in\mathbb{R}[x]$, then $f(3)=3^3-2\cdot 3^2+2=11\in\mathbb{R}$.
(b) If $f(x)=[3]+[1]x+[-3]x^4\in\mathbb{Z}_5[x]$, then $f([2])=[3]\oplus([1]\odot[2])\oplus([-3]\odot[2]^4)=[3]\oplus[2]\oplus([2]\odot[1])=[7]=[2]\in\mathbb{Z}_5$.

Remainder Theorem. *If $f(x)\in F[x]$, and $c\in F$, then the remainder in the division of $f(x)$ by $x-c$ is $f(c)$.*

PROOF. Because $\deg(x-c)=1$, the remainder in the division of $f(x)$ by $x-c$ must be either 0 or of degree 0. Thus, for some $q(x)\in F[x]$,

$$f(x) = (x-c)q(x) + r, \qquad \text{with } r\in F.$$

This yields

$$f(c) = (c-c)q(c) + r = r. \quad \square$$

Example 41.4. (a) Divide $f(x)=x^3-2x^2+2\in\mathbb{R}[x]$ by $x-3$. The quotient is x^2+x+3 and the remainder is 11:

$$f(x) = (x-3)(x^2+x+3) + 11.$$

Also, $f(3)=11$, as we saw in Example 41.3(a).

(b) Divide $f(x)=[3]+[1]x+[-3]x^4 \in \mathbb{Z}_5[x]$ by $[1]x+[-2]$. The quotient is $[2]x^3+[4]x^2+[3]x+[2]$ and the remainder is $[2]$:

$$f(x) = ([1]x+[-2])([2]x^3+[4]x^2+[3]x+[2])+[2].$$

Also, $f([2])=[2]$, as we saw in Example 41.3(b).

If $f(x)$, $g(x) \in F[x]$, with $g(x) \neq 0$, then $f(x)$ is *divisible* by $g(x)$ over F if $f(x)=g(x)q(x)$ for some $q(x) \in F[x]$. Thus $f(x)$ is divisible by $g(x)$ if the remainder in the division of $f(x)$ by $g(x)$ is zero. If $f(x)$ is divisible by $g(x)$ (over F), then we also say that $g(x)$ is a *factor* of $f(x)$ (over F).

Factor Theorem. *If $f(x) \in F[x]$ and $c \in F$, then $x-c$ is a factor of $f(x)$ iff $f(c)=0$.*

PROOF. This is an immediate corollary of the Remainder Theorem. □

An element $c \in F$ is called a *root* (or *zero*) of a polynomial $f(x) \in F[x]$ if $f(c)=0$. By the Factor Theorem, c is a root of $f(x)$ iff $x-c$ is a factor of $f(x)$.

PROBLEMS

41.1. For each pair of polynomials $f(x)$ and $g(x)$, determine $q(x)$ and $r(x)$, the quotient and remainder in the division of $f(x)$ by $g(x)$.
(a) $f(x)=x^3+x+1$, $g(x)=x-1$, both in $\mathbb{Q}[x]$.
(b) $f(x)=x^4-1$, $g(x)=-x^2+2$, both in $\mathbb{Q}[x]$.
(c) $f(x)=x^3-3$, $g(x)=x^4+x$, both in $\mathbb{Z}_3[x]$.
(d) $f(x)=x^2+2$, $g(x)=x-1$, both in $\mathbb{Z}_3[x]$.
(e) $f(x)=3x^4+2x^2-1$, $g(x)=2x^2+4x$, both in $\mathbb{Z}_5[x]$.

41.2. Verify that $\deg f_1(x) < \deg f(x)$ in the proof of the Division Algorithm.

41.3. Use the Remainder Theorem to determine the remainder when $2x^5-3x^3+2x+1 \in \mathbb{R}[x]$ is divided by $x-2 \in \mathbb{R}[x]$.

41.4. Use the Remainder Theorem to determine the remainder when $2x^5-3x^3+2x+1 \in \mathbb{R}[x]$ is divided by $x+3 \in \mathbb{R}[x]$.

41.5. What is the remainder when $2x^5-3x^3+2x+1 \in \mathbb{Z}_7[x]$ is divided by $x-2 \in \mathbb{Z}_7[x]$?

41.6. What is the remainder when $ix^9+3x^7+x^6-2ix+1 \in \mathbb{C}[x]$ is divided by $x+i \in \mathbb{C}[x]$?

41.7. Use the Factor Theorem to answer each of the following questions.
(a) Is $x-3 \in \mathbb{Q}[x]$ a factor of $3x^3-9x^2-7x+21 \in \mathbb{Q}[x]$?
(b) Is $x+2 \in \mathbb{R}[x]$ a factor of $x^3+8x^2+6x-8 \in \mathbb{R}[x]$?
(c) For which $k \in \mathbb{Q}$ is $x-1$ a factor of $x^3+2x^2+x+k \in \mathbb{Q}[x]$?
(d) Is $x-2 \in \mathbb{Z}_5[x]$ a factor of $2x^5-3x^4-4x^3+3x \in \mathbb{Z}_5[x]$?
(e) For which $k \in \mathbb{C}$ is $x+i$ a factor of $ix^9+3x^7+x^6-2ix+k \in \mathbb{C}[x]$?

41.8. Find all odd primes p for which $x-2$ is a factor of $x^4+x^3+x^2+x$ in $\mathbb{Z}_p$.

41.9. Prove that for each prime p there is an infinite field of characteristic p. (*Suggestion*: Consider the field of quotients of $\mathbb{Z}_p[x]$. Fields of quotients are discussed in Section 24.)

41.10. Construct an example to show that the Division Algorithm is not true if F is replaced by the integral domain $\mathbb{Z}$. (Compare Problem 41.11.)

41.11. Prove that if the Division Algorithm is true for polynomials over an integral domain D, then D is a field. (Compare Problem 41.10.)

41.12. Use the Factor Theorem to construct a single polynomial $f(x) \in \mathbb{Z}_5[x]$ such that every element of $\mathbb{Z}_5[x]$ is a root of $f(x)$.

41.13. Prove that if p is a prime, then each element of $\mathbb{Z}_p$ is a root of $x^p - x$. (*Suggestion*: If $[a] \neq [0]$, then $[a]^{p-1} = [1]$ because $\mathbb{Z}_p^{\#}$ is a group with respect to $\odot$ by the corollary of Theorem 20.1.)

41.14. (Lagrange's interpolation formula). Assume that $a_0, a_1, \ldots, a_n$ are distinct elements of a field F, and that $b_0, b_1, \ldots, b_n \in F$. Let

$$f(x) = \sum_{k=0}^{n} \frac{b_k(x-a_0)\cdots(x-a_{k-1})(x-a_{k+1})\cdots(x-a_n)}{(a_k-a_0)\cdots(a_k-a_{k-1})(a_k-a_{k+1})\cdots(a_k-a_n)}. \quad (41.1)$$

Verify that $f(a_j) = b_j$ for $0 \leq j \leq n$. (This shows that there exists a polynomial of degree at most n that takes on given values at $n+1$ distinct given points of F. Problem 50.5 will show that such a polynomial is unique.)

41.15. (a) Use Problem 41.14 to write a polynomial $f(x) \in \mathbb{R}[x]$ such that $f(1)=2$, $f(2)=3$, and $f(3)=-1$.

(b) Use Problem 41.14 to write a polynomial $f(x) \in \mathbb{Z}_5[x]$ such that $f([1]) = [2]$, $f([2]) = [3]$, and $f([3]) = [-1]$.

SECTION 42. FACTORIZATION OF POLYNOMIALS

The next theorem is a direct parallel of Theorem 38.1.

Theorem 42.1. *If $a(x)$ and $b(x)$ are polynomials over a field F, not both the zero polynomial, then there is a unique monic polynomial $d(x)$ over F such that*

(a) $d(x)|a(x)$ and $d(x)|b(x)$, and

(b) if $c(x)$ is a polynomial such that $c(x)|a(x)$ and $c(x)|b(x)$, then $c(x)|d(x)$.

The polynomial $d(x)$ in the theorem is called the *greatest common divisor* of $a(x)$ and $b(x)$. Just as in the case of the integers, the existence of the greatest common divisor will be shown using the *Euclidean Algorithm*—this time for polynomials.

PROOF. First assume $b(x) \neq 0$. By the Division Algorithm there are unique polynomials $q_1(x)$ and $r_1(x)$ such that

$$a(x) = b(x)q_1(x) + r_1(x), \quad \text{with } r_1(x) = 0 \quad \text{or} \quad \deg r_1(x) < \deg b(x).$$

If $r_1(x)=0$, then $b(x)|a(x)$; thus $d(x)=b(x)$ satisfies parts (a) and (b). If $r_1(x)\neq 0$, then we apply the Division Algorithm repeatedly, just as in the proof of Theorem 38.1:

$$\begin{aligned}
a(x)&=b(x)q_1(x)+r_1(x), && \deg r_1(x)<\deg b(x)\\
b(x)&=r_1(x)q_2(x)+r_2(x), && \deg r_2(x)<\deg r_1(x)\\
r_1(x)&=r_2(x)q_3(x)+r_3(x), && \deg r_3(x)<\deg r_2(x)\\
&\ \ \vdots\\
r_{k-2}(x)&=r_{k-1}(x)q_k(x)+r_k(x), && \deg r_k(x)<\deg r_{k-1}(x)\\
r_{k-1}(x)&=r_k(x)q_{k+1}(x).
\end{aligned}$$

Here we must eventually get the zero polynomial as a remainder because $\deg r_1(x)>\deg r_2(x)>\deg r_3(x)>\cdots$. Let $r_k(x)$ denote the last nonzero remainder. The proof that $r_k(x)$ satisfies both of the requirements (a) and (b) for $d(x)$ is similar to the proof of Theorem 38.1 (Problem 42.3). If the leading coefficient of $r_k(x)$ is r, then $r^{-1}\cdot r_k(x)$ is a *monic* polynomial satisfying (a) and (b).

If $b(x)=0$, and a_m is the leading coefficient of $a(x)$, then $a_m^{-1}\cdot a(x)$ is a monic polynomial satisfying both requirements (a) and (b) for $d(x)$. Thus we have proved the existence of a greatest common divisor.

The proof of uniqueness relies on the requirement that the greatest common divisor be monic. It is similar to the uniqueness proof in Theorem 38.1 and is left as an exercise (Problem 42.4). □

Example 42.1. Here the Euclidean Algorithm is applied to compute the greatest common divisor of $a(x)=x^4-x^3-x^2+1$ and $b(x)=x^3-1$, considered as polynomials over the field of rationals.

$$\begin{aligned}
x^4-x^3-x^2+1&=(x^3-1)(x-1)+(-x^2+x)\\
x^3-1&=(-x^2+x)(-x-1)+(x-1)\\
-x^2+x&=(x-1)(-x)
\end{aligned}$$

Therefore, the greatest common divisor is $x-1$.

The next theorem follows from the proof of Theorem 42.1 in the same way that Theorem 38.2 follows from the proof of Theorem 38.1 (Problem 42.5).

Theorem 42.2. *If $a(x)$ and $b(x)$ are polynomials over a field F, not both the zero polynomial, and $d(x)$ is their greatest common divisor, then there exist* *polynomials* $u(x)$ *and* $v(x)$ *over* F *such that*

$$d(x) = a(x)u(x) + b(x)v(x).$$

Example 42.2. Consider Example 42.1. From the first equation there,

$$-x^2 + x = (x^4 - x^3 - x^2 + 1) - (x^3 - 1)(x - 1).$$

Using this with the second equation, we get

$$\begin{aligned} x - 1 &= (x^3 - 1) - (-x^2 + x)(-x - 1) \\ &= (x^3 - 1) - [(x^4 - x^3 - x^2 + 1) - (x^3 - 1)(x - 1)](-x - 1) \\ &= (x^4 - x^3 - x^2 + 1)(x + 1) + (x^3 - 1)[1 + (x - 1)(-x - 1)] \\ &= (x^4 - x^3 - x^2 + 1)(x + 1) + (x^3 - 1)(-x^2 + 2). \end{aligned}$$

Thus $d(x) = x - 1 = a(x)u(x) + b(x)v(x)$ for $u(x) = x + 1$ and $v(x) = -x^2 + 2$.

Two polynomials $f(x)$ and $g(x)$ over a field F are said to be *associates* if $f(x) = c \cdot g(x)$ for some nonzero element c of F. For example, $2x^2 - 1$ and $6x^2 - 3 = 3(2x^2 - 1)$ are associates over $\mathbb{Q}$. Notice that each nonzero polynomial has precisely one monic polynomial among its associates. Notice also that each polynomial of degree at least one has two obvious sets of divisors: its associates and the polynomials of degree zero. If a polynomial of degree at least one has no other divisors, then it is said to be *irreducible* (or *prime*). Thus, if $f(x) = g(x)h(x)$, and $f(x)$ is irreducible, then one of $g(x)$ and $h(x)$ is of degree zero and the other is an associate of $f(x)$. If a polynomial is not irreducible, then it is said to be *reducible*.

The property of being irreducible depends on the field F. For example, $x^2 - 2$ is irreducible over the field of rational numbers, but not over the field of real numbers: $x^2 - 2 = (x + \sqrt{2})(x - \sqrt{2})$. Problems 42.12 through 42.15 have to do with showing whether polynomials are reducible or irreducible. The irreducible polynomials over a field F play the same role for $F[x]$ that the prime numbers do for the ring of integers. Here is the first indication of that. (Compare Lemma 39.2.)

Corollary. *If* F *is a field,* $a(x), b(x), p(x) \in F[x], p(x)$ *is irreducible, and* $p(x) | a(x)b(x)$*, then* $p(x) | a(x)$ *or* $p(x) | b(x)$.

Proof. If $p(x) \nmid a(x)$, then the greatest common divisor of $p(x)$ and $a(x)$ is e, the polynomial of degree zero with the coefficient the unity of F. Thus, if $p(x) \nmid a(x)$, then by Theorem 42.2 there are polynomials $u(x)$ and $v(x)$ such that $e = u(x)p(x) + v(x)a(x)$. Multiplication of both sides of this equation by $b(x)$ leads to $b(x)=u(x)p(x)b(x)+v(x)a(x)b(x)$. Because $p(x)|p(x)$ and $p(x)|a(x)b(x)$, we conclude that $p(x)|[u(x)p(x)b(x)+v(x)a(x)b(x)]$, and therefore $p(x)|b(x)$. Thus if $p(x)\nmid a(x)$, then $p(x)|b(x)$, which proves the corollary. □

Corollary. *If $p(x), a_1(x), a_2(x), \ldots, a_n(x)$ are polynomials over F, with $p(x)$ irreducible and $p(x)|a_1(x)a_2(x)\cdots a_n(x)$, then $p(x)|a_i(x)$ for some $i(1 \leq i \leq n)$.*

PROOF. Use the preceding corollary and induction on n (Problem 42.10). □

Unique Factorization Theorem. *Each polynomial of degree at least one over a field F can be written as an element of F times a product of monic irreducible polynomials over F, and, except for the order in which these irreducible polynomials are written, this can be done in only one way.*

PROOF. Let S denote the set of those polynomials over F that are of degree at least one and that *cannot* be written as stated. We shall prove that S is empty. If not, then by the Least Integer Principle (applied to the set of degrees of polynomials in S) there is a polynomial of least positive degree in S; let $a(x)$ denote such a polynomial and assume $\deg a(x)=n$. Then $a(x)$ is not irreducible, and so it can be factored as $a(x)=a_1(x)a_2(x)$, where $1<\deg a_1(x)<n$ and $1<\deg a_2(x)<n$. By the choice of $a(x), a_1(x)$, and $a_2(x)$, we know $a_1(x)\notin S$ and $a_2(x)\notin S$. Therefore, $a_1(x)$ and $a_2(x)$ can each be written as an element of F times a product of monic irreducible polynomials, and so the same is true of $a(x)=a_1(x)a_2(x)$. This contradicts the fact that $a(x)\in S$, and we therefore conclude that S must be empty, as stated.

The proof of the last part of the theorem is similar to the last part of the proof of the Fundamental Theorem of Arithmetic (Section 39) and is left as an exercise (Problem 42.11). □

Example 42.3. The polynomial $3x^4-3x^2-6$ can be factored as

$$3(x^2-2)(x^2+1) \qquad \text{in } \mathbb{Q}[x],$$

$$3(x+\sqrt{2})(x-\sqrt{2})(x^2+1) \qquad \text{in } \mathbb{R}[x],$$

and

$$3(x+\sqrt{2})(x-\sqrt{2})(x+i)(x-i) \qquad \text{in } \mathbb{C}[x].$$

Each factor is irreducible in its context.

PROBLEMS

42.1. Use the Euclidean Algorithm to compute the greatest common divisors of the following pairs of polynomials over $\mathbb{Q}$.
(a) x^3-3x^2+3x-2 and x^2-5x+6
(b) x^4+3x^2+2 and x^5-x
(c) x^3+x^2-2x-2 and $x^4-2x^3+3x^2-6x$

42.2. (a) to (c) Express each greatest common divisor in Problem 42.1 as a linear combination of the two given polynomials (as in Theorem 42.2).

42.3. Prove that the polynomial $r_k(x)$ in the proof of Theorem 42.1 satisfies both of the requirements (a) and (b) of the theorem.

42.4. Prove the uniqueness of $d(x)$ in Theorem 42.1.

42.5. Prove Theorem 42.2. (The remark preceding it suggests the method.)

42.6. Explain why each nonzero polynomial has precisely one monic polynomial among its associates.

42.7. (a) Determine the monic associate of $2x^3-x+1\in\mathbb{Q}[x]$.
(b) Determine the monic associate of $1+x-ix^2\in\mathbb{C}[x]$.
(c) Determine the monic associate of $2x^5-3x^2+1\in\mathbb{Z}_7[x]$.

42.8. Verify that $x^3-3\in\mathbb{Z}_7[x]$ is irreducible. (*Suggestion*: Use the Factor Theorem.)

42.9. Write $x^3+3x^2+3x+4\in\mathbb{Z}_5[x]$ as a product of irreducible polynomials.

42.10. Write the proof of the second corollary of Theorem 42.2.

42.11. Complete the proof of the Unique Factorization Theorem.

42.12. (a) Prove that $(x-1)|f(x)$ in $\mathbb{Z}_2[x]$ iff $f(x)$ has an even number of nonzero coefficients.
(b) Prove that if $\deg f(x)>1$, then $f(x)$ is irreducible over $\mathbb{Z}_2$ iff $f(x)$ has constant term 1 and an odd number of nonzero coefficients.
(c) Determine all irreducible polynomials of degree 4 or less over $\mathbb{Z}_2$.
(d) Write each polynomial of degree 3 over $\mathbb{Z}_2$ as a product of irreducible factors.

42.13. (a) By counting the number of distinct possibilities for $(x-a)(x-b)$, verify that there are $p(p+1)/2$ monic reducible polynomials of degree 2 over $\mathbb{Z}_p$ (p a prime).
(b) How many monic irreducible polynomials of degree 2 over $\mathbb{Z}_p$ are there?

42.14. State and prove a theorem establishing the existence of a unique *least common multiple* for every pair of polynomials, not both the zero polynomial, over a field F. (Compare Problem 39.10 and Theorem 42.1.)

42.15. (Eisenstein's irreducibility criterion) *Assume that p is a prime, $f(x)=a_0+a_1x+\cdots+a_nx^n\in\mathbb{Z}[x]$, $p|a_i$ $(0\le i\le n-1)$, $p^2\nmid a_0$, and $p\nmid a_n$. Then $f(x)$ is irreducible in $\mathbb{Z}[x]$.* Give an indirect proof of this by justifying each of the following statements.
(a) Assume that $f(x)=(b_0+b_1x+\cdots+b_ux^u)(c_0+c_1x+\cdots+c_vx^v)$. Then p does not divide both b_0 and c_0.
(b) But p divides one of b_0 and c_0. Assume that $p\nmid b_0$ and $p|c_0$.
(c) Since $p\nmid c_v$ (why?), not all c_j are divisible by p. Let k be the smallest integer such that $p\nmid c_k$ and $p|c_j$ for $0\le j<k$.

(d) Because $a_k = b_k c_0 + b_{k-1} c_1 + \cdots + b_0 c_k$, we can conclude that $p | b_0 c_k$, which is a contradiction.

(For applications, see Problems 42.16 and 42.17.)

42.16. Use Eisenstein's irreducibility criterion (Problem 42.15) to show that each of the following polynomials is irreducible in $\mathbb{Z}[x]$.

(a) $x^3 + 6x^2 + 3x + 3$

(b) $x^5 - 5x^3 + 15$

42.17. Use Eisenstein's irreducibility criterion (Problem 42.15) to prove that if p is a prime then the polynomial

$$f(x) = (x^p - 1)/(x - 1) = x^{p-1} + x^{p-2} + \cdots + x + 1$$

is irreducible in $\mathbb{Z}[x]$. (*Suggestion*: Replace x by $y+1$, and let $g(y)$ denote the result. Show that $g(y)$ is irreducible in $\mathbb{Z}[y]$, and explain why this implies that $f(x)$ is irreducible in $\mathbb{Z}[x]$. The polynomial $f(x)$ is called a *cyclotomic polynomial*; if p is odd, its roots are the imaginary pth roots of unity.)

SECTION 43. UNIQUE FACTORIZATION DOMAINS

The similarity between the factorization theorems for integers and polynomials has been emphasized throughout this chapter. We shall be concerned now—and again in Section 47—with the question of whether these theorems can be generalized to other rings. This question was first studied in the early- to mid-1800s, and it is important to know that it was studied out of necessity: progress in solving several notable problems in number theory depended on understanding factorization in rings of numbers other than the integers. Examples 43.1 and 43.2 will illustrate the kinds of rings that were involved; then Section 47 will give a more detailed account of the relevant history. Two points can be made now, however: first, the ideas we are to meet here and in Section 47 came not merely from abstraction for abstraction's sake, but from attempts to solve specific problems; second, these ideas, and the problems that brought them into focus, gave rise to algebraic number theory, one of the most interesting and challenging branches of modern mathematics.

The proofs in this section are similar to proofs already given for $\mathbb{Z}$ and $F[x]$, and will be left as exercises. Except for Section 47, the remainder of the book will be independent of this section.

Suppose that D is an integral domain. If $a, b \in D$, with $b \neq 0$, then a is *divisible* by b if $a = bc$ for some $c \in D$. If a is divisible by b, we also say that b *divides* a, or that b is a *factor* of a, and we write $b | a$.

An element of D is called a *unit* if it divides the unity, e, of D. The units of $\mathbb{Z}$ are 1 and -1. If F is a field, then the units of $F[x]$ are the polynomials of degree zero (that is, the nonzero constant polynomials).

Elements a and b of D are called *associates* if $a | b$ and $b | a$. It can be proved that a and b are associates iff $a = bu$ for some unit u of D (Problem

43.11). Elements $a,b \in \mathbb{Z}$ are associates iff $a = \pm b$. Polynomials $f(x), g(x) \in F[x]$ are associates iff $g(x) = c \cdot f(x)$ for some $c \in F$.

An element in an integral domain D is *irreducible* if it is not a unit of D and its only divisors in D are its associates and the units of D. The irreducible elements of $\mathbb{Z}$ are the primes and their negatives. The irreducible elements of $F[x]$ are the irreducible polynomials of $F[x]$.

Definition. An integral domain D is a *unique factorization domain* provided that

(a) if $a \in D$, $a \neq 0$, and a is not a unit, then a can be written as a product of irreducible elements of D, and

(b) if $a \in D$ and

$$a = p_1 p_2 \cdots p_s = q_1 q_2 \cdots q_t,$$

where each p_i and each q_j is irreducible, then $s = t$ and there is a permutation π of $\{1,2,\ldots,s\}$ such that p_i and $q_{\pi(i)}$ are associates for $1 \leq i \leq s$.

The ring of integers is a unique factorization domain (Section 39); so is the ring $F[x]$ of polynomials over any field F (Section 42). Following is an example of an integral domain that is not a unique factorization domain. We shall see in Section 47 that there is a historical connection between examples of this type and attempts to prove Fermat's Last Theorem (which will be stated in Section 47).

Example 43.1. Let $\mathbb{Z}[\sqrt{-5}\,] = \{a + b\sqrt{-5} \mid a,b \in \mathbb{Z}\}$. It can be verified that $\mathbb{Z}[\sqrt{-5}\,]$ is a subring of $\mathbb{C}$ (Problem 43.2). Therefore, since $\mathbb{Z}[\sqrt{-5}\,]$ contains 1, it is an integral domain. In proving that $\mathbb{Z}[\sqrt{-5}\,]$ is not a unique factorization domain, we shall use the mapping N from $\mathbb{Z}[\sqrt{-5}\,]$ to the set of nonnegative integers defined by

$$N(a + b\sqrt{-5}\,) = |a + b\sqrt{-5}\,|^2 = a^2 + 5b^2.$$

This mapping N is called a *norm*. If $z, w \in \mathbb{Z}[\sqrt{-5}\,]$, then

(a) $N(z) \geq 0$,
(b) $N(z) = 0$ iff $z = 0$, and
(c) $N(zw) = N(z)N(w)$ (Problem 43.3).

To determine the units of $\mathbb{Z}[\sqrt{-5}\,]$, we first observe that if $zw = 1$, then $N(z)N(w) = N(zw) = N(1) = 1$. Therefore, if $z = a + b\sqrt{-5}$ is a unit, then $N(z) = a^2 + 5b^2 = 1$, so that $a = \pm 1$ and $b = 0$. Thus the units of $\mathbb{Z}[\sqrt{-5}\,]$ are ± 1. It follows that an element of $\mathbb{Z}[\sqrt{-5}\,]$ has two associates, itself and its negative.

Now observe that

$$9 = 3 \cdot 3 = (2+\sqrt{-5})(2-\sqrt{-5}),$$

and that $9, 3, 2 \pm \sqrt{-5} \in \mathbb{Z}[\sqrt{-5}]$. If we show that 3 and $2 \pm \sqrt{-5}$ are irreducible in $\mathbb{Z}[\sqrt{-5}]$, then we will have shown that $\mathbb{Z}[\sqrt{-5}]$ is not a unique factorization domain, because by the preceding paragraph 3 is not an associate of either $2+\sqrt{-5}$ or $2-\sqrt{-5}$. We shall prove that 3 is irreducible, and leave the proof for $2 \pm \sqrt{-5}$ as an exercise (Problem 43.4).

Assume that $3 = zw$, with $z, w \in \mathbb{Z}[\sqrt{-5}]$. Then $N(z)N(w) = N(zw) = N(3) = 9$; hence $N(z)$ is either 1, 3, or 9. If $N(z) = 1$, then z is a unit, and if $N(z) = 9$, then $N(w) = 1$ and w is a unit. Therefore, if 3 is to have a factor that is neither an associate of 3 nor a unit, then that factor must have norm 3. But $a^2 + 5b^2 \neq 3$ for all integers a and b, and so $\mathbb{Z}[\sqrt{-5}]$ has no element of norm 3. Thus 3 is irreducible in $\mathbb{Z}[\sqrt{-5}]$.

If you examine the proofs of unique factorization in $\mathbb{Z}$ and $F[x]$, you will see that they both rely on the Division Algorithm. We shall now see that essentially the same method can work in other cases. Specifically, Theorem 43.1 will show that an integral domain is a unique factorization domain if it satisfies the following definition.

Definition. An integral domain D is a *Euclidean domain* if for each nonzero element $a \in D$ there exists a nonnegative integer $d(a)$ such that
(a) if a and b are nonzero elements of D, then $d(a) \leq d(ab)$, and
(b) if $a, b \in D$, with $b \neq 0$, then there exist elements $q, r \in D$ such that $a = bq + r$, with $r = 0$ or $d(r) < d(b)$.

The ring of integers is a Euclidean domain with $d(a) = |a|$. If F is a field, then $F[x]$ is a Euclidean domain with $d(f(x)) = \deg f(x)$. The following example, first introduced by Gauss, has historical interest which will be discussed in Section 47.

Example 43.2. Let $\mathbb{Z}[i] = \{a + bi \mid a, b \in \mathbb{Z}\}$. The elements of $\mathbb{Z}[i]$ are called *Gaussian integers*. It is easy to verify that $\mathbb{Z}[i]$ is an integral domain, and we shall prove that it is a Euclidean domain with respect to $d(a+bi) = |a+bi|^2 = a^2 + b^2$.

If z and w are nonzero elements of $\mathbb{Z}[i]$, then

$$d(z) = |z|^2 \leq |z|^2|w|^2 = |zw|^2 = d(zw).$$

Thus d satisfies condition (a) of the definition.

To verify condition (b), assume that $z, w \in \mathbb{Z}[i]$ with $w \neq 0$. Then $zw^{-1} \in \mathbb{C}$, and in fact $zw^{-1} = a + bi$ with $a, b \in \mathbb{Q}$. Let m and n be integers such that $|a - m| \leq \frac{1}{2}$ and $|b - n| \leq \frac{1}{2}$. Then

$$zw^{-1} = a + bi = m + ni + \left[(a-m)+(b-n)i\right],$$

and

$$\begin{aligned} z &= (m+ni)w + \left[(a-m)+(b-n)i\right]w \\ &= qw + r, \end{aligned}$$

where $q = m + ni$ and $r = [(a-m)+(b-n)i]w$. Here $r \in \mathbb{Z}[i]$ because $qw \in \mathbb{Z}[i]$ and $z \in \mathbb{Z}[i]$. It is now sufficient to show that $d(r) < d(w)$:

$$\begin{aligned} d(r) &= d[(a-m)+(b-n)i]d(w) \\ &= [(a-m)^2 + (b-n)^2]d(w) \\ &\leq [(\tfrac{1}{4}) + (\tfrac{1}{4})]d(w) \\ &< d(w). \end{aligned}$$

It can be shown that the units in $\mathbb{Z}[i]$ are ± 1 and $\pm i$ (Problem 43.5). Notice that 2 can be factored in $\mathbb{Z}[i]$ as $2 = i(1-i)^2$, where i is a unit and $1 - i$ is irreducible.

Because we have proved that $\mathbb{Z}[i]$ is a Euclidean domain, we can conclude that it is a unique factorization domain because of the following theorem.

Theorem 43.1. *Every Euclidean domain is a unique factorization domain.*

The proof of Theorem 43.1 is similar to the proofs of unique factorization in the special cases $\mathbb{Z}$ and $F[x]$. It can be carried out by first working through proofs of each of the following lemmas. The details will be left as exercises, but just by reading the statements of these results you can get an understanding of how ideas of factorization extend beyond the most familiar examples.

Lemma 43.1. *If D is an integral domain and $a, b \in D$, then a and b are associates iff $a = bu$ for some unit u of D.*

Definition. If a and b are elements of an integral domain D, not both zero, then an element $d \in D$ is a *greatest common divisor* of a and b provided that
(a) $d | a$ and $d | b$, and
(b) if $c \in D$, $c | a$, and $c | b$, then $c | d$.

A pair of elements need not have a *unique* greatest common divisor by this definition. For example, both ± 2 are greatest common divisors of 4 and 6 in $\mathbb{Z}$. And $x+1$, $2x+2$, and $(\frac{1}{2})x+(\frac{1}{2})$ are all greatest common divisors of x^2-1 and x^2+2x+1 in $\mathbb{Q}[x]$. The uniqueness of greatest common divisors in $\mathbb{Z}$ (Theorem 38.1) depended in part on the requirement of being positive. Similarly, the uniqueness of greatest common divisors in $F[x]$ (Theorem 42.1) depended in part on the requirement of being monic. For more general discussions of factorization we must sacrifice these strong forms of uniqueness and use the definition above. However, Lemma 43.3 will show that even in the general case we do not lose all uniqueness. Before stating that, however, we state the following important fact.

Lemma 43.2. *Any two nonzero elements a and b of a Euclidean domain D have a greatest common divisor d in D, and $d = ar + bs$ for some $r, s \in D$.*

Lemma 43.3. *If d_1 and d_2 are both greatest common divisors of elements a and b in an integral domain D, then d_1 and d_2 are associates in D.*

Lemma 43.4. *Assume that a and b are nonzero elements of a Euclidean domain D. Then $d(a) = d(ab)$ iff b is a unit in D.*

Lemma 43.5. *Assume that a, b, and c are nonzero elements of a Euclidean domain D with unity e. If a and b have greatest common divisor e, and $a|bc$, then $a|c$.*

Lemma 43.6. *If a, b, and p are nonzero elements of a Euclidean domain D, and p is irreducible and $p|ab$, then $p|a$ or $p|b$.*

In Section 47 we shall see how the ideas in this section are connected with other ideas about rings that are, on the surface, unrelated to factorization. For references for the history related to this chapter see the notes at the end of Chapter XII.

PROBLEMS

Assume that D is an integral domain throughout these problems.

43.1. Sketch a proof, along the lines of Example 43.1, that $\mathbb{Z}[\sqrt{-3}\,]$ is not a unique factorization domain. [*Suggestion*: $4 = 2 \cdot 2 = (1+\sqrt{-3}\,)(1-\sqrt{-3}\,)$.]

43.2. Prove that $\mathbb{Z}[\sqrt{-5}\,]$ is subring of $\mathbb{C}$ (Example 43.1).

43.3. Verify the properties (a), (b), and (c) of the norm in Example 43.1.
43.4. Prove that $2 \pm \sqrt{-5}$ are irreducible in $\mathbb{Z}[\sqrt{-5}]$ (Example 43.1).
43.5. Show that the units in $\mathbb{Z}[i]$ are ± 1 and $\pm i$.
43.6. Explain why every field is a unique factorization domain.
43.7. Prove that every field is a Euclidean domain.
43.8. (a) What are the units of $\mathbb{Z}[x]$?
(b) Prove or disprove that $\mathbb{Z}[x]$ is a unique factorization domain.
(c) Prove or disprove that $\mathbb{Z}[x]$ is a Euclidean domain with $d(f(x)) = \deg f(x)$.
43.9. Prove that an element a of a Euclidean domain is a unit iff $d(a) = 1$.
43.10. Prove that if $a \in D$ and $a \neq 0$, then a is a greatest common divisor of a and 0.
43.11. Prove Lemma 43.1.
43.12. Prove Lemma 43.2.
43.13. Prove Lemma 43.3.
43.14. Prove Lemma 43.4.
43.15. Prove Lemma 43.5.
43.16. Prove Lemma 43.6.
43.17. Prove Theorem 43.1.
43.18. Prove that if $a, u \in D$, and u is a unit, then $u \mid a$.
43.19. Prove that the units of D form a group with respect to the multiplication in D.

CHAPTER XII

Quotient Rings

The first two sections of this chapter will show that the definitions and basic theorems for ring homomorphisms and quotient rings directly parallel those for group homomorphisms and quotient groups. The third section develops some facts about polynomial rings that will be used later in studying questions about fields and polynomial equations. In the last section of the chapter we shall see that ideals, which are the kernels of ring homomorphisms, arise in a natural way in the study of factorization in rings.

SECTION 44. HOMOMORPHISMS OF RINGS. IDEALS

Definition. If R and S are rings, then a mapping $\theta: R \rightarrow S$ is a (ring) *homomorphism* if

$$\theta(a+b) = \theta(a) + \theta(b)$$

and

$$\theta(ab) = \theta(a)\theta(b)$$

for all $a, b \in R$.

Thus a ring isomorphism (Section 21) is a ring homomorphism that is one-to-one and onto. In the conditions $\theta(a+b)=\theta(a)+\theta(b)$ and $\theta(ab)=\theta(a)\theta(b)$, the operations on the left in each case are those of R, and the operations on the right are those of S. Because of the first of these conditions, a ring homomorphism $\theta: R \rightarrow S$ is necessarily a group homomorphism from the additive group of R to the additive group of S. It follows that any statement about group homomorphisms translates into some statement about ring homomorphisms. For example, if $\theta: R \to S$ is a ring homomorphism, then $\theta(0_R) = 0_S$ and $\theta(-a) = -\theta(a)$ for all $a \in R$.

Example 44.1. The mapping $\theta: \mathbb{Z} \to \mathbb{Z}_n$ defined by $\theta(a) = [a]$ was seen to be a homomorphism of additive groups in Example 28.1. But also $\theta(ab) = [ab] = [a] \odot [b]$ for all $a, b \in \mathbb{Z}$ so that θ is a ring homomorphism.

Example 44.2. Let R and S be rings, and let $R \times S$ be the direct sum of R and S (Example 18.6). The mappings $\pi_1: R \times S \to R$ and $\pi_2: R \times S \to S$ defined by $\pi_1((r,s)) = r$ and $\pi_2((r,s)) = s$ are ring homomorphisms (Problem 44.3).

Definition. If $\theta: R \to S$ is a ring homomorphism, then *Ker* θ, the *kernel* of θ, is the set of all elements $r \in R$ such that $\theta(r) = 0_S$.

Thus the kernel of a ring homomorphism θ is just the kernel of θ thought of as a homomorphism of the additive groups of the rings. Having observed this, we know at once that Kerθ is a subgroup of the additive group of R. Kernels of ring homomorphisms are, in fact, subrings. But just as kernels of group homomorphisms are special among all subgroups, being normal, kernels of ring homomorphisms are special among all subrings—they are ideals, in the following sense.

Definition. A subring I of a ring R is an *ideal* of R if $ar \in I$ and $ra \in I$ for all $a \in I$ and all $r \in R$.

The important point here is that in the products ar and ra, r can be any element in R; r is not restricted to I. Notice that $\mathbb{Z}$ is a subring of $\mathbb{Q}$, but $\mathbb{Z}$ is not an ideal of $\mathbb{Q}$. Other examples of subrings that are not ideals are given in the problems.

Theorem 44.1. *Let $\theta: R \to S$ be a ring homomorphism.*

(*a*) *The image of θ is a subring of S.*
(*b*) *The kernel of θ is an ideal of R.*
(*c*) *θ is one-to-one iff Ker $\theta = \{0_R\}$.*

PROOF. (a) Because θ is an additive group homomorphism, it follows from Section 28 (after Example 28.4) that $\theta(R)$, the image of θ, is a subgroup of the additive group of S. Thus it suffices to prove that $\theta(R)$ is

closed with respect to multiplication. To this end, assume that $s_1, s_2 \in \theta(R)$. Then $\theta(r_1)=s_1$ and $\theta(r_2)=s_2$ for some $r_1, r_2 \in R$, and $\theta(r_1 r_2)=\theta(r_1)\theta(r_2)=s_1 s_2$. Therefore $s_1 s_2 \in \theta(R)$, as required.
(b) We have already observed that $\operatorname{Ker}\theta$ is a subgroup of the additive group of R. On the other hand, if $a \in \operatorname{Ker}\theta$ and $r \in R$, then $\theta(ar)=\theta(a)\theta(r)=0\cdot\theta(r)=0$ and $\theta(ra)=\theta(r)\theta(a)=\theta(r)\cdot 0=0$. This proves that $\operatorname{Ker}\theta$ is an ideal of R.
(c) This is a direct consequence of the last part of Theorem 28.1; simply change to additive notation. □

Example 44.3. The kernel of the homomorphism $\theta: \mathbb{Z} \to \mathbb{Z}_n$ in Example 44.1 is the set (ideal) consisting of all multiples of the integer n. The kernel of the homomorphism $\pi_1: R \times S \to R$ in Example 44.2 is $\{(0,s) | s \in S\}$; this ideal of $R \times S$ is isomorphic to S (Problem 44.3).

The ideals in the first part of the preceding example, the kernels of the homomorphisms $\theta: \mathbb{Z} \to \mathbb{Z}_n$, belong to an important general class defined as follows. Let R be a commutative ring with unity e, and let $a \in R$; then (a) will denote the set of all multiples of a by elements of R:

$$(a) = \{ra | r \in R\}.$$

We shall verify that (a) is an ideal of R. First, (a) is a subgroup of the additive group of R:

(i) If $r, s \in R$, so that $ra, sa \in (a)$, then $ra + sa = (r+s)a \in (a)$.
(ii) $0 = 0a \in R$.
(iii) The negative of an element ra in (a) is $(-r)a$, and it is also in (a).

Second, if $ra \in (a)$ and $s \in R$, then $s(ra) = (sr)a \in (a)$. Thus (a) is an ideal of R, as claimed.

Ideals of the form (a) are called *principal ideals*. The ideal (a) is the smallest ideal of R containing a. Every ideal of $\mathbb{Z}$ is a principal ideal (Problem 44.13). We shall prove in Theorem 46.3 that if F is a field, then every ideal of the polynomial ring $F[x]$ is a principal ideal. Problem 44.14 gives an example of an ideal that is not principal.

Some rings, such as $\mathbb{Z}$, have many ideals. At the other extreme are rings having no ideals except the two obvious ones, (0) and the ring itself. The next example shows that any field has this property.

Example 44.4. If F is a field, then F has no ideals other than (0) and F. *Proof*: Assume that I is an ideal of F and that $I \neq (0)$; we shall prove that $I = F$. Let $a \in I, a \neq 0$. Then a has an inverse in F because F is a field. If e

is the unity of F, then $e=a\cdot a^{-1}\in I$ because I is an ideal. But now if r is any element of F, then $r=e\cdot r\in I$, again because I is an ideal. This proves that $I=F$. □

Problem 44.11 gives a partial converse to the statement proved in Example 44.4.

PROBLEMS

44.1. Which of the following mappings are ring homomorphisms? Determine the kernel of each mapping that is a homomorphism.
(a) $\theta:\mathbb{Z}\to\mathbb{Z}$ by $\theta(a)=3a$
(b) $\theta:\mathbb{Z}\to\mathbb{Z}$ by $\theta(a)=a^2$
(c) $\theta:\mathbb{Z}_6\to\mathbb{Z}_3$ by $\theta([a])=[a]$
(d) $\theta:\mathbb{C}\to\mathbb{R}$ by $\theta(z)=|z|$
(e) $\theta:\mathbb{C}\to\mathbb{C}$ by $\theta(z)=iz$
(f) $\theta:M(2,\mathbb{Z})\to M(2,\mathbb{Z})$ by $\theta\left(\begin{bmatrix} a & b \\ c & d\end{bmatrix}\right)=\begin{bmatrix} a & c \\ b & d\end{bmatrix}$

44.2. Verify that if R and S are rings, and $\theta:R\to S$ is defined by $\theta(r)=0_S$ for each $r\in R$, then θ is a homomorphism.

44.3. (a) Prove that the mapping $\pi_1:R\times S\to R$ defined in Example 44.2 is a (ring) homomorphism.
(b) Prove that $\mathrm{Ker}\,\pi_1\approx S$. (See Example 44.3.)

44.4. Prove that a homomorphic image of a commutative ring is commutative.

44.5. Prove or disprove that if a ring R has a unity, then every homomorphic image of R has a unity.

44.6. Prove that if R is a ring, $\alpha:\mathbb{Q}\to R$ and $\beta:\mathbb{Q}\to R$ are ring homomorphisms, and $\alpha(a)=\beta(a)$ for each $a\in\mathbb{Z}$, then $\alpha=\beta$ [that is, $\alpha(a)=\beta(a)$ for each $a\in\mathbb{Q}$].

44.7. (a) Determine the smallest subgroup containing $\frac{1}{2}$, in the additive group of $\mathbb{Q}$.
(b) Determine the smallest subgroup containing $\frac{1}{2}$, in the multiplicative group of $\mathbb{Q}$.
(c) Determine the smallest subring of $\mathbb{Q}$ containing $\frac{1}{2}$.
(d) Determine the smallest ideal of $\mathbb{Q}$ containing $\frac{1}{2}$.
(e) Determine the smallest subfield of $\mathbb{Q}$ containing $\frac{1}{2}$.

44.8. Every subring of $\mathbb{Z}$ is an ideal of $\mathbb{Z}$. Why?

44.9. Prove that the constant polynomials in $\mathbb{Z}[x]$ form a subring that is not an ideal.

44.10. Prove that if R is a commutative ring with unity, and $a\in R$, then (a) is the smallest ideal of R containing a. (Compare Theorem 7.3.)

44.11. Prove that if R is a commutative ring with unity, and R has no ideals other than (0) and R, then R is a field. (*Suggestion*: Use Problem 44.10 to show that each nonzero element has a multiplicative inverse.)

44.12. Determine all of the ideals of $\mathbb{Z}_{12}$.

44.13. Prove that if I is an ideal of $\mathbb{Z}$, then either $I=(0)$ or $I=(n)$, where n is the least positive integer in I. (See Problem 10.18.)

44.14. (An ideal that is not a principal ideal.)

(a) Let I denote the set of all polynomials in $\mathbb{Z}[x]$ that have an even number as the constant term. Prove that I is an ideal of $\mathbb{Z}[x]$.

(b) Prove that I is not a principal ideal of $\mathbb{Z}[x]$.

44.15. Prove or disprove that if $\theta: R \to S$ is a ring homomorphism, then the image of θ is an ideal of S. (*Suggestion*: See Problem 44.9.)

44.16. Prove that if F is a field, R is a ring, and $\theta: F \to R$ is a ring homomorphism, then either θ is one-to-one or $\theta(a) = 0$ for all $a \in F$.

44.17. The *center* of a ring is the subring defined by $\{c \in R \mid cr = rc$ for every $r \in R\}$. (See Problems 19.19, 19.20, and 19.21.) Prove that the center of $M(2, \mathbb{Z})$ consists of all matrices of the form

$$\begin{bmatrix} a & 0 \\ 0 & a \end{bmatrix}, \qquad a \in \mathbb{Z}.$$

Verify that this is not an ideal of $M(2, \mathbb{Z})$. (Therefore the center of a ring need not be an ideal.)

44.18. A subring I of a ring R is a *left ideal* of R if $ra \in I$ for all $r \in R$ and all $a \in I$. A *right ideal* is defined similarly.

(a) Verify that the set of all matrices of the form

$$\begin{bmatrix} a & 0 \\ b & 0 \end{bmatrix} \qquad (a, b \in \mathbb{Z})$$

is a left but not a right ideal of $M(2, \mathbb{Z})$.

(b) Find a right ideal of $M(2, \mathbb{Z})$ that is not a left ideal.

(c) Verify that if R is a ring and $a \in R$, then $\{r \in R \mid ra = 0\}$ is a left ideal of R. Determine this left ideal for $R = M(2, \mathbb{Z})$ and

$$a = \begin{bmatrix} 0 & 1 \\ 0 & 0 \end{bmatrix}.$$

(d) Verify that if R is a ring and $a \in R$, then $\{r \in R \mid ar = 0\}$ is a right ideal of R. Determine this right ideal for R and a as in part (c).

SECTION 45. QUOTIENT RINGS

In this section we shall discuss quotient rings, the ring analogues of quotient groups, and use them to prove that every ideal of a ring is a kernel of some homomorphism. We shall also prove the ring version of the Fundamental Homomorphism Theorem.

If I is an ideal of a ring R, then I is a subgroup of the additive group of R, and it is even normal. (The additive group of R is Abelian, and every subgroup of an Abelian group is normal.) Therefore, we can talk about the quotient group R/I. Because the elements of R/I are the cosets of I formed relative to addition, these elements will be written in the form $I + r$ $(r \in R)$. The next theorem shows that R/I can be made into a ring in a very natural way. The construction is merely a generalization of that used for $\mathbb{Z}_n$ in Example 18.2: there, $\mathbb{Z}_n$ can be thought of as $\mathbb{Z}/(n)$ for (n) the ideal consisting of all integral multiples on n.

Theorem 45.1. *Let I be an ideal of a ring R, and let R/I denote the set of all right cosets of I considered as a subgroup of the additive group of R. For $I+a\in R/I$ and $I+b\in R/I$, let $(I+a)+(I+b)=I+(a+b)$ and $(I+a)(I+b)=I+ab$. With these operations R/I is a ring, called the* quotient ring *of R by I.*

PROOF. Because R/I is a group with respect to addition (by Theorem 29.1 in additive notation) it suffices to verify the properties that involve multiplication. The first step in this is to verify that the multiplication on R/I is well-defined. Thus assume that $I+a_1=I+a_2$ and $I+b_1=I+b_2$; we must show that $I+a_1b_1=I+a_2b_2$. From $I+a_1=I+a_2$ we have $a_1=n_1+a_2$ for some $n_1\in I$. Also, from $I+b_1=I+b_2$ we have $b_1=n_2+b_2$ for some $n_2\in I$. This implies that $a_1b_1=(n_1+a_2)(n_2+b_2)=n_1n_2+n_1b_2+a_2n_2+a_2b_2$. But $n_1n_2+n_1b_2+a_2n_2\in I$ because $n_1,n_2\in I$ and I is an ideal of R. Therefore, a_1b_1 has the form $a_1b_1=n_3+a_2b_2$ with $n_3\in I$, so that $I+a_1b_1=I+a_2b_2$, as required.

Here is a verification of one of the distributive laws. Assume that $a,b,c\in R$. Then $(I+a)[(I+b)+(I+c)]=(I+a)[I+(b+c)]=I+a(b+c)=I+(ab+ac)=(I+ab)+(I+ac)=(I+a)(I+b)+(I+a)(I+c)$. The verification of associativity of multiplication, as well as the other distributive law, will be left as an exercise (Problem 45.2). □

If R is commutative, and I is any ideal in R, then R/I will be commutative (Problem 45.3). If R has a unity e, then $I+e$ will be a unity for R/I (Problem 45.4). On the other hand, R/I will not necessarily be an integral domain when R is an integral domain. For instance, $\mathbb{Z}$ is an integral domain, but $\mathbb{Z}_6$ is not (Example 19.2). More will be said about the properties of particular quotient rings in the next section.

Theorem 45.2. *If R is a ring and I is an ideal of R, then the mapping $\eta:R\to R/I$ defined by*

$$\eta(a) = I + a \qquad \text{for each } a\in R$$

is a homomorphism of R onto R/I, and $\operatorname{Ker}\eta=I$.

PROOF. This proof is similar to that of Theorem 29.2 (Problem 45.5). □

As in the case of groups, the mapping $\eta:R\to R/I$ in Theorem 45.2 is called the *natural homomorphism* of R onto R/I. Notice that Theorem 45.2 shows that every ideal is a kernel. The next theorem shows that every homomorphic image of a ring R is isomorphic to a quotient ring of R.

Fundamental Homomorphism Theorem for Rings. *Let R and S be rings, and let $\theta: R \to S$ be a homomorphism from R onto S with $\operatorname{Ker}\theta = I$. Then the mapping $\phi: R/I \to S$ defined by*

$$\phi(I+a) = \theta(a) \qquad \text{for each } I+a \in R/I$$

is an isomorphism of R/I onto S. Therefore

$$R/I \approx S.$$

PROOF. If the Fundamental Homomorphism Theorem for groups, and its proof (Section 30), are changed into additive notation, then the only thing that remains to be shown is that ϕ preserves multiplication. To do this, let $I+a \in R/I$ and $I+b \in R/I$. Then $\phi((I+a)(I+b)) = \phi(I+ab) = \theta(ab) = \theta(a)\theta(b) = \phi(I+a)\phi(I+b)$, as required. □

PROBLEMS

45.1. Define $\theta: \mathbb{Z}_{12} \to \mathbb{Z}_4$ by $\theta([a]_{12}) = [a]_4$ for each $[a]_{12} \in \mathbb{Z}_{12}$.
- (a) Verify that θ is well-defined.
- (b) Verify that θ is a ring homomorphism.
- (c) Use the Fundamental Homomorphism Theorem for Rings to explain why $\mathbb{Z}_{12}/([4]) \approx \mathbb{Z}_4$.
- (d) Construct Cayley tables for the ring operations on $\mathbb{Z}_{12}/([4])$. (Compare Problem 29.3.)

45.2. (a) Prove that multiplication is associative in every quotient ring R/I.
(b) Verify the distributive law

$$[(I+a)+(I+b)](I+c) = (I+a)(I+c)+(I+b)(I+c)$$

for every quotient ring. (See the end of the proof of Theorem 45.1.)

45.3. Prove that if R is commutative and I is an ideal of R, then R/I is commutative.

45.4. Prove that if R has a unity e, then $I+e$ is a unity for R/I.

45.5. Prove Theorem 45.2.

45.6. An ideal P of a commutative ring R is a *prime* ideal if $P \neq R$ and, for all $a, b \in R$, $ab \in P$ implies that $a \in P$ or $b \in P$. Prove that if n is a positive integer, then (n) is a prime ideal of $\mathbb{Z}$ iff n is a prime.

45.7. Assume that R is a commutative ring with unity and $P \neq R$ is an ideal of R. Prove that P is a prime ideal iff R/P is an integral domain. (Prime ideals are defined in Problem 45.6.)

45.8. (First Isomorphism Theorem for Rings.) If I and J are ideals of a ring R, then $I+J$ is defined to be $\{a+b \mid a \in I \text{ and } b \in J\}$. Verify that $I+J$ is an ideal of R, J is an ideal of $I+J$, $I \cap J$ is an ideal of I, and

$$\frac{I}{I \cap J} \approx \frac{I+J}{J}.$$

(*Suggestion*: See Problem 30.14.)

45.9. (Second Isomorphism Theorem for Rings.) Prove that if I and J are ideals of a ring R, with $J \subseteq I$, then I/J is an ideal of R/J, and $(R/J)/(I/J) \approx R/I$.
(*Suggestion*: See Problem 30.15.)

SECTION 46. QUOTIENT RINGS OF $F[X]$

We know that a polynomial over a field F need not have a root in F. In Chapter VI it was shown how this problem could be overcome in special cases by appropriate field extensions: the real numbers produced a root for $x^2-2 \in \mathbb{Q}[x]$, and the complex numbers produced a root for $x^2+1 \in \mathbb{R}[x]$. We shall see that by using quotient rings the problem can be solved in general. Specifically, we shall see that if $f(x)$ is any nonconstant polynomial over any field F, then $f(x)$ has a root in some extension of F; and that extension can be chosen to be (isomorphic to) a quotient ring of $F[x]$. The proof of this will be given in the next chapter, and will make use of the facts about quotient rings developed in this section.

Theorem 46.1. *Assume that F is a field and that $p(x) \in F[x]$. Then $F[x]/(p(x))$ is a field iff $p(x)$ is irreducible over F.*

PROOF. Let I denote the principal ideal $(p(x))$ throughout the proof. Suppose first that $p(x)$ is reducible over F, say $p(x)=a(x)b(x)$ with both $a(x)$ and $b(x)$ of degree less than that of $p(x)$. We shall prove that in this case $F[x]/I$ is not a field. The degree of any nonzero polynomial in I must be at least as great as $\deg p(x)$; thus $a(x) \notin I$ and $b(x) \notin I$. Therefore, $I+a(x)$ and $I+b(x)$ are both nonzero elements of $F[x]/I$. But $(I+a(x))(I+b(x))=I+a(x)b(x)=I+p(x)=I$, the zero element of $F[x]/I$. We conclude that $F[x]/I$ has divisors of zero so that $F[x]/I$ is not a field (it is not even an integral domain). This proves that if $F[x]/I$ is a field, then $p(x)$ must be irreducible.

Suppose now that $p(x)$ is irreducible. Problem 46.1 asks you to show that $F[x]/I$ is commutative and that $I+e$ is a unity for $F[x]/I$ (where e is the unity of F). Thus it suffices to prove that each nonzero element of $F[x]/I$ has a multiplicative inverse in $F[x]/I$. Assume that $I+f(x)$ is nonzero. Then $f(x) \notin I$, which means that $f(x)$ is not a multiple of $p(x)$ in $F[x]$. Because $p(x)$ is irreducible, this implies that $p(x)$ and $f(x)$ have greatest common divisor e (Problem 46.2). Therefore, by Theorem 42.2, $e=p(x)u(x)+f(x)v(x)$ for some $u(x), v(x) \in F[x]$. It follows that $e-f(x)v(x)=p(x)u(x) \in I$, and hence $I+e=I+f(x)v(x)=(I+f(x))(I+v(x))$. This shows that $I+v(x)$ is a multiplicative inverse of $I+f(x)$. $\square$

The following theorem is helpful when working with quotient rings $F[x]/(p(x))$, whether $p(x)$ is irreducible or not.

Theorem 46.2. *Assume that F is a field, $p(x)=a_0+a_1x+\cdots+a_nx^n$ is a polynomial of degree n over F, and I is the ideal $(p(x))$ of $F[x]$. Then each element of $F[x]/I$ can be expressed uniquely in the form*

$$I+\left(b_0+b_1x+\cdots+b_{n-1}x^{n-1}\right), \qquad \text{with } b_0,b_1,\ldots,b_{n-1}\in F. \quad (46.1)$$

Moreover, $\{I+b \mid b\in F\}$ is a subfield of $R[x]/I$ isomorphic to F.

PROOF. If $I+f(x)\in F[x]/I$, then by the Division Algorithm $f(x)=p(x)q(x)+r(x)$ for some $q(x),r(x)\in F[x]$ with $r(x)=0$ or $\deg r(x)<\deg p(x)$. Since $f(x)-r(x)=p(x)q(x)\in I$, we have $I+f(x)=I+r(x)$; therefore, each element of $F[x]/I$ can be expressed in at least one way in the form (46.1). On the other hand, if

$$I+\left(b_0+b_1x+\cdots+b_{n-1}x^{n-1}\right)=I+\left(c_0+c_1x+\cdots+c_{n-1}x^{n-1}\right),$$

then

$$(b_0-c_0)+(b_1-c_1)x+\cdots+(b_{n-1}-c_{n-1})x^{n-1}\in I$$

so that

$$p(x) \text{ divides } \left[(b_0-c_0)+(b_1-c_1)x+\cdots+(b_{n-1}-c_{n-1})x^{n-1}\right].$$

This implies that

$$(b_0-c_0)+(b_1-c_1)x+\cdots+(b_{n-1}-c_{n-1})x^{n-1}=0,$$

because $\deg p(x)=n>n-1$. Therefore

$$b_0=c_0,\ b_1=c_1,\ldots,b_{n-1}=c_{n-1}.$$

This proves uniqueness.

It is now easy to verify that $b\mapsto I+b$ is a one-to-one ring homomorphism, and this will prove the last part of the theorem (Problem 46.3). □

Notice that the proof shows that if $f(x)\in F[x]$, and if $f(x)=p(x)q(x)+r(x)$ with $r(x)=0$ or $\deg r(x)<\deg p(x)$, then $I+f(x)=I+r(x)$. This is important when computing in $F[x]/(p(x))$, as we shall see in the following example and again in Chapter XIII.

Example 46.1. We shall now show how the preceding ideas can be used to construct the field of complex numbers from the field of real numbers. This will be a special case of what is to come in the next chapter. Let

$p(x)=1+x^2$, which is irreducible over $\mathbb{R}$. Also let $I=(1+x^2)$. By Theorem 46.1, $\mathbb{R}[x]/I$ is a field. In fact, $\mathbb{R}[x]/I \approx \mathbb{C}$, as we shall now verify.

Each element of $\mathbb{R}[x]/I$ can be written uniquely as $I+(a+bx)$ with $a,b \in \mathbb{R}$, by Theorem 46.2. Define $\theta : \mathbb{R}[x]/I \to \mathbb{C}$ by

$$\theta(I+(a+bx)) = a + bi.$$

That θ is one-to-one and onto, and preserves addition, is left as an exercise (Problem 46.4). To verify that θ preserves multiplication, we first write

$$\theta[(I+(a+bx))(I+(c+dx))] = \theta[I+(ac+(ad+bc)x+bdx^2)].$$

To continue, we must determine $u,v \in \mathbb{R}$ such that

$$I+(ac+(ad+bc)x+bdx^2) = I+(u+vx).$$

This is done by using the remark following the proof of Theorem 46.2: divide $ac+(ad+bc)x+bdx^2$ by $1+x^2$; the remainder will be $u+vx$. The result is

$$ac + (ad+bc)x + bdx^2 = (1+x^2)bd + (ac-bd) + (ad+bc)x,$$

so that $u+vx=(ac-bd)+(ad+bc)x$. Therefore

$$\begin{aligned}\theta[(I+(a+bx))(I+(c+dx))] &= \theta[I+((ac-bd)+(ad+bc)x)]\\ &= (ac-bd)+(ad+bc)i\\ &= (a+bi)(c+di)\\ &= \theta(I+(a+bx))\theta(I+(c+dx)).\end{aligned}$$

This proves that θ preserves multiplication. Thus $\mathbb{R}[x]/I \approx \mathbb{C}$, as claimed.

With the notation $a+bi$ $(a,b \in \mathbb{R})$ for the elements of $\mathbb{C}$, the number i is a root of $1+x^2$. To verify the corresponding statement with $\mathbb{R}[x]/I$ in place of $\mathbb{C}$, we must check that $1+z^2=0$ for some element z of $\mathbb{R}[x]/I$. Because of the isomorphism $b \mapsto I+b$ from Theorem 46.2, in this equation $(1+z^2=0)$ 1 should be interpreted as $I+1$, and 0 should be interpreted as $I+0$. Because $\theta(I+x)=i$, one root must be $z=I+x \in \mathbb{R}[x]/I$. To check this directly, write

$$\begin{aligned}(I+1)+(I+x)^2 &= (I+1)+(I+x^2)\\ &= I+(1+x^2)\\ &= I+0.\end{aligned}$$

The next theorem shows that Theorem 46.2 covers all quotient rings of $F[x]$.

Theorem 46.3. *If F is a field, then every ideal of the polynomial ring $F[x]$ is a principal ideal.*

PROOF. Let I denote an ideal of $F[x]$. If $I=\{0\}$, then I is the principal ideal (0). Assume $I\neq\{0\}$, and let $g(x)$ denote any polynomial of least degree among the nonzero polynomials in I. We shall show that $I=(g(x))$; certainly $I\supseteq(g(x))$. Let $f(x)\in I$. By the Division Algorithm there are polynomials $q(x),r(x)\in F[x]$ such that

$$f(x) = g(x)q(x) + r(x) \text{ with } r(x) = 0 \text{ or } \deg r(x) < \deg g(x).$$

Then $f(x)\in I$ and $g(x)q(x)\in I$ and so $r(x)=f(x)-g(x)q(x)\in I$. Thus $r(x)=0$, for otherwise $r(x)\in I$ and $\deg r(x)<\deg g(x)$, contradicting the choice of $g(x)$ as a polynomial of least degree in I. Therefore, $f(x)=g(x)q(x)\in(g(x))$ so that $I=(g(x))$. □

Notice that the proof of Theorem 46.3 shows that $I=(g(x))$ for $g(x)$ any polynomial of least degree among the nonzero polynomials in I.

PROBLEMS

46.1. Prove that if F is a field with unity e, and I is an ideal of $F[x]$, then $F[x]/I$ is commutative with unity $I+e$.

46.2. Prove that if F is a field and $p(x),f(x)\in F[x]$, with $p(x)$ irreducible and $p(x)\nmid f(x)$, then $p(x)$ and $f(x)$ have greatest common divisor e, where e is the unity of F. (Compare Problem 38.8.)

46.3. Verify that the mapping $b\mapsto I+b$ in the proof of Theorem 46.2 is a one-to-one ring homomorphism.

46.4. Prove that the mapping θ in Example 46.1 is one-to-one and onto, and preserves addition.

46.5. Prove that if F is a subfield of a field E, and $c\in E$, then $\theta: F[x]\to E$ defined by $\theta(f(x))=f(c)$ is a homomorphism.

46.6. If θ is any homomorphism of the type in Problem 46.5, then $\operatorname{Ker}\theta=(b(x))$ for some $b(x)\in F[x]$, because $F[x]$ is a principal ideal domain. Determine such a polynomial $b(x)$ for each of the following examples. (*Suggestion*: Use the Factor Theorem.)

(a) $\theta:\mathbb{Q}[x]\to\mathbb{Q}$ by $\theta(f(x))=f(0)$

(b) $\theta:\mathbb{Q}[x]\to\mathbb{Q}$ by $\theta(f(x))=f(3)$

(c) $\theta:\mathbb{Q}[x]\to\mathbb{R}$ by $\theta(f(x))=f(\sqrt{2}\,)$

(d) $\theta:\mathbb{R}[x]\to\mathbb{C}$ by $\theta(f(x))=f(i)$

(e) $\theta:\mathbb{R}[x]\to\mathbb{C}$ by $\theta(f(x))=f(-i)$

46.7. Suppose that F is a field and $f(x),g(x)\in F[x]$. Prove that $(f(x))=(g(x))$ iff $f(x)$ and $g(x)$ are associates. (Associates are defined in Section 42.)

46.8. Prove that if F is a field and I is an ideal of $F[x]$, then there is a unique monic polynomial $m(x)\in F[x]$ such that $I=(m(x))$.

46.9. (a) Prove or disprove that if $(f(x))=(g(x))$, then $\deg f(x)=\deg g(x)$.

(b) Prove or disprove that if $\deg f(x)=\deg g(x)$, then $(f(x))=(g(x))$.

46.10. True or false: If $f(x)\in(g(x))$ and $\deg f(x)=\deg g(x)$, then $(f(x))=(g(x))$.

46.11. The proof of Theorem 46.3 uses the Least Integer Principle. Where?
46.12. Determine all of the prime ideals of $F[x]$, where F is a field. (See Problems 45.6 and 45.7.)

SECTION 47. FACTORIZATION AND IDEALS

The theme in Chapters X and XI was factorization; the theme in this chapter has been the mutually equivalent ideas of ring homomorphism, ideal, and quotient ring. In this section we shall see that these apparently unrelated themes are, in fact, not unrelated. Specifically, we shall see in Theorem 47.2 that information about the ideals in a ring can often tell us when there is unique factorization in the ring. Then, by looking at some penetrating discoveries from nineteenth-century number theory, we shall see that in many rings factorization can most appropriately be studied by considering "products of ideals" in the ring, rather than just products of elements of the ring. This section is designed to show the relation between some general ideas connecting number theory and rings—details and proofs will be omitted or left to the problems. You should be familiar with Sections 43 and 44, and Theorem 46.3. We begin with a key definition.

Definition. An integral domain in which every ideal is a principal ideal is called a *principal ideal domain.*

Examples of principal ideal domains include the ring of integers (Problem 44.13), and the ring $F[x]$ of polynomials over any field F (Theorem 46.3). Recall that both $\mathbb{Z}$ and $F[x]$ are Euclidean domains; therefore, the fact that they are also principal ideal domains is a special case of the following theorem.

Theorem 47.1. *Every Euclidean domain is a principal ideal domain.*

The proof of Theorem 47.1 is similar to the proof of Theorem 46.3, and will be left as an exercise (Problem 47.1). The converse of Theorem 47.1 is false; that is, not every principal ideal domain is a Euclidean domain. An example is given by the ring of all complex numbers of the form $a+b(1+\sqrt{-19})/2$ for $a,b\in\mathbb{Z}$. (See [2] for a discussion of this.)

The following theorem gives a direct link between factorization and ideals.

Theorem 47.2. *Every principal ideal domain is a unique factorization domain.*

A proof of Theorem 47.2 can be constructed by working through Problems 47.2 through 47.8; they provide a thorough test of your grasp of the ideas in the last few chapters. The converse of Theorem 47.2, like the converse of Theorem 47.1, is false. For example, $\mathbb{Z}[x]$ is a unique factorization domain, but it is not a principal ideal domain: the set of all polynomials that have an even number as constant term forms an ideal that is not principal (Problems 43.8 and 44.14).

Here is a summary of the relationships between several important classes of integral domains.

$$\begin{matrix}\text{Euclidean} \\ \text{domains}\end{matrix} \subset \begin{matrix}\text{principal ideal} \\ \text{domains}\end{matrix} \subset \begin{matrix}\text{unique factorization} \\ \text{domains}\end{matrix} \subset \begin{matrix}\text{integral} \\ \text{domains}\end{matrix} \tag{47.1}$$

Each class is contained in, but different from, the class that follows it. Problem 47.9 asks you to place a number of specific examples in the smallest possible class of this sequence.

The extension of factorization theorems to rings other than the integers first arose in the last century, from two different sources. The first was in work on biquadratic reciprocity, and the second was in work on Fermat's Last Theorem. Here, briefly, is what was involved.

If the congruence $x^2 \equiv a \pmod{m}$ has a solution, then a is said to be a *quadratic residue* of m; if there is no solution, then a is said to be a *quadratic nonresidue* of m. (For example: 1 and 4 are quadratic residues of 5; 2 and 3 are quadratic nonresidues of 5.) A. M. Legendre (1752–1833) introduced the following convenient notation for working with quadratic residues: If p is an odd prime and $p \nmid a$, then the symbol (a/p) is defined by

$$(a/p) = \begin{cases} 1 & \text{if } a \text{ is a quadratic residue of } p \\ -1 & \text{if } a \text{ is a quadratic nonresidue of } p. \end{cases}$$

(Thus $(1/5) = (4/5) = 1$ and $(2/5) = (3/5) = -1$.)

The most famous theorem about quadratic residues is the *law of quadratic reciprocity*: If p and q are distinct odd primes, then

$$(p/q)(q/p) = (-1)^{\left(\frac{1}{4}\right)(p-1)(q-1)}.$$

Another way to say this is as follows: If either $p \equiv 1 \pmod{4}$ or $q \equiv 1 \pmod{4}$, then p is a quadratic residue of q iff q is a quadratic residue of p; if both $p \equiv 3 \pmod{4}$ and $q \equiv 3 \pmod{4}$, then p is a quadratic residue of q iff q is a quadratic nonresidue of p. The law of quadratic reciprocity was stated by Euler and Legendre, but Gauss gave the first complete proof. Gauss then searched for an equally comprehensive law for biquadratic residues: a is a *biquadratic residue* of m if $x^4 \equiv a \pmod{m}$ has a solution. He found such a law: the *law of biquadratic reciprocity*. The detailed statement of this law is not important here. What is important is that in order to solve this problem about biquadratic residues, Gauss stepped outside of $\mathbb{Z}$ —the ring in which the problem was posed—and worked in the larger ring $\mathbb{Z}[i]$, now known as the ring of Gaussian integers (Example 43.2). In the process he proved that unique factorization holds in $\mathbb{Z}[i]$. Gauss also used similar ideas to develop a theory of cubic reciprocity (replacing x^4 by x^3).

Now we turn to the connection between factorization and Fermat's Last Theorem. There are infinitely many triples (x,y,z) of integers, called *Pythagorean triples*, such that $x^2+y^2=z^2$. (Only solutions with nonzero integers are of interest, of course. See Problem 47.11.) Many such triples were known to the Babylonians as early as 1600 B.C., and they were also studied by the Greek mathematician Diophantus in his book *Arithmetica* (c. A.D. 250). In 1637, in a marginal note of a copy of *Arithmetica*, Fermat wrote that there are no solutions in positive integers of $x^n+y^n=z^n$ if $n>2$. This became known as Fermat's Last Theorem, even though Fermat gave no proof. In fact, it is very unlikely that Fermat had a proof. In any event, no one has given a proof since, even though many outstanding mathematicians have tried. By the 1840s, mathematicians who worked on Fermat's Last Theorem were concentrating on those cases where n is an odd prime, and several realized the usefulness of factoring the left side of $x^p+y^p=z^p$ as

$$x^p + y^p = (x+y)(x+uy)\cdots(x+u^{p-1}y),$$

where u is an imaginary pth root of unity. This revealed the problem as one of factorization: Are there nonzero integers x, y, and z such that

$$z^p = (x+y)(x+uy)\cdots(x+u^{p-1}y)?$$

Because u is imaginary, this took them outside of $\mathbb{Z}$. Specifically, they were faced with questions about factorization in the ring $\mathbb{Z}[u]$, where u is a solution of

$$u^{p-1} + u^{p-2} + \cdots + u + 1 = 0. \tag{47.2}$$

At least one erroneous proof of Fermat's Last Theorem rested on the mistaken assumption that unique factorization holds in $\mathbb{Z}[u]$. Although unique factorization holds in $\mathbb{Z}[u]$ for many values of p [with u a solution of (47.2)], it does not hold for all values. The smallest value for which unique factorization fails is $p=23$. Although the ring $\mathbb{Z}[\sqrt{-5}\,]$ is not of the type being considered here, the failure of unique factorization in it shows what can happen (Example 43.1).

The German mathematician Ernst Kummer (1810–1893) was able to overcome the failure of unique factorization in part by inventing new "ideal numbers." By using these ideal numbers, along with the elements in the original ring $\mathbb{Z}[u]$, Kummer was able to prove the impossibility of $x^p + y^p = z^p$ for many values of p, thereby adding considerably to what was known about Fermat's Last Theorem. Although the problem is still unsettled, it is known that at least there are no solutions for $n \leq 100{,}000$.

From the modern point of view, the proper setting for questions about factorization in the rings considered by Gauss and Kummer is algebraic number theory. This theory was created by Richard Dedekind (1831–1916) in the 1870s. An *algebraic number* is any complex number that is a solution of an equation

$$a_n z^n + \cdots + a_1 z + a_0 = 0, \tag{47.3}$$

where the coefficients are in $\mathbb{Z}$. It can be shown that the set of all algebraic numbers forms a field. An *algebraic integer* is any solution of an equation of the form (47.3) where the coefficients are in $\mathbb{Z}$ and $a_n = 1$. The set of all algebraic integers forms an integral domain. This integral domain of algebraic integers contains many other integer domains, and some of these, such as $\mathbb{Z}[i]$, are unique factorization domains, while others, such as $\mathbb{Z}[\sqrt{-5}\,]$, are not.

One of the basic problems of algebraic number theory is to determine just which rings of algebraic integers are unique factorization domains. Theorem 47.2 gives a partial answer: if every ideal is principal, then factorization is unique. By replacing products of numbers by products of ideals, Dedekind was able to construct a theory of unique factorization for all rings of algebraic numbers. If I and J are ideals of a ring R, then the product IJ is defined to be the ideal generated by all products ab for $a \in I$, $b \in J$. An ideal $I \neq R$ is said to be *prime* if $ab \in I$ implies that $a \in I$ or $b \in I$. The fundamental theorem of ideal theory states that every nonzero ideal in a ring of algebraic integers can be factored uniquely as a product of prime ideals. In particular, if a denotes any algebraic integer, then the principal ideal (a) can be factored uniquely as a product of prime ideals. With these ideas it is possible to develop theorems for the set of ideals in a general ring of algebraic integers that are similar to the theorems for the set of elements in every unique factorization domain.

The preceding discussion shows part of what grew from the problems studied by Gauss and Kummer. These ideas form the basis for much of modern algebraic number theory, which has come to be seen as the appropriate place to study many of the more difficult questions about ordinary integers. These ideas also provided many of the problems that have shaped modern ring theory, including the abstract theory of ideals; this part of algebra owes a great deal to Emmy Noether (1882–1935) and her students, and is discussed in Chapter 10 of E. T. Bell's *Development of Mathematics* (reference [1] from Chapter VI).

PROBLEMS

47.1. Prove Theorem 47.1.

47.2. Assume that D is an integral domain and that $a,b \in D$.

(a) Prove that $a|b$ iff $(b) \subseteq (a)$.

(b) Prove that a and b are associates iff $(b)=(a)$.

47.3. If a and b are elements of a principal ideal domain D, not both zero, then a and b have a greatest common divisor in D. Prove this by justifying each of the following statements.

(a) If I is defined by $I=\{ax+by|x,y \in D\}$, then I is an ideal of D.

(b) $I=(d)$ for some $d \in D$.

(c) An element d, chosen as in part (b), is a greatest common divisor of a and b.

47.4. Prove that if a and b are elements of a principal ideal domain D with unity e, and a and b have greatest common divisor e, then $e=ar+bs$ for some $r,s \in D$. (See Problem 47.3.)

47.5. Prove that if a, b, and p are nonzero elements of a principal ideal domain D, and p is irreducible and $p|ab$, then $p|a$ or $p|b$. (See Problem 47.4.)

47.6. Prove that if D is a principal ideal domain, $a_1,a_2,\ldots,a_n,p \in D$, and p is irreducible and $p|a_1a_2 \cdots a_n$, then $p|a_i$ for some i. (See Problem 47.5.)

47.7. If D is a principal ideal domain, $a \in D$, and a is not a unit of D, then a can be written as a product of irreducible elements of D. Give an indirect proof of this by justifying each of the following statements.

(a) Assume that a cannot be written as a product of irreducible elements. Then a is not irreducible, and so $a=a_1b_1$ where neither a_1 nor b_1 is a unit.

(b) Either a_1 or b_1 cannot be written as a product of irreducible elements; assume that a_1 cannot be written as a product of irreducible elements.

(c) Since $a_1|a$, Problem 47.2 implies that $(a) \subseteq (a_1)$.

(d) Since b_1 is not a unit, Problem 47.2 implies that $(a) \neq (a_1)$. Therefore $(a) \subset (a_1)$.

(e) Repeat parts (a) through (d) with a_1 in place of a to deduce the existence of an element $a_2 \in D$ such that $(a) \subset (a_1) \subset (a_2)$.

(f) This can be done repeatedly, yielding a sequence

$$(a) \subset (a_1) \subset (a_2) \subset \cdots$$

of ideals of D. Let $I=\{x \in D | x \in (a_i) \text{ for some } a_i\}$, and prove that I is an ideal of D.

(g) Because I is an ideal, $I=(c)$ for some $c \in I$. Therefore, $c \in (a_k)$ for some k, and then $I=(c) \subseteq (a_k)$. This contradicts the fact that all of the inclusions $(a_i) \subset (a_{i+1})$ in part (f) are strict inequalities.

(h) This completes the proof.

47.8. Use Problems 47.6 and 47.7 to prove Theorem 47.2.

47.9. Consider the four classes of rings in (47.1). Determine the smallest class to which each of the following rings belongs.

(a) $\mathbb{Z}$ (b) $\mathbb{Z}_5$ (c) $\mathbb{Q}$ (d) $\mathbb{Z}_5[x]$ (e) $\mathbb{Z}[i]$ (f) $\mathbb{Z}[\sqrt{-5}\,]$
(g) $\mathbb{Z}[x]$ (h) $\mathbb{Q}[x]$

47.10. (a) Which integers $k \in \{1,2,\ldots,10\}$ are quadratic residues of 11?
(b) Is 11 a quadratic residue of 7?
(c) Verify the law of quadratic reciprocity for $p=7$ and $q=11$.

47.11. Prove that if m and n are positive integers with $m>n$, and $x=m^2-n^2$, $y=2mn$, $z=m^2+n^2$, then (x,y,z) is a Pythagorean triple.

NOTES ON CHAPTER XII

Reference [1] is an excellent source for the detailed history of many of the ideas in Section 47; also see references [1] and [3] from Chapter VI. Reference [3] below is an elementary introduction to algebraic number theory.

Although we have considered the impact of number theory on algebra, we have not considered the impact of either geometry or topology. For both of the latter, see the expository articles under *algebraic structures, algebraic geometry*, and *algebraic topology* in the bibliography by Gaffney and Steen, listed in the Introduction.

1. Edwards, H. M., *Fermat's Last Theorem: A Genetic Introduction to Algebraic Number Theory*, Springer-Verlag, New York, 1977.
2. Motzkin, T., *The Euclidean Algorithm*, Bulletin of the American Mathematical Society, **55** (1949) 1142–1146.
3. Pollard, H., *Theory of Algebraic Numbers* (Carus Monograph No. 9, The Mathematical Association of America) John Wiley, New York, 1950.

CHAPTER XIII

Field Extensions

Fields are most conveniently analyzed by studying how they are built up from smaller subfields. Recall that a field E is an extension of a field F if E contains F, or, more generally, a subfield isomorphic to F. We know from Sections 21 and 24 that any field of prime characteristic p is an extension of $\mathbb{Z}_p$, and any field of characteristic 0 is an extension of $\mathbb{Q}$. Therefore, if we were to determine all extensions of $\mathbb{Z}_p$ and $\mathbb{Q}$, we would have determined all fields. Although that is asking too much, it is nonetheless true that a lot *is* known about field extensions. In the first section we shall look at how to construct extensions that will solve a particular kind of problem, namely, that of providing roots for polynomials; the extension of $\mathbb{R}$ to $\mathbb{C}$ to obtain a root for $1+x^2$ is a special case. We shall also introduce some basic definitions about field extensions that will be used in Chapters XIV and XV. In the second section of the chapter ideas from the first section will be used to explain how to construct all finite fields. This chapter depends heavily on Section 46.

SECTION 48. ADJOINING ROOTS

Let E be an extension field of a field F, and let S be a subset of E. There is at least one subfield of E containing both F and S, namely, E itself. The intersection of all the subfields of E that contain both F and S is a subfield of E (Problem 48.1); it will be denoted $F(S)$. If $S \subseteq F$, then $F(S)=F$. If $S=\{a_1,a_2,\ldots,a_n\}$, then $F(S)$ will be denoted $F(a_1,a_2,\ldots,a_n)$. For example, $\mathbb{R}(i)=\mathbb{C}$. The field $F(S)$ will consist of all the elements of E that can be obtained from F and S by repeated applications of the operations of

E—addition, multiplication, and the taking of additive and multiplicative inverses (Problem 48.3).

If $E = F(a)$ for some $a \in E$, then E is said to be a *simple extension* of F. We can classify the simple extensions of F by making use of $F[x]$, the ring of polynomials in the indeterminate x over F, and $F[a] = \{a_0 + a_1 a + \cdots + a_n a^n | a_0, a_1, \ldots, a_n \in F\}$, the ring of all polynomials in a. The difference between $F[x]$ and $F[a]$ is that two polynomials in $F[x]$ can be equal only if coefficients on like powers of x are equal, whereas in some cases two polynomials in $F[a]$ can be equal without coefficients on like powers of a being equal. For example,

$$1 + 3\sqrt{2} = -1 + 3\sqrt{2} + \sqrt{2}^{\,2} \qquad \text{in } \mathbb{Q}[\sqrt{2}\,],$$

but

$$1 + 3x \neq -1 + 3x + x^2 \qquad \text{in } \mathbb{Q}[x].$$

For E an extension of F, and $a \in E$, define

$$\theta : F[x] \to F[a]$$

by

$$\theta(a_0 + a_1 x + \cdots + a_n x^n) = a_0 + a_1 a + \cdots + a_n a^n.$$

It can be verified that θ is a ring homomorphism (Problem 46.5). Therefore, by the Fundamental Homomorphism Theorem for Rings, $F[x]/I \approx F[a]$, where $I = \operatorname{Ker}\theta$ is an ideal of $F[x]$. Notice that I consists precisely of those polynomials having a as a root, because $f(x) \in I$ iff $\theta(f(x)) = f(a) = 0$. By Theorem 46.3 every ideal of $F[x]$ is a principal ideal, so that $I = (p(x))$ for some $p(x) \in F[x]$. Because $F[a] \subseteq E$, and E has no zero divisors, $F[a]$ cannot have zero divisors. Therefore, $F[x]/(p(x))$ cannot have zero divisors, so either $p(x) = 0$ or $p(x)$ is irreducible over F (Problem 48.4). We shall consider these two cases separately.

If $p(x) = 0$, then $I = \{0\}$ and $F[x] \approx F[a]$. The field of quotients of $F[x]$ can be thought of as the set of all quotients $f(x)/g(x)$ with $f(x), g(x) \in F[x]$ and $g(x) \neq 0$ (Problem 48.8). Also, $F(a)$ consists of all "quotients" $f(a)g(a)^{-1}$ with $f(a), g(a) \in F[a]$ and $g(a) \neq 0$. It follows that $F(a)$ is isomorphic to the field of quotients of $F[x]$. In this case $F(a)$ is said to be a *simple transcendental extension* of F, and a is said to be *transcendental* over F.

Example 48.1. It can be shown—but not easily—that the real number π is not a root of any polynomial with rational coefficients [1]. Therefore, $\mathbb{Q}(\pi)$ is a simple transcendental extension of $\mathbb{Q}$, and π is transcendental over $\mathbb{Q}$. The elements of $\mathbb{Q}(\pi)$ are the rational expressions

$$\frac{a_0+a_1\pi+a_2\pi^2+\cdots+a_n\pi^n}{b_0+b_1\pi+b_2\pi^2+\cdots+b_m\pi^m}$$

for $a_0,a_1,\ldots,a_n,b_0,b_1,\ldots,b_m\in\mathbb{Q}$.

Now consider the case $p(x)\neq 0$. As we have observed, in this case $p(x)$ is irreducible over F. Therefore, by Theorem 46.1, $F[x]/(p(x))$ is a field so that $F[a]$, being isomorphic to $F[x]/(p(x))$, is also a field. It follows that $F(a)=F[a]$. In this case $F(a)$ is said to be a *simple algebraic extension* of F, and a is said to be *algebraic* over F. The next theorem says more about fields $F[x]/(p(x))$.

Theorem 48.1. *Assume that F is a field, and that $p(x)\in F[x]$ is irreducible over F. Then $F[x]/(p(x))$ is a field extension of F, and $p(x)$ has a root in $F[x]/(p(x))$.*

PROOF. As before, let $I=(p(x))$. Theorem 46.1 shows that $F[x]/I$ is a field. In Theorem 46.2 we showed that by identifying each element $I+b\in F[x]/I$ with $b\in F$, we obtain a subfield of $F[x]/I$ that is isomorphic to F. Therefore $F[x]/I$ is an extension of F.

Assume that $p(x)=a_0+a_1x+\cdots+a_nx^n$, and let α denote the element $I+x\in F[x]/I$. Then

$$\begin{aligned} p(\alpha) &= a_0+a_1(I+x)+\cdots+a_n(I+x)^n \\ &= I+(a_0+a_1x+\cdots+a_nx^n) \\ &= I+p(x) \\ &= I. \end{aligned}$$

But I is the zero of $F[x]/I$. Thus α is a root of $p(x)$ in $F[x]/I$. □

Corollary. *If F is a field, and $f(x)$ is a polynomial of positive degree over F, then $f(x)$ has a root in some extension of F.*

PROOF. If $f(x)$ is irreducible over F, then Theorem 48.1 applies directly. Otherwise, $f(x)$ has some irreducible factor $p(x)\in F[x]$, and then $f(x)=p(x)q(x)$ for some $q(x)\in F[x]$. Apply Theorem 48.1 to $p(x)$. This gives a root α for $p(x)$ in $F[x]/(p(x))$, and this α is also a root for $f(x)$ because $f(\alpha)=p(\alpha)q(\alpha)=0\cdot q(\alpha)=0$. □

An important fact about Theorem 48.1 is that it shows how to construct a field having a root for $p(x)$ beginning with only F and $p(x)$; we need not have an extension before we begin. Example 46.1 illustrated this for the case of the polynomial x^2+1, irreducible over $\mathbb{R}$. Here is another example.

Example 48.2. The polynomial x^2-2 is irreducible over $\mathbb{Q}$. By Theorem 48.1, $\mathbb{Q}[x]/(x^2-2)$ is an extension of $\mathbb{Q}$, and it contains a root for x^2-2. The point being stressed here is that this requires no previous knowledge of $\mathbb{R}$, which we know to contain the root $\sqrt{2}$ of x^2-2.

Let $I=(x^2-2)$. Theorem 46.2 tells us that each element of $\mathbb{Q}[x]/I$ can be expressed uniquely in the form $I+(a+bx)$ with $a,b\in\mathbb{Q}$. The mapping $\theta:\mathbb{Q}[x]/I\to\mathbb{Q}(\sqrt{2})$ defined by

$$\theta(I+(a+bx)) = a+b\sqrt{2}$$

is an isomorphism. A direct proof of this is similar to that in Example 46.1, and is left as an exercise (Problem 48.9). The calculation of a typical product in $\mathbb{Q}[x]/I$ follows (see the remark after Theorem 46.2):

$$\begin{aligned}[I+(1-2x)][I+(2+x)] &= I+(2-3x-2x^2)\\ &= I+(-2-3x),\end{aligned}$$

because

$$2-3x-2x^2 = (x^2-2)(-2)+(-2-3x).$$

The corresponding calculation in $\mathbb{Q}(\sqrt{2})\subseteq\mathbb{R}$ would be

$$\begin{aligned}(1-2\sqrt{2})(2+\sqrt{2}) &= 2-3\sqrt{2}-2(\sqrt{2})^2\\ &= -2-3\sqrt{2}.\end{aligned}$$

We close with two more important ideas relating to field extensions. If $f(x)\in F[x]$, then an extension field E of F such that $E=F(a_1,a_2,\ldots,a_n)$ and $f(x)=a(x-a_1)(x-a_2)\cdots(x-a_n)\in F[x]$ is called a *splitting* (or *root*) *field* of $f(x)$ over F. In other words, a splitting field of $f(x)$ is just large enough to contain F and the roots of $f(x)$. It can be proved that such a field does exist (Problem 48.10). Moreover, it is uniquely determined by F and $f(x)$, in the sense that if E_1 and E_2 are splitting fields of $f(x)$ over F, then $E_1\approx E_2$. The splitting field of x^2-2 over $\mathbb{Q}$ is $\mathbb{Q}(\sqrt{2})$.

If F is a field, and P is the intersection of all the subfields of F, then P is a subfield of F. If F has characteristic 0, then $P\approx\mathbb{Q}$. If F has prime characteristic p, then $P\approx\mathbb{Z}_p$. (See Problem 48.11.) In either case P is called the *prime subfield* of F.

PROBLEMS

Assume that F is a field throughout these problems.

48.1. Prove that if E is a field and T is a subset of E, then the intersection of all the subfields of E that contain T is a subfield of E. (See Problem 20.11.) [If F is a subfield of E and S is a subset of E, and $T = F \cup S$, then the subfield obtained here is $F(S)$.]

48.2. Prove that if $a + bi$ is imaginary, then $\mathbb{R}(a+bi) = \mathbb{C}$.

48.3. Prove the last statement in the first paragraph of this section.

48.4. Prove that if $F[x]/(p(x))$ has no zero divisors, then either $p(x) = 0$ or $p(x)$ is irreducible over F.

48.5. Show that every element of $\mathbb{Q}(\sqrt{5})$ can be written in the form $a + b\sqrt{5}$, with $a, b \in \mathbb{Q}$.

48.6. Show that every element of $\mathbb{Q}(\sqrt[3]{5})$ can be written in the form $a + b\sqrt[3]{5} + c\sqrt[3]{25}$, with $a, b, c \in \mathbb{Q}$.

48.7. Prove that $\mathbb{Q}(\sqrt{2}) \not\approx \mathbb{Q}(\sqrt{3})$.

48.8. Prove that the set of all quotients $f(x)/g(x)$ with $f(x), g(x) \in F[x]$, and $g(x) \neq 0$, is a field with respect to the usual operations from elementary algebra. Also explain why this field is isomorphic to the field of quotients of $F[x]$.

48.9. Prove that the mapping θ in Example 48.2 is an isomorphism.

48.10. Prove that there is a splitting field for each $p(x) \in F[x]$. (*Suggestion*: Use the corollary of Theorem 48.1.)

48.11. Prove that the prime subfield of F is isomorphic to $\mathbb{Q}$ if F has characteristic 0, and is isomorphic to $\mathbb{Z}_p$ if F has characteristic p.

SECTION 49. FINITE FIELDS

In Section 20 we proved that $\mathbb{Z}_n$ is a field iff n is a prime. This established the existence of a field of order p for each prime p, but it left open the question of whether there are fields of other finite orders. We settled this in a special case with Example 20.1, which gave Cayley tables defining operations for a field of order 4. It can be proved that there is a finite field of order q iff q is a power of a prime. We shall prove half of this in Theorem 49.1: The order of a finite field must be a power of a prime. We shall then construct fields of order 8 and 9; these constructions will make use of theorems that we have proved about polynomials and field extensions. We shall close with some remarks about what is needed regarding polynomials and field extensions to prove the existence of a field of *each* prime power order. The proof of Theorem 49.1 will make use of the idea of the dimension of a vector space. This is reviewed in Appendix D; if it is not familiar, simply pass on to the examples that follow Theorem 49.1.

Assume that E is a field extension of F. We can think of E as a vector space over F in the following way: the vectors are the elements of E; the scalars are the elements of F; and the "product" of a scalar $a \in F$ and a vector $b \in E$ is simply the product ab in the field E. Problem 49.1 asks for a proof that this does yield a vector space. The dimension of this vector space is called the *degree* of E over F, and is denoted $[E:F]$. For example, $[\mathbb{C}:\mathbb{R}]=2$, because $\{1,i\}$ is a basis for $\mathbb{C}$ over $\mathbb{R}$: each element of $\mathbb{C}$ can be written uniquely as $a \cdot 1 + b \cdot i$ with $a,b \in \mathbb{R}$.

Theorem 49.1. *If F is a finite field, then the order of F is p^n, where p is the (prime) characteristic of F and n is a positive integer.*

PROOF. The characteristic of F cannot be 0, so it must be a prime, which we denote by p. The prime subfield of F can be identified with $\mathbb{Z}_p$, and F will be a vector space over this prime subfield. If $[F:\mathbb{Z}_p]=n$, with $\{b_1,b_2,\ldots,b_n\}$ a basis for F over $\mathbb{Z}_p$, then each element of F can be written uniquely as a linear combination

$$a_1b_1 + a_2b_2 + \cdots + a_nb_n \tag{49.1}$$

with $a_1,a_2,\ldots,a_n \in \mathbb{Z}_p$. The choices for $a_1,a_2,\ldots,a_n$ can be made independently, so that there are p^n total linear combinations (49.1). Thus the order of F is p^n. □

To construct finite fields we make use of the ideas in Sections 46 and 48. If $p(x) \in \mathbb{Z}_p[x]$, and $p(x)$ is irreducible and of degree n over $\mathbb{Z}_p$, then $\mathbb{Z}_p[x]/(p(x))$ will be a field by Theorem 46.1. Its order will be p^n, because, by Theorem 46.2, each element of the field can be expressed uniquely in the form (46.1). Here are two examples. To simplify notation in these examples, *we shall write a in place of $[a]$ for the elements of Z_n*; addition and multiplication of coefficients will be performed modulo n, of course. Also, we shall denote elements of $\mathbb{Z}_n[x]/(p(x))$ by $b_0+b_1x+\cdots+b_{n-1}x^{n-1}$ rather than $I+(b_0+b_1x+\cdots+b_{n-1}x^{n-1})$.

Example 49.1. The polynomial $1+x+x^2 \in \mathbb{Z}_2[x]$ is irreducible over $\mathbb{Z}_2$ (Problem 49.2). Therefore, $\mathbb{Z}_2[x]/(1+x+x^2)$ is a field of order $2^2=4$. Tables 49.1 and 49.2 are the Cayley tables for the field operations; they are followed by several examples of calculations of entries in the tables.

Table 49.1

+	0	1	x	$1+x$
0	0	1	x	$1+x$
1	1	0	$1+x$	x
x	x	$1+x$	0	1
$1+x$	$1+x$	x	1	0

Table 49.2

$\cdot$	0	1	x	$1+x$
0	0	0	0	0
1	0	1	x	$1+x$
x	0	x	$1+x$	1
$1+x$	0	$1+x$	1	x

Remember that computations with the coefficients are to be performed using the rules of $\mathbb{Z}_2$. Thus, for example, $x+(1+x)=1+2x=1+0x=1$. Also, an element $f(x)$ really denotes $I+f(x)$; therefore, the simplifications afforded by the remarks after Theorem 46.2 apply: if $f(x)=(1+x+x^2)q(x)+r(x)$, then $I+f(x)=I+r(x)$. Hence $x(1+x)=x+x^2=1$ in the table, because

$$x+x^2=(1+x+x^2)\cdot 1+1 \qquad \text{in } \mathbb{Z}_2[x].$$

Also, $(1+x)(1+x)=1+2x+x^2=1+x^2=x$, because

$$1+x^2=(1+x+x^2)\cdot 1+x \qquad \text{in } \mathbb{Z}_2[x].$$

Example 49.2. The polynomial $1+x^2$ is irreducible over $\mathbb{Z}_3$ (Problem 49.3). Therefore, $\mathbb{Z}_3[x]/(1+x^2)$ is a field of order $3^2=9$. Tables 49.3 and 49.4 are the Cayley tables for the operations; here are several examples of the necessary calculations.

The elements are added as any polynomials, with the coefficients reduced by the rules in $\mathbb{Z}_3$. For example,

$$(1+2x)+(2+2x)=3+4x=0+x=x.$$

Simplifications of products are made as in Example 49.1, with $I=(p(x))=(1+x^2)\subset\mathbb{Z}_3[x]$ in this case. Thus

$$(1+2x)(2+2x)=2+6x+4x^2=2+x^2 \qquad \text{in } \mathbb{Z}_3[x],$$

Table 49.3

$+$	0	1	2	x	$2x$	$1+x$	$1+2x$	$2+x$	$2+2x$
0	0	1	2	x	$2x$	$1+x$	$1+2x$	$2+x$	$2+2x$
1	1	2	0	$1+x$	$1+2x$	$2+x$	$2+2x$	x	$2x$
2	2	0	1	$2+x$	$2+2x$	x	$2x$	$1+x$	$1+2x$
x	x	$1+x$	$2+x$	$2x$	0	$1+2x$	1	$2+2x$	2
$2x$	$2x$	$1+2x$	$2+2x$	0	x	1	$1+x$	2	$2+x$
$1+x$	$1+x$	$2+x$	x	$1+2x$	1	$2+2x$	2	$2x$	0
$1+2x$	$1+2x$	$2+2x$	$2x$	1	$1+x$	2	$2+x$	0	x
$2+x$	$2+x$	x	$1+x$	$2+2x$	2	$2x$	0	$1+2x$	1
$2+2x$	$2+2x$	$2x$	$1+2x$	2	$2+x$	0	x	1	$1+x$

Table 49.4

$\cdot$	0	1	2	x	$2x$	$1+x$	$1+2x$	$2+x$	$2+2x$
0	0	0	0	0	0	0	0	0	0
1	0	1	2	x	$2x$	$1+x$	$1+2x$	$2+x$	$2+2x$
2	0	2	1	$2x$	x	$2+2x$	$2+x$	$1+2x$	$1+x$
x	0	x	$2x$	2	1	$2+x$	$1+x$	$2+2x$	$1+2x$
$2x$	0	$2x$	x	1	2	$1+2x$	$2+2x$	$1+x$	$2+x$
$1+x$	0	$1+x$	$2+2x$	$2+x$	$1+2x$	$2x$	2	1	x
$1+2x$	0	$1+2x$	$2+x$	$1+x$	$2+2x$	2	x	$2x$	1
$2+x$	0	$2+x$	$1+2x$	$2+2x$	$1+x$	1	$2x$	x	2
$2+2x$	0	$2+2x$	$1+x$	$1+2x$	$2+x$	x	1	2	$2x$

and

$$2+x^2=(1+x^2)\cdot 1+1 \qquad \text{in } \mathbb{Z}_3[x],$$

so that

$$I+(2+x^2)=I+1 \qquad \text{and} \qquad (1+2x)(2+2x)=1$$

in the table.

Assume now that F is a field of order p^n. Because F has field characteristic p, $pa=0$ for each $a\in F$, and so each nonzero element of F has order p in the additive group of F. Our remarks about a finite field as a vector space over $\mathbb{Z}_p$ can be used to show that, *as additive groups*, $F\approx\mathbb{Z}_p\times\mathbb{Z}_p\times\cdots\times\mathbb{Z}_p$ (n copies of $\mathbb{Z}_p$; see Problem 49.4). The multiplicative group of F is of order $p^n - 1$; therefore, by a corollary of Lagrange's theorem, $a^{p^n-1}=1$ for each $a \in F$. (As in Examples 49.1 and 49.2 we are using 1 for the unity of F.) It can be proved that the multiplicative group of F will have at least one element of order $p^n - 1$. This means that *the multiplicative group of a finite field is cyclic*. In Example 49.1, either x or $1+x$ is a generator for the multiplicative group of F. In Example 49.2, either $1+x$ or $1+2x$ or $2+x$ or $2+2x$ is a generator.

Because $a^{p^n-1}=1$ for each $a \in F$, it follows that each $a \in F$ is a root of $x^{p^n} - x \in \mathbb{Z}_p[x]$. It can be shown that F is a splitting field for $x^{p^n} - x$ over $\mathbb{Z}_p$. By the remarks at the end of Section 48 we know that this splitting field is essentially unique. It has order p^n and is called the *Galois field* of order p^n; it is commonly denoted by $GF(p^n)$. By the existence and uniqueness of splitting fields we can say that *there is a field of each prime power order* p^n, *and all fields of order* p^n *are isomorphic*.

In practice, Galois fields are constructed from irreducible polynomials of degree n, as in our two examples. Such polynomials always exist, although there is no general rule that produces them.

PROBLEMS

49.1. Prove that if E is a field extension of F, then E is a vector space over F with respect to the operations described in the second paragraph of this section.

49.2. Verify that $1+x+x^2 \in \mathbb{Z}_2[x]$ is irreducible over $\mathbb{Z}_2$. (*Suggestion*: Use the Factor Theorem, from Section 41.)

49.3. Verify that $1+x^2 \in \mathbb{Z}_3[x]$ is irreducible over $\mathbb{Z}_3$. (*Suggestion*: Use the Factor Theorem, from Section 41.)

49.4. Assume that F is a finite field of order p^n, and let $\{b_1, b_2, \ldots, b_n\}$ be a basis for F as a vector space over $\mathbb{Z}_p$. If $a \in F$, then a can be written uniquely in the form of Equation (49.1); define $\theta: F \to \mathbb{Z}_p \times \mathbb{Z}_p \times \cdots \times \mathbb{Z}_p$ by

$$\theta(a) = (a_1, a_2, \ldots, a_n).$$

Prove that θ is an isomorphism of additive groups.

49.5. The first step in the proof of Theorem 49.1 states that the characteristic of a finite field cannot be 0. Give a reason.

49.6. (a) Verify that $1+x+x^3 \in \mathbb{Z}_2[x]$ is irreducible over $\mathbb{Z}_2$.
(b) Construct addition and multiplication tables for the field $\mathbb{Z}_2[x]/(1+x+x^3)$.

49.7. (a) Verify that $1+x+x^4 \in \mathbb{Z}_2[x]$ is irreducible over $\mathbb{Z}_2$.
(b) Find a generator for the (cyclic) multiplicative group of the field $\mathbb{Z}_2[x]/(1+x+x^4)$.

49.8. Construct addition and multiplication tables for the ring $\mathbb{Z}_2[x]/(x^2)$. Give two proofs that the ring is not a field, one based on Theorem 46.1 and another based directly on the tables that you construct.

NOTES ON CHAPTER XIII

General references for the topics in this chapter are the same as those listed in the notes at the end of Chapter XIV. The following book contains proofs that π and e are transcendental.

1. Niven, I., *Irrational Numbers* (Carus Monograph No. 11, The Mathematical Association of America), John Wiley, New York, 1956.

CHAPTER XIV

Polynomial Equations

The desire to answer questions about roots of polynomials was the most important factor in the early history of modern algebra. This chapter contains some of the basic facts and history concerning these questions.

SECTION 50. ROOTS OF A POLYNOMIAL

Recall that an element c of a field F is a root of a polynomial $f(x) \in F[x]$ if $f(c)=0$. By the Factor Theorem (Section 41), $f(c)=0$ iff $x-c$ is a factor of $f(x)$. If $(x-c)^m$ divides $f(x)$, but no higher power of $x-c$ divides $f(x)$, then c is called a root of *multiplicity* m. When we count the number of roots of a polynomial, each root of multiplicity m is counted m times. For example, $x^3-x^2-x+1=(x-1)^2(x+1)$ has 1 as a root of multiplicity two, and -1 as a root of multiplicity one; it has no other root. Thus we say that this polynomial has three roots.

Theorem 50.1. *A polynomial $f(x)$ of degree $n \geq 1$ over a field F has at most n roots in F.*

PROOF. The proof will be by induction on n. If $n=1$, then $f(x)=a_0+a_1x$, with $a_1 \neq 0$, and the only root is $-a_1^{-1}a_0$. Thus assume that $n>1$, and assume the theorem true for polynomials of degree less than n. If $f(x)$ has no root, we are through. If c is a root, then by the Factor Theorem $f(x)=(x-c)f_1(x)$ for some $f_1(x) \in F[x]$, and deg $f_1(x)=n-1$. By the induction hypothesis $f_1(x)$ has at most $n-1$ roots in F. It will follow that

$f(x)$ has at most n roots in F if $f(x)$ has no roots in F except c and the roots of $f_1(x)$. But this is so because if $a \in F$ then $f(a)=(a-c)f_1(a)$, so that $f(a)=0$ only if $a-c=0$ or $f_1(a)=0$ (because F has no divisors of zero). Thus $f(a)=0$ only if $a=c$ or a is a root of $f_1(x)$. □

We have seen that a polynomial over a field may have no roots in that field. For example, x^2-2 and x^2+1 have no roots in the field of rationals. However, the Fundamental Theorem of Algebra (Section 26) ensures that each polynomial over the complex numbers has at least one complex root. In fact, we can prove more, as in the following corollary:

Corollary. *Each polynomial of degree $n \geq 1$ over the field $\mathbb{C}$ of complex numbers has n roots in $\mathbb{C}$.*

PROOF. We shall induct on n. Assume that $f(x)=a_0+a_1x+\cdots+a_nx^n \in \mathbb{C}[x]$. If $n=1$, then $f(x)$ has one root, namely, $-a_1^{-1}a_0$. Assume the corollary to be true for all polynomials of degree less than n. If $n>1$, then $f(x)$ has a root, say c, by the Fundamental Theorem of Algebra. Thus, by the Factor Theorem, $f(x)=(x-c)f_1(x)$ for some $f_1(x) \in \mathbb{C}[x]$, and $\deg f_1(x)=n-1$. By the induction hypothesis, $f_1(x)$ has $n-1$ roots in $\mathbb{C}$. Each of these is clearly a root of $f(x)$, as is c, and hence $f(x)$ has *at least* n roots in $\mathbb{C}$. By Theorem 50.1, $f(x)$ has *at most* n roots in $\mathbb{C}$. Thus $f(x)$ has exactly n roots in $\mathbb{C}$. □

PROBLEMS

50.1. (a) Construct a polynomial over $\mathbb{R}$ having 2 as a root of multiplicity one and 3 as a root of multiplicity two.
(b) Construct a polynomial over $\mathbb{C}$ having i as a root of multiplicity two and $-i$ as a root of multiplicity one.

50.2. Construct a polynomial over $\mathbb{Z}_5$ having [3] as a root of multiplicity two and [1] as a root of multiplicity one.

50.3. Prove that $x^2-1 \in \mathbb{Z}_{12}[x]$ has four roots in $\mathbb{Z}_{12}$. Does this contradict Theorem 50.1? Why?

50.4. Prove that if Q denotes the division ring of quarternions introduced in Problem 26.16, then $x^2+1 \in Q[x]$ has infinitely many roots in Q. (Here 1 denotes the unity of Q, that is, the 2×2 identity matrix.)

50.5. Assume that $a_0, a_1, \ldots, a_n$ are distinct elements of a field F, that $b_0, b_1, \ldots, b_n \in F$, and that $f(x), g(x) \in F[x]$ are of degree n or less and satisfy

$$f(a_j) = b_j = g(a_j) \qquad \text{for } 0 \leq j \leq n.$$

Prove that $f(x)=g(x)$ in $F[x]$. (Compare Problem 41.14.)

50.6. Assume that F is a field, $f(x) \in F[x]$, and E is an extension of F. Prove that an element c is a multiple root of $f(x)$ *iff* $f(c) = f'(c) = 0$. [*Multiple root* means root of multiplicity greater than one. The polynomial $f'(x)$ is the formal derivative of $f(x)$, defined in Problem 40.12. *Suggestion*: Use the Division Algorithm to write $f(x) = (x-c)^2 q(x) + r(x)$, where $r(x)$ is linear. Verify that $f(c) = r(c)$ and $f'(c) = r'(c)$. Next, explain why $f(x) = (x-c)^2 q(x) + (x-c)f'(c) + f(c)$. Finally, use the latter equation to examine when $(x-c)^2 | f(x)$.]

SECTION 51. RATIONAL ROOTS. CONJUGATE ROOTS

The first theorem in this section gives information about the *rational roots* of a polynomial with *integral coefficients*; the second gives information about the *complex roots* of a polynomial with *real coefficients*.

Theorem 51.1. *Let $f(x) = a_0 + a_1 x + \cdots + a_n x^n$ be a polynomial with integral coefficients. If r/s is a rational root of $f(x)$, and $(r,s) = 1$, then $r | a_0$ and $s | a_n$.*

Example 51.1. The rational roots of $f(x) = 2 - 3x - 8x^2 + 12x^3$ will be represented by fractions with numerators chosen from the divisors of 2, and denominators chosen from the divisors of 12. It can be verified that $\frac{1}{2}, -\frac{1}{2}$, and $\frac{2}{3}$ are roots of $f(x)$. Theorem 51.1 gives the same possibilities for the rational roots of $g(x) = 12x^3 - 3x + 2$ as for the rational roots of $f(x)$. But $g(x)$ has no rational root (Problem 51.4).

PROOF OF THEOREM 51.1. If r/s is a root of $f(x)$, then $a_0 + a_1(r/s) + \cdots + a_n(r/s)^n = 0$. Therefore $a_0 s^n + a_1 r s^{n-1} + \cdots + a_n r^n = 0$. Solving this first for $a_0 s^n$, and then for $a_n r^n$, we get

$$a_0 s^n = -\left[a_1 r s^{n-1} + \cdots + a_n r^n\right] \tag{51.1}$$

and

$$a_n r^n = -\left[a_0 s^n + \cdots + a_{n-1} r^{n-1} s\right]. \tag{51.2}$$

Because r is a divisor of the right side of (51.1), $r | a_0 s^n$. But $(r,s) = 1$, and so $(r, s^n) = 1$, and therefore $r | a_0$ by Lemma 39.1. Similarly, because s is a divisor of the right side of (51.2), $s | a_n r^n$. This implies that $s | a_n$ because $(s, r^n) = 1$, again by Lemma 39.1. □

Corollary. *A rational root of a monic polynomial $a_0 + a_1 x + \cdots + x^n$ with integral coefficients must be an integer and a divisor of a_0.*

PROOF. Apply Theorem 51.1 with $a_n = 1$. □

Before stating the next theorem, we first recall from elementary algebra that each quadratic equation $ax^2 + bx + c = 0$ has two solutions, given by

$$x = \frac{-b \pm \sqrt{b^2 - 4ac}}{2a}. \tag{51.3}$$

The number $b^2 - 4ac$ is called the *discriminant* of $ax^2 + bx + c$. Assume that a, b, and c are real. Then the two solutions in (51.3) will be real and equal if $b^2 - 4ac = 0$, real and unequal if $b^2 - 4ac > 0$, and imaginary and unequal if $b^2 - 4ac < 0$. In the latter case the two solutions will be complex conjugates of each other (that is, of the form $u + vi$ and $u - vi$, where $u, v \in \mathbb{R}$). Notice that the conjugate of a real number u is u itself. Thus we can say in any case that if a complex number $u + vi$ is a solution of a quadratic equation with real coefficients, then its conjugate $u - vi$ is also a solution. This is a special case of the next theorem.

Theorem 51.2. *If $a + bi$ is a root of a polynomial $f(x)$ with real coefficients, then its complex conjugate $a - bi$ is also a root of $f(x)$.*

PROOF. We assume that $b \neq 0$; otherwise the theorem is trivially true. In light of the Factor Theorem, we know that $x - (a + bi)$ is a factor of $f(x)$, and it suffices to establish that $x - (a - bi)$ is also a factor of $f(x)$. We shall do the latter by showing that $[x - (a + bi)][x - (a - bi)] = [(x - a) - bi][(x - a) + bi] = (x - a)^2 + b^2$ is a factor of $f(x)$.

Since $(x - a)^2 + b^2 = x^2 - 2ax + (a^2 + b^2)$ is a polynomial with real coefficients, we can divide it into $f(x)$, applying the Division Algorithm for polynomials over $\mathbb{R}$. This yields unique $q(x), r(x) \in \mathbb{R}[x]$ such that

$$f(x) = \left[(x - a)^2 + b^2\right]q(x) + r(x), \qquad \text{with } r(x) = 0 \text{ or } \deg r(x) < 2.$$

If $r(x) \neq 0$, then $r(x)$ must have the form $cx + d$. Then for $x = a + bi$ we have

$$0 = f(a + bi) = r(a + bi) = c(a + bi) + d = (ca + d) + cbi. \tag{51.4}$$

(Here we have used the fact that $(x - a)^2 + b^2 = 0$ for $x = a + bi$ because $(x - a)^2 + b^2 = [x - (a + bi)][x - (a - bi)]$.) But (51.4), in turn, implies that $ca + d = 0$ and $cb = 0$. Because $b \neq 0$, we conclude that $c = 0$; and then $d = 0$. Therefore $r(x) = 0$, and $[x - (a + bi)][x - (a - bi)]$ is a factor of $f(x)$. □

Corollary. *A polynomial over the field $\mathbb{R}$ of real numbers is irreducible over R iff it is linear (first degree), or quadratic (second degree) with a negative discriminant.*

PROOF. Any linear polynomial is irreducible, and any quadratic polynomial over $\mathbb{R}$ with a negative discriminant is irreducible over $\mathbb{R}$ because its roots are imaginary. It now suffices to prove the converse.

Assume that $f(x) \in \mathbb{R}[x]$ and that $f(x)$ is irreducible and nonlinear; we shall prove that $f(x)$ must be quadratic with a negative discriminant.

We can think of $f(x)$ as a polynomial over $\mathbb{C}$; as such, it has a root, say $a+bi$, in $\mathbb{C}$. If $b=0$, then $f(a)=0$ so that $x-a$ is a factor of $f(x)$, contradicting that $f(x)$ is nonlinear and irreducible over R. Thus $b \neq 0$. The proof of the preceding theorem shows that $(x-a)^2+b^2$, which is in $\mathbb{R}[x]$, must be a factor of $f(x)$. Since $f(x)$ is irreducible, it follows that $f(x)$ must be an element of $\mathbb{R}$ times $(x-a)^2+b^2$, and thus, in particular, $f(x)$ must be quadratic. The discriminant of $f(x)$ must be negative in this case because its roots are imaginary ($b \neq 0$). □

PROBLEMS

51.1. Prove that if $f(x) \in \mathbb{R}[x]$ has an imaginary root of multiplicity two, then $\deg f(x) \geq 4$.

51.2. Give an example of a quadratic polynomial over $\mathbb{C}$ that has an imaginary root of multiplicity two. (Compare Problem 51.1.)

51.3. Give an example of a polynomial over $\mathbb{Z}_2$ that is irreducible and of degree 3. (Compare the corollary of Theorem 51.2.)

51.4. Prove that $12x^3-3x+2$ has no rational root.

51.5. Find all rational roots of each of the following polynomials over $\mathbb{Q}$.
(a) $4x^3-7x-3$
(b) $2-11x+17x^2-6x^3$
(c) $2x^3-x^2+8x-4$

51.6. Write each of the following polynomials over $\mathbb{Q}$ as a product of factors that are irreducible over $\mathbb{Q}$.
(a) x^3-x^2-5x+5
(b) $3x^3-2x^2+3x-2$
(c) x^3-2x^2+2x

51.7. (a) to (c). Repeat Problem 51.6 using factors that are irreducible over $\mathbb{R}$.

51.8. (a) to (c). Repeat Problem 51.6 using factors that are irreducible over $\mathbb{C}$.

SECTION 52. AN INTRODUCTION TO GALOIS THEORY

The challenge of solving polynomial equations

$$a_n x^n + \cdots + a_1 x + a_0 = 0 \tag{52.1}$$

has been one of the most important in the history of algebra. Methods for solving first- and second-degree equations go back to the early Egyptians and Babylonians. In the sixteenth century, Italian mathematicians (del Ferro, Tartaglia, and Ferrari) succeeded in solving cubic (third-degree) and quartic (fourth-degree) equations. Their solutions give formulas for writing the roots in terms of the coefficients $a_0, a_1, \ldots, a_n$, much as the quadratic formula (51.3) for the solutions of second-degree equations. It then became a challenge to do the same for equations of degree higher than four. Specifically, the problem, as it eventually came to be interpreted, was to show that it is possible to express the roots of an equation (52.1) in terms of the coefficients $a_0, a_1, \ldots, a_n$ using only addition, subtraction, multiplication, division, and the extraction of roots, each applied only finitely many times. When the roots can be so expressed, it is said that the equation is *solvable by radicals*.

In 1770–1771 Lagrange took the first steps in settling this problem by introducing methods for studying polynomial equations that in reality involved group theory. Lagrange sensed that the key to understanding these equations and their roots was related to the effect, on the original equations and on certain related equations, of permutations of the roots of the equation.

Early in the nineteenth century Paulo Ruffini (Italian) and N. H. Abel (Norwegian), drawing on the ideas introduced by Lagrange, showed that, in fact, there are equations of each degree higher than four that are *not* solvable by radicals. The complete answer to the question of which equations are solvable by radicals was finally given by Evariste Galois, another French mathematician, in the mid-nineteenth century. Galois was able to show that a group can be associated with each polynomial equation (52.1) in such a way that the property of solvability by radicals is directly related to a corresponding property of the group. This group property is called *solvability* and is defined as follows: A group G is *solvable* if it contains a finite sequence of subgroups

$$G = G_0 \rhd G_1 \rhd G_2 \rhd \cdots \rhd G_n = \{e\},$$

with each subgroup normal in the next, as indicated, such that each of the successive quotient groups G_{i-1}/G_i is Abelian. What Galois proved is that an equation is solvable by radicals iff its associated group (now called its *Galois group*) is solvable. The point of the earlier work by Ruffini and Abel was that for each degree higher than four there exist polynomial equations whose Galois groups are not solvable.

The work by Galois became the basis for what is now known as *Galois theory*. Today this theory has to do with the study of groups of automorphisms of field extensions; theorems concerning solvability by radicals are merely a special part of the general theory. We shall now give a very brief sketch of the parts of Galois theory relating to solvability by radicals. *All fields are assumed to be subfields of the field of complex numbers.*

We begin by recalling some facts from Section 48. If F is a field, and $a \in \mathbb{C}$, then $F(a)$ denotes the smallest subfield of $\mathbb{C}$ that contains both F

and a. Also, in general, if F is a field and S is a subset of $\mathbb{C}$, then $F(S)$ denotes the smallest subfield of $\mathbb{C}$ that contains both F and S. The field $F(S)$ is the intersection of all subfields of $\mathbb{C}$ that contain both F and S. If $f(x) \in F[x]$, then an extension field E of F such that $E = F(a_1, \ldots, a_n)$ and $f(x) = a(x - a_1) \cdots (x - a_n) \in E[x]$ is called a *splitting* (or *root*) *field* of $f(x)$ over F. In other words, a splitting field of $f(x)$ is just large enough to contain F and the roots of $f(x)$. It can be proved that such a field exists, and that it is uniquely determined by F and $f(x)$.

Next we need the concept of field automorphism. An *automorphism* of a field E is a one-to-one mapping σ of E onto E such that $\sigma(a+b) = \sigma(a) + \sigma(b)$ and $\sigma(ab) = \sigma(a)\sigma(b)$ for all $a, b, \in E$ (for example, $\sigma : \mathbb{C} \to \mathbb{C}$ defined by $\sigma(x+yi) = x - yi$). If F is a subfield of a field E, then G_F will denote the set of all automorphisms σ of E such that $\sigma(a) = a$ for all $a \in F$. It is easy to show that G_F (with composition as operation) is a group (Problem 52.2); it is called the *Galois group* of E over F. If E is the splitting field of a polynomial $f(x) \in F[x]$, then G_F is called the *Galois group* of $f(x)$ over F. The fundamental theorem of Galois theory establishes a correspondence between the subgroups of G_F and the subfields of E that contain F. This correspondence is used to deduce properties of E from properties of G_F.

A polynomial equation $f(x) = 0$, with $f(x) \in F[x]$, is said to be *solvable by radicals* over F if the splitting field E of $f(x)$ over F can be embedded in a field D such that

$$F = F_0 \subset F_1 \subset \cdots \subset F_k = D,$$

where $F_j = F_{j-1}(a_{j-1})$ for $1 \le j \le k$ with each a_{j-1} a root of an equation

$$x^{n_j} - c_{j-1} = 0 \qquad \text{with } c_{j-1} \in F_{j-1}.$$

Now we can state the basic theorem on solvability by radicals: *An equation $f(x) = 0$ is solvable by radicals over F iff G_F, the Galois group of $f(x)$ over F, is a solvable group.*

The fact that there exist equations of each degree greater than four that are not solvable by radicals can now be shown by establishing these additional facts: (a) for each $n \ge 5$ there is an equation of degree n over the field of rational numbers whose Galois group is S_n, and (b) if $n \ge 5$, then S_n is not solvable.

PROBLEMS

52.1. Prove that the set of all automorphisms of a field E is a group with respect to composition.

52.2. Prove that if F is a subfield of field E, and G_F is the set of all automorphisms σ of E such that $\sigma(a) = a$ for each $a \in F$, then G_F is a subgroup of the group of automorphisms of E.

52.3. Show that the group S_3 is solvable.

52.4. Prove that if the groups A and B are solvable, then so is $A \times B$.

52.5. Prove that $\mathbb{Q}(\sqrt{2}, i) = \mathbb{Q}(\sqrt{2} + \sqrt{2}\, i)$.

52.6. Prove that if a is a primitive nth root of unity, then the splitting field of $x^n - 1$ over $\mathbb{Q}$ is $\mathbb{Q}(a)$. (*Primitive root* is defined in Problem 27.7.)

NOTES ON CHAPTER XIV

References [1] through [7] are general works on modern algebra. Each of these references contains more extensive discussions of fields and polynomial equations than this book.

The early editions of [7], which were written in the 1930s, have had a major influence on most other textbooks in modern algebra. Reference [1] has been widely used since the appearance of its first edition in 1941. References [3], [5], and [6] are graduate-level textbooks; reference [4] is written at a level between those books and this one. Reference [2] is a well-known series treating algebra as well as other parts of mathematics; although originally published in French, some volumes are available in English translation.

1. Birkhoff, G., and S. MacLane, *A Survey of Modern Algebra*, 3rd ed., Macmillan, New York, 1965.
2. Bourbaki, N., *Éléments de mathématiques*, Hermann, Paris.
3. Goldhaber, J., and G. Ehrlich, *Algebra*, Macmillan, New York, 1970.
4. Herstein, I. N., *Topics in Algebra*, 2nd ed., John Wiley, New York, 1975.
5. Hungerford, T. W., *Algebra*, Holt, Rinehart and Winston, New York, 1974.
6. Jacobson, N., *Basic Algebra I*, Freeman, San Francisco, 1974.
7. Van der Waerden, B. I., *Algebra*, 7th ed., 2 vols., Ungar, New York, 1970.

The following two books are specialized, treating fields and polynomial equations.

8. Artin, E., *Galois Theory*, 2nd ed., University of Notre Dame, Notre Dame, 1944.
9. Gaal, L., *Classical Galois Theory with Examples*, 2nd ed., Chelsea, New York, 1973.

CHAPTER XV

Geometric Constructions

In the last chapter we saw that in working to solve certain polynomial equations by radicals, mathematicians in the seventeenth and eighteenth centuries had tried to do the impossible. The story in this chapter is similar: in working to carry out certain geometric constructions using only the straightedge and compass, mathematicians in ancient Greece also had tried to do the impossible. What makes this story especially interesting is that although the constructions were to be geometric, the proofs of their impossibility involve algebra. In fact, the key algebraic concepts are the same as those used to analyze solvability by radicals.

After showing why geometric construction problems are equivalent to problems in algebra, we shall determine which of these algebraic problems have solutions. We shall then show how these results apply to the constructions originally attempted by the Greeks.

SECTION 53. THREE FAMOUS PROBLEMS

In the fifth century B.C., early in the history of Greek geometry, three problems began to attract increasing attention.

I. The duplication of the cube.
II. The trisection of an arbitrary angle.
III. The quadrature of the circle.

Each involved the construction of one geometrical segment from another, using only an (unmarked) straightedge and a compass. With the first the problem was to construct the edge of a cube having twice the volume of a given cube; with the second the problem was to show that any angle could be trisected; and with the third the problem was to construct the side of a square having the same area as a circle of given radius.

It must be stressed that these problems are concerned only with the question of whether the constructions can, in theory, be carried out in a finite number of steps using only a straightedge and compass. For practical purposes there is no reason to restrict the tools to a straightedge and compass, and the constructions can be carried out to any desired degree of accuracy by other means.

As a first step in analyzing the three problems, let us rephrase each of them using numbers. In I, if the edge of the given cube is taken as the unit of length, and the edge of the required cube is denoted by x, then the volumes of the two cubes are 1 cubic unit and x^3 cubic units, respectfully. Thus I can be rephrased as follows:

> **I′.** *Given a segment of length* 1, *construct a segment of length* x *with* $x^3 = 2$.

Ultimately, we shall show that the construction II is impossible in general by showing that an angle of 60° cannot be trisected. (Some angles can, in fact, be trisected. But the problem in II is whether they all can be trisected, and one example will suffice to prove otherwise.) Thus we restrict attention now to an angle of 60°. It is easy to show that an angle can be constructed iff a segment the length of its cosine can be constructed from a segment of unit length (Problem 53.1; here and elsewhere we can assume a segment of unit length as given). It can be shown with elementary trigonometry (Problem 53.2) that if A is any angle, then

$$\cos A = 4\cos^3\left(\frac{A}{3}\right) - 3\cos\left(\frac{A}{3}\right). \tag{53.1}$$

If we take $A = 60°$, and let $x = \cos(A/3)$, this simplifies to

$$8x^3 - 6x - 1 = 0.$$

Thus the problem of trisecting a 60° angle can be rephrased like this:

> **II′.** *Given a segment of length* 1, *construct a segment of length* x *with* $8x^3 - 6x - 1 = 0$.

For III, we can take the radius of the given circle as the unit of length. Then the area of the circle is π square units, and the problem becomes that of constructing a segment of length x such that $x^2 = \pi$. So III can be rephrased as follows:

> **III′.** *Given a segment of length* 1, *construct a segment of length* x *with* $x^2 = \pi$.

With I′, II′, and III′, each of the original problems becomes a problem of whether a certain multiple of a given length can be constructed. The key in each case is in the following fact, which will be proved in the next section:

> *A necessary and sufficient condition that a segment of length x can be constructed with a straightedge and compass, beginning with a segment of length* 1, *is that x can be obtained from* 1 *by a finite number of rational operations and square roots.*

(The rational operations referred to here are addition, subtraction, multiplication, and division.) After we have proved this fact, the only remaining questions will be whether the solutions of the equations in I′, II′, and III′ satisfy the specified condition; those questions will be considered in Section 55.

Now assume that a unit of length has been given. Just what this length is can vary from problem to problem, but, whatever it is, we shall regard it as fixed until the end of the next section. We can construct perpendicular lines with a straightedge and compass, and so we do that, and then mark off the given unit of length on each of the perpendicular lines (axes), to establish coordinates for points in the plane of the axes. Following the usual conventions, we label the axes with x and y, directed as shown in Figure 53.1.

The points O, P, and Q will be called *constructible points*, as will any other points that can be obtained by starting with the points O, P, and Q and making repeated use of a straightedge and compass. In order to determine which points are constructible, we must introduce two further notions. A *constructible line* is a line (in the xy-plane) passing through two constructible points. A *constructible circle* is a circle (in the xy-plane) having its center at a constructible point, and its radius equal to the distance between two constructible points. Thus a point other than O or P or Q is constructible if it is a point of intersection of two constructible lines, of two constructible circles, or of a constructible line and a constructible circle. Finally, a *constructible number* is a number that is either the

Figure 53.1

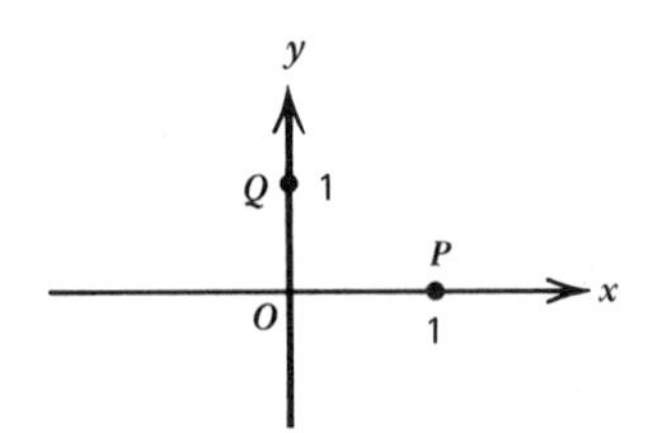

distance between two constructible points or the negative of such a distance. (This definition depends on the chosen unit of length, of course.)

Example 53.1. The unit circle is constructible: its center is at point O and its radius is OP. The line bisecting the first and third quadrants is constructible, because the lines OP and OQ are constructible, and the bisector of an angle between two constructible lines is constructible (Problem 53.4). Therefore, point R (Figure 53.2) is constructible, being the intersection of a constructible line and a constructible circle. Also, $\sqrt{2}$ is a constructible number, because it is the distance between P and Q.

Example 53.2. A point is constructible iff each of its coordinates is a constructible number. Because, given constructible numbers a and b, the points (a,O) and (O,b) are constructible in one step; and then the point (a,b) is constructible as a point of intersection of two circles, one with center at (a,O) and radius $|b|$, the other with center at (O,b) and radius $|a|$. Conversely, given a constructible point S with coordinates (a,b), we can construct lines through S perpendicular to each coordinate axis; the points of intersection of these lines with the coordinate axes have coordinates (a,O) and (O,b), so that a and b are constructible.

We are primarily interested in constructible numbers, but we must get at those by working with constructible points, lines, and circles. Let $\mathbb{K}$ denote the set of all constructible numbers. This is a subset of $\mathbb{R}$. In fact, by the following theorem, $\mathbb{K}$ is a sub*field* of $\mathbb{R}$.

Theorem 53.1. *The set $\mathbb{K}$ of constructible numbers is a field.*

Proof. We are assuming that 1, a unit of length, has been given. Thus it suffices to prove that if $a \in \mathbb{K}$ and $b \in \mathbb{K}$, then $a+b \in \mathbb{K}$, $a-b \in \mathbb{K}$, $ab \in \mathbb{K}$,

Figure 53.2

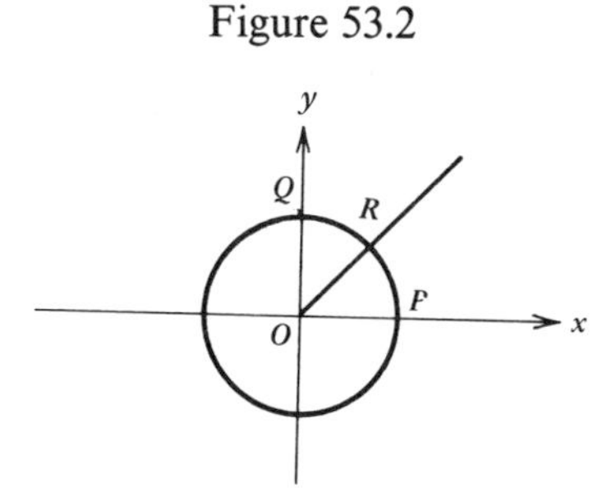

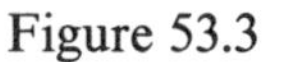
Figure 53.3

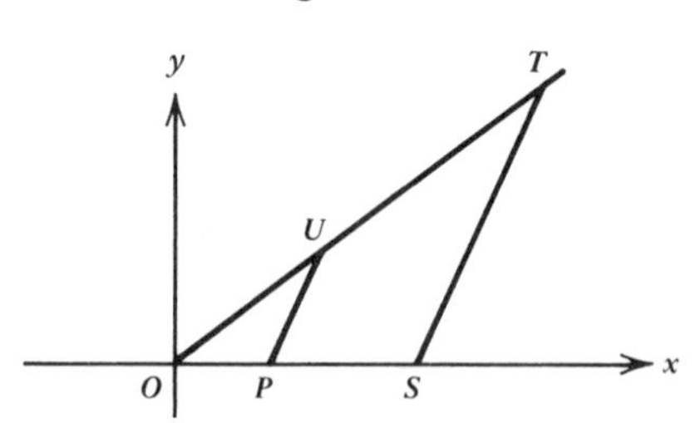

and, if $b \neq 0$ as well, then $a/b \in \mathbb{K}$. The proof for a/b will be given here; the other parts will be left to the problems.

Assume that a and b are constructible numbers, with $b \neq 0$. It will suffice to prove that $a/b \in \mathbb{K}$ for $a > 0$ and $b > 0$. As before, assume OP of unit length (Figure 53.3). The point S, b units from O on the x-axis, is constructible; and the point T, a units from O on the line $x = y$, is constructible (Problem 53.5). The line through P parallel to ST is constructible (Problem 53.6); let this line intersect OT at U. The point U is constructible, and, because triangles OPU and OST are similar, $OU/OP = OT/OS$. Therefore, the distance between O and U is a/b, so that $a/b \in \mathbb{K}$. □

Being a subfield of $\mathbb{R}$, the field $\mathbb{K}$ must contain all rational numbers. In the next section we shall determine what other numbers are in $\mathbb{K}$.

PROBLEMS

All constructions are to involve only an (unmarked) straightedge and a compass.

53.1. Show how to construct an angle if a segment the length of its cosine and a segment of unit length are given.

53.2. Derive the trigonometric identity (53.1).

53.3. Show how to construct a perpendicular to a given line at a given point on the line.

53.4. Show how to construct the bisector of a given angle.

53.5. Explain why the point T is constructible in the proof of Theorem 53.1.

53.6. Explain why the line through P parallel to ST is constructible in the proof of Theorem 53.1.

53.7. Prove that if a and b are constructible numbers, then so are $a+b$, $a-b$, and ab.

SECTION 54. CONSTRUCTIBLE NUMBERS

The next lemma will be used in proving Theorem 54.1, which will characterize constructible numbers.

Lemma 54.1. *Let F be a subfield of $\mathbb{R}$.*
(*a*) *If a line contains two points whose coordinates are in F, then the line has an equation $ax+by+c=0$ for some $a,b,c\in F$.*
(*b*) *If a circle has its radius in F and its center at a point whose coordinates are in F, then the circle has an equation $x^2+y^2+dx+ey+f=0$ for some $d,e,f\in F$.*

PROOF. (a) An equation for the line through (x_1,y_1) and (x_2,y_2) is

$$\frac{y-y_1}{x-x_1}=\frac{y_2-y_1}{x_2-x_1}.$$

This can be put in the form $ax+by+c=0$ with each of the coefficients in F because each is a rational combination of x_1, y_1, x_2, and y_2.
(b) Similar to part (a), beginning with

$$(x-x_1)^2+(y-y_1)^2=r^2$$

with $x_1,y_1,r\in F$. □

If F is a subfield of $\mathbb{R}$ and $a\in\mathbb{R}$, then $F(a)$ is the smallest subfield of $\mathbb{R}$ containing both F and a (see Section 48). For example, $\mathbb{Q}(\sqrt{2})$ consists of all real numbers $c+d\sqrt{2}$ with $c,d\in\mathbb{Q}$. Notice that each number in $\mathbb{Q}(\sqrt{2})$ is constructible, because $\mathbb{Q}\subset\mathbb{K}$ and $\sqrt{2}\in\mathbb{K}$ (Example 53.1). Each number in $\mathbb{Q}(\sqrt{2})(\sqrt{3})=\mathbb{Q}(\sqrt{2},\sqrt{3})$ is also constructible, because $\sqrt{3}\in\mathbb{K}$ (see Problem 54.2). We shall now prove that all constructible numbers belong to fields that are built up in this way.

Theorem 54.1. *A real number r is constructible iff there is a finite sequence $\mathbb{Q}=F_0\subset F_1\subset\cdots\subset F_k$ of subfields of $\mathbb{R}$ with $r\in F_k$, where $F_j=F_{j-1}(\sqrt{a_{j-1}})$ and a_{j-1} is a positive number in F_{j-1} for $1\le j\le k$.*

PROOF. Assume first that r is constructible. Then r can be constructed in a finite number of steps beginning with 1, which is in $\mathbb{Q}$. In light of Lemma 54.1, it suffices to prove that if F is any subfield of $\mathbb{K}$, and a real number r is a coordinate of an intersection point of lines or circles having equations with coefficients in F, then $r\in F(\sqrt{a})$ for some positive real number $a\in F$.

It is easy to see that in solving two simultaneous linear equations $a_1x+b_1y+c_1=0$ and $a_2x+b_2y+c_2=0$, with coefficients in F, we shall not be led outside F. Thus, if r is a coordinate of the intersection of the corresponding lines, then $r \in F$, which is sufficient.

If y is eliminated from $ax+by+c=0$ and $x^2+y^2+dx+ey+f=0$, the result is a quadratic equation $Ax^2+Bx+C=0$ with $A,B,C \in F$. Therefore, the x-coordinates of the intersection of the corresponding line and circle are given by

$$x = \frac{-B \pm \sqrt{B^2-4AC}}{2A}.$$

If these are real numbers, then $x \in F(\sqrt{a}\,)$ for $a=B^2-4AC>0$, as required. The case of the y-coordinate is similar.

The coordinates of the intersection points of two circles, $x^2+y^2+d_1x+e_1y+f_1=0$ and $x^2+y^2+d_2x+e_2y+f_2=0$, can be obtained by first subtracting, thereby getting $(d_1-d_2)x+(e_1-e_2)y+(f_1-f_2)=0$, and then using this equation with that of the first circle. The result is similar to that in the case of a line and a circle.

Now assume that r is an element of a field F_k as described in the theorem. We know that each number in $\mathbb{Q}$ is constructible, and so in order to prove that r is constructible it suffices to prove that if each number in a subfield F of $\mathbb{R}$ is constructible, and if a is a positive number in F, then each number in $F(\sqrt{a}\,)$ is constructible. Because each number in $F(\sqrt{a}\,)$ is of the form $c+d\sqrt{a}$ for $c,d \in F$, and because products and sums of constructible numbers are constructible, it suffices to prove that $\sqrt{a}$ is constructible. But it is an elementary construction problem to show that $\sqrt{a}$ is constructible from 1 and a (Problem 54.3). □

PROBLEMS

54.1. Carefully fill in the details in the proof of Lemma 54.1.
54.2. Prove that $\sqrt{3} \in \mathbb{K}$, without using Theorem 54.1.
54.3. Prove that $\sqrt{a}$ is constructible from 1 and a.
54.4. Prove that $\sqrt[4]{2} \in \mathbb{K}$.

SECTION 55. IMPOSSIBLE CONSTRUCTIONS

We shall now use the characterization of constructible numbers given in Theorem 54.1 to prove the impossibility of duplication of the cube and trisection of a 60° angle. We shall also indicate, without proof, why quadrature of the circle is impossible. In settling the first two problems we shall need the following theorem.

Theorem 55.1. *If a cubic equation with rational coefficients has no rational root, then it has no constructible root.*

The following two lemmas will be used in the proof of the theorem.

Lemma 55.1. *If the roots of* ax^3+bx^2+cx+d *are* r_1, r_2, *and* r_3, *then* $r_1+r_2+r_3=-b/a$.

PROOF. The polynomials ax^3+bx^2+cx+d and $a(x-r_1)(x-r_2)(x-r_3)=ax^3-a(r_1+r_2+r_3)x^2+a(r_1r_2+r_1r_3+r_2r_3)x-ar_1r_2r_3$ have the same roots and the same leading coefficients, and therefore the coefficients on all like powers of x must be equal (Problem 55.1). Thus, in particular, $b=-a(r_1+r_2+r_3)$. □

Lemma 55.2. *If* F *is a subfield of* $\mathbb{R}$, *and a cubic polynomial* $ax^3+bx^2+cx+d\in F[x]$ *has a root* $p+q\sqrt{r}$, *where* $p,q,r\in F$, $r>0$, *and* $\sqrt{r}\notin F$, *then* $p-q\sqrt{r}$ *is also a root.*

PROOF. The proof is similar to that of Theorem 51.2 and is left as an exercise (Problem 55.2). □

PROOF OF THEOREM 55.1. Assume that $ax^3+bx^2+cx+d\in\mathbb{Q}[x]$ has no rational root, but that it does have a root in $\mathbb{K}$. We shall show that this leads to a contradiction. Let $u\in\mathbb{K}$ be a root of the given cubic, and assume u chosen so that the corresponding field F_k in Theorem 54.1 satisfies the condition that k is minimal with respect to the condition of containing such a root. Then $k\neq 0$ because $u\notin\mathbb{Q}=F_0$. Therefore, $u=p+q\sqrt{r}$ for some $p,q,r\in F_{k-1}$, $\sqrt{r}\notin F_{k-1}$. By Lemma 55.2, $p-q\sqrt{r}$ is also a root of the given cubic. By Lemma 55.1, the sum of the roots of the cubic is $-b/a$, so that if v is the third root, then $-b/a=v+(p+q\sqrt{r})+(p-q\sqrt{r})$, and $v=-b/a-2p$. But $a,b,p\in F_{k-1}$, and hence $v\in F_{k-1}$, contradicting the minimality of k. □

Impossibility of the duplication of the cube.

From I′ in Section 53, the question is whether x^3-2 has a constructible root. Theorem 51.1 can be used to show that x^3-2 has no rational root, and therefore, by Theorem 55.1, x^3-2 has no constructible root.

Impossibility of the trisection of an arbitrary angle.

From II′ in Section 53, if a 60° angle could be trisected, then $8x^3-6x-1$ would have a constructible root. But Theorem 51.1 can be used to show that $8x^3-6x-1$ has no rational root, and therefore, again by Theorem 55.1, $8x^3-6x-1$ has no constructible root.

Discussion of the quadrature of the circle.

It can be proved from Theorem 54.1 that any constructible number is a root of a polynomial with integral coefficients; any root of such a polynomial is called an *algebraic number*. Thus from III′ in Section 53 the impossibility of the quadrature of the circle would follow if it were shown that $\sqrt{\pi}$ is not algebraic. This was done by the German mathematician C. L. F. Lindemann, in 1882. The proof requires methods that are not algebraic, however, and thus it will not be given here. (See reference [1] of Chapter XIII.)

PROBLEMS

55.1. Prove that if two polynomials over $\mathbb{C}$ have the same degree, the same roots, and the same leading coefficients, then the coefficients on all like powers of x must be equal. (*Suggestion:* Subtract one of the polynomials from the other. The result is a polynomial of smaller degree than that of the given polynomial, but with as many roots as the given polynomial. Apply Theorem 50.1.)

55.2. Prove Lemma 55.2.

55.3. Prove that neither x^3-2 nor $8x^3-6x-1$ has a rational root.

55.4. Prove the following generalization of Lemma 55.1: If the roots of $a_nx^n + \cdots + a_1x + a_0$ are $r_1, r_2, \ldots, r_n$, then

$$r_1 + r_2 + \cdots + r_n = -a_{n-1}/a_n$$

and

$$r_1r_2\cdots r_n = (-1)^n a_0/a_n.$$

NOTES ON CHAPTER XV

For more details concerning the history of geometric construction problems, see either the book listed below or the books by Boyer and May listed in the Introduction.

1. Klein, F., *Famous Problems of Elementary Geometry*, Chelsea, New York, 1955.

CHAPTER XVI

Algebraic Coding

Occasional errors are inevitable when information is transmitted from one place to another, as from a satellite to the earth, or when information is stored and then retrieved, as in the magnetic storage of a computer. These errors are likely to occur randomly, caused by such things as lightning or circuitry failure. The task for coding theory is to minimize the effect of these errors. This is done by introducing redundancy into the information before it is transmitted or stored, in such a way as to help detect when errors occur or even to correct errors that occur. Interestingly, the techniques used in doing this have been mostly algebraic, involving finite fields and ideas from linear algebra. The study of these algebraic techniques is the domain of algebraic coding theory. A thorough introduction to this theory would draw on much of the preceding material in this book. The treatment here will be less thorough than that, to make it more easily accessible and of manageable length; the only prerequisites are parts of Chapters I to V, material from linear algebra that is reviewed in Appendix D, and some elementary facts about probability that are used in the last section (and can be ignored by taking Theorem 59.1 on faith).

SECTION 56. INTRODUCTION

Example 56.1. Suppose that a message consisting of a single binary digit, either 0 or 1, is to be sent from one place to another. There is some probability, however small, that 1 will be received if 0 is sent, and that 0 will be received if 1 is sent. One way to minimize the chance of error is to send the message three times: send 000 in place of 0, and 111 in place of 1. If 000 is received, it is reasonable to assume that 000 was sent. If either 100, 010, or 001 is received, it is still reasonable to assume that 000 was sent, because the probability of one error is smaller than the probability of two errors. (This is true if the probability of an error for any one symbol is small, an assumption we shall make throughout.) If either 111, 011, 101, or 110 is received, the assumption would be that 111 was sent. The correspondence (mapping) $0 \mapsto 000$, $1 \mapsto 111$ is called a *code*. This code will detect if two or less errors have occurred; this will be the case if any message other than 000 or 111 is received. The code cannot detect when three errors have occurred. Any message subjected to a single error will be interpreted correctly; this is not true for double errors.

In general, there will be several steps in transmitting information, as indicated schematically in Figure 56.1. We are not concerned with the engineering problems involved in this process, but rather with the idealized problem of the design of efficient codes. (Notice that this is not the same as constructing "secret" codes.) Our interest is in the *encoder* and the *decoder*. The encoder will take information from the source and produce a *code*, which will be a sequence of 0's and 1's, or more generally elements from any specified finite field. The decoder will receive this code after it has been subjected to possible alteration during transmittal or storage. The decoder must be designed to make a decision as to what message was sent based on what message is received, or perhaps to reject the message altogether if it differs substantially from anything that might have been sent.

The codes we shall consider are called *block codes*. Any such code consists of a sequence of *code words* (or *blocks*). The code words are themselves sequences of *digits*, which will be elements from some predetermined finite field F. If the digits are from $\mathbb{Z}_2$, the code is called a *binary code*. When working with $\mathbb{Z}_p$, we shall follow the practice, which we began in Chapter XI, of omitting brackets from elements of $\mathbb{Z}_p$, writing $+$ instead of $\oplus$, using juxtaposition or $\cdot$ instead of $\odot$, and relying on the context to remind us that all calculations are to be done modulo n. In $\mathbb{Z}_5$, for example, $3(4+3)=3\cdot 2=6=1$.

All code words will have the same length, n, called the *code length* (or *block length*) of the code. Code words of length n can be thought of as elements of F^n, the vector space of n-tuples over the field F. These words will be written as in Example 56.1; for example, 010 rather than $(0,1,0)$.

Figure 56.1

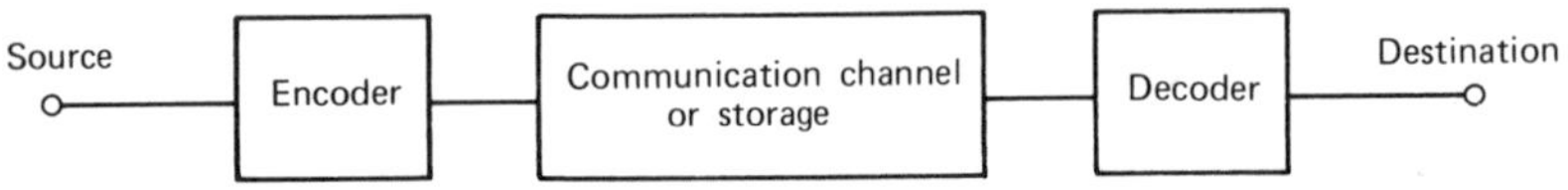

Source → Encoder → Communication channel or storage → Decoder → Destination

Only code words will leave the encoder. (In Example 56.1, the code words are 000 and 111.) But any word of length n can be a *received word*, that is, a word received by the decoder. (In Example 56.1, there are eight possible received words.) The decision-making problem of the decoder is, given a received word, to either assign it a code word or declare it to be an error. The solution chosen for this problem can be conveniently presented by means of a *decoding table*, as in the following example.

Example 56.2. Table 56.1 is a decoding table for a code with code words 10001, 11100, 01011, and 00110. If any one of these four code words is received, it will be decoded as itself. If any other word is received, it will be decoded as the code word heading the column in which the received word appears. For instance, 01110 would be decoded as 00110. Each received word that is the result of an error in a single digit will be decoded correctly by the table—in each column these are the five words above the dotted line. Some received words that result from errors in two or more digits will also be decoded correctly—these are below the dotted line. However, some received words that are the result of errors in two digits will be decoded incorrectly; for example, if 11100 is sent and then received as 10110, it will be incorrectly decoded as 00110. If we arrange to have "error" signaled whenever the received word is not a code word, then all errors in one or two digits will be detected.

The *distance* between two words is the number of positions in which they differ. The distance between 10001 and 01011 is 3, for example. The smallest distance between all pairs of code words of a code will be called the *minimum distance* of the code. In Example 56.2, the minimum distance is 3; that is why all errors in one or two digits can be detected—such errors cannot change one code word into a different code word.

Table 56.1

code word	10001	11100	01011	00110
received words with one error	10000	11101	01010	00111
	10011	11110	01001	00100
	10101	11000	01111	00010
	11001	10100	00011	01110
	00001	01100	11011	10110
received words with two errors	00000	11010	01101	10111
	00101	11111	01000	10010

We shall often use the following decoding rule: To decode an n-tuple, choose the code word closest to it in terms of the distance defined above; if this offers more than one choice, choose arbitrarily. This is called the *maximum-likelihood decoding rule*; under certain mild restrictions it maximizes the probability of correct decoding. (This will be discussed in Section 59.) Table 56.1 is based on this rule.

Notice that a code will *detect* patterns of t or fewer errors if its minimum distance is $t+1$, because in this case a code word will not be changed into a different code word through errors in t or fewer of its digits. Also, a code will *correct* patterns of t or fewer errors if its minimum distance is $2t+1$, because in this case any word resulting from t or fewer errors will be decoded correctly, since no word is within distance t of more than one code word. This shows the desirability of having a code with code words as far apart as possible. One of the most effective methods for designing such codes makes use of finite fields and ideas from linear algebra, and will be discussed in the next section.

PROBLEMS

56.1. Construct a decoding table for the code in Example 56.1. What is the minimum distance for the code?

56.2. Use the maximum-likelihood decoding rule to construct a decoding table for a binary code with code words 0000 and 1111. If 0000 is sent, which received words that result from errors in two digits will be decoded incorrectly?

56.3. Consider Example 56.2, and assume that the code word 10001 is sent. Which received words that result from errors in two digits will be decoded incorrectly?

56.4. Use the maximum-likelihood decoding rule to construct a decoding table for a binary code with code words 1001, 0101, 0010, 1110. What is the minimum distance for the code?

56.5. (a) Find three code words for a binary code of length 3 such that the minimum distance of the code is 2. (b) Is there a binary code with length 4 and three code words such that the minimum distance of the code is 3?

56.6. Denote the distance between two words x and y by $d(x,y)$. [Example: $d(10001, 01011)=3$.] Verify that

(a) $d(x,y)=0$ iff $x=y$,

(b) $d(x,y)=d(y,x)$, and

(c) $d(x,z)\leq d(x,y)+d(y,z)$ for all words x,y,z of a code.

56.7. Prove that a code with minimum distance d can correct all patterns of s or fewer errors and detect all patterns of t or fewer errors ($t\geq s$) if $t+s<d$. [*Suggestion*: For each word x, let $S_r(x)$ denote the *sphere* consisting of all words y such that $d(x,y)\leq r$. Draw a picture representing two spheres whose centers are code words a minimum distance apart. Problem 56.6 will support your geometrical intuition.]

SECTION 57. LINEAR CODES

The words in our codes have been arbitrary subsets of F^n, the vector space of n-tuples over a finite field F. One additional requirement—that the set of code words form a *subspace* of F^n—provides the bridge between the problems of coding and the techniques of matrix theory and linear algebra. The following definition singles out the resulting codes for fields of prime order.

Definition. An (n,k) *linear code* over $\mathbb{Z}_p$ is a code whose code words form a k-dimensional subspace of $\mathbb{Z}_p^n$.

If $\{w_1, w_2, \ldots, w_k\}$ is a basis for the subspace of code words, then this subspace is the set of all linear combinations

$$a_1 w_1 + a_2 w_2 + \cdots + a_k w_k,$$

with $a_1, a_2, \ldots, a_k \in \mathbb{Z}_p$. There are p choices for each a_i $(1 \le i \le k)$, so there are p^k code words in an (n,k) linear code over $\mathbb{Z}_p$. Hereafter, we shall consider only linear codes, and our examples will even be restricted to linear *binary codes*, that is, linear codes over $\mathbb{Z}_2$. The code in Example 56.2 is not a linear code, because the code words do not form a subspace of $\mathbb{Z}_2^5$.

Example 57.1. The set $\{10001, 11100, 01011\}$ is a basis for a (5, 3) linear binary code. The set (subspace) of code words is

$$\{00000, 10001, 11100, 01011, 01101, 11010, 10111, 00110\}.$$

A $k \times n$ matrix whose rows form a basis for an (n,k) linear code is called a *generator matrix* for the code. The basis in Example 57.1 gives the generator matrix

$$G = \begin{bmatrix} 1 & 0 & 0 & 0 & 1 \\ 1 & 1 & 1 & 0 & 0 \\ 0 & 1 & 0 & 1 & 1 \end{bmatrix}. \tag{57.1}$$

The code words come from forming all linear combinations of the rows of a generator matrix. These linear combinations can be obtained by multiplying the generator matrix on the left by the $p^k \times k$ matrix whose rows are the elements of $\mathbb{Z}_p^k$. For Example 57.1, an appropriate matrix is

$$A = \begin{bmatrix} 0 & 0 & 0 \\ 1 & 0 & 0 \\ 0 & 1 & 0 \\ 0 & 0 & 1 \\ 1 & 1 & 0 \\ 1 & 0 & 1 \\ 0 & 1 & 1 \\ 1 & 1 & 1 \end{bmatrix}.$$

Then

$$AG = \begin{bmatrix} 0 & 0 & 0 & 0 & 0 \\ 1 & 0 & 0 & 0 & 1 \\ 1 & 1 & 1 & 0 & 0 \\ 0 & 1 & 0 & 1 & 1 \\ 0 & 1 & 1 & 0 & 1 \\ 1 & 1 & 0 & 1 & 0 \\ 1 & 0 & 1 & 1 & 1 \\ 0 & 0 & 1 & 1 & 0 \end{bmatrix}$$

The rows of AG are the code words in the same order as they are listed in Example 57.1. Notice that a generator matrix completely determines the code and, in general, will be much more compact than the list of all code words.

Another way to specify an (n,k) linear code is by giving an $(n-k)\times n$ matrix H whose rows form a basis for the null space of the generator matrix G. (The null space of a matrix is discussed in Appendix D.) This matrix H has the property that

$$GH' = 0, \tag{57.2}$$

where H' denotes the transpose of H. An element $w \in \mathbb{Z}_p^n$ is in the null space of G iff

$$Gw' = 0. \tag{57.3}$$

For the generator matrix in (57.1), $w = (a_1, a_2, \ldots, a_5)$ is in the null space iff

$$\begin{aligned} a_1 \qquad\qquad\quad + a_5 &= 0 \\ a_1 + a_2 + a_3 \qquad\quad &= 0 \\ a_2 \qquad + a_4 + a_5 &= 0 \end{aligned} \tag{57.4}$$

In this case a_4 and a_5 can be assigned arbitrarily, and the equations in (57.4) will then determine a_1, a_2, a_3. The null space is {00000, 01110, 11001, 10111}, and we can use

$$H = \begin{bmatrix} 0 & 1 & 1 & 1 & 0 \\ 1 & 1 & 0 & 0 & 1 \end{bmatrix}.$$

The subspace of code words can be recovered from H, because an element $v \in \mathbb{Z}_p^n$ is a code word iff

$$vH' = 0, \tag{57.5}$$

or

$$Hv' = 0. \tag{57.6}$$

The conditions (57.3) and (57.6) exhibit the fact that the row space of H is the null space of G iff the row space of G is the null space of H. The matrix H is called a *parity-check matrix* for the code. The reason for this name can be seen from looking at (57.5) for H as in our example: if we write $v=(b_1,b_2,\ldots,b_5)$, then (57.5) is equivalent to the system of equations

$$\begin{aligned} b_2+b_3+b_4 \qquad &=0\\ b_1+b_2 \qquad\qquad +b_5&=0. \end{aligned}$$

The first equation says that an even number of the middle three components of $(b_1,b_2,\ldots,b_5)$ must be zero; this is a parity check on the middle three components. The second equation is a parity check on the first, second, and fifth components.

The row space of a generator matrix G will not be changed if elementary row operations are performed on G. If columns of G are interchanged, the change in the code given by G will amount only to a permutation of the components of each code word. If one generator matrix can be obtained from another by a combination of elementary row operations and column permutations, then the codes corresponding to the two matrices are said to be *equivalent*.

Theorem 57.1. *Every (n,k) linear code is equivalent to a code with a generator matrix of the form*

$$G=\begin{bmatrix} 1 & 0 & \cdots & 0 & p_{1,1} & \cdots & p_{1,n-k}\\ 0 & 1 & \cdots & 0 & p_{2,1} & \cdots & p_{2,n-k}\\ & & & & \cdots & & \\ 0 & 0 & \cdots & 1 & p_{k,1} & \cdots & p_{k,n-k} \end{bmatrix}=[I_k|P]. \qquad (57.7)$$

This is called the standard generator matrix *for the code.*

PROOF. By elementary row operations a generator matrix for the given code can be put in row-reduced echelon form (Appendix D). The row space has dimension k so that each row of this row-reduced echelon matrix will have a nonzero entry. By appropriate permutation of the columns the matrix can be put in the form (57.7). □

One of the simplest ways to form an (n,k) linear code over $\mathbb{Z}_p$ is to begin with the elements of $\mathbb{Z}_p^k$ and transform them into elements of $\mathbb{Z}_p^n$ through right multiplication by a standard generator matrix. That is, with $v=(a_1,a_2,\ldots,a_k)\in\mathbb{Z}_p^k$, and G as in (57.7), we associate the code word

$$vG=(a_1,a_2,\ldots,a_k,b_1,b_2,\ldots,b_{n-k}),$$

where

$$b_j = \sum_{i=1}^{k} a_i p_{ij} \qquad (1 \le j \le n-k).$$

A code of this type is called a *systematic code*. The first k components of each code word are called the *information symbols*, and the last $n-k$ components are called the *check symbols*. By Theorem 57.1 every linear code is equivalent to a systematic code. The next theorem shows how to get a parity-check matrix for a systematic code.

Theorem 57.2. *The parity-check matrix for a code with generator matrix G as in* (57.7) *is*

$$H = \begin{bmatrix} -p_{1,1} & \cdots & -p_{k,1} & 1 & 0 & \cdots & 0 \\ -p_{1,2} & \cdots & -p_{k,2} & 0 & 1 & \cdots & 0 \\ & \cdots & & & & \cdots & \\ -p_{1,n-k} & \cdots & -p_{k,n-k} & 0 & 0 & \cdots & 1 \end{bmatrix} = [-P'|I_{n-k}].$$

PROOF. It is easy to verify that $GH'=0$. Therefore, the row space of H is contained in the null space of G. But the rows of H are linearly independent; thus dim(row space H)$=n-k$. Therefore, the row space of H must be all of the null space of G, because dim(null space G)$=n-$dim(row space G)$=n-k$. □

Example 57.2. The standard generator matrix arising from the generator matrix G in (57.1) is

$$\begin{bmatrix} 1 & 0 & 0 & 0 & 1 \\ 0 & 1 & 0 & 1 & 1 \\ 0 & 0 & 1 & 1 & 0 \end{bmatrix}.$$

An element $(a_1, a_2, a_3) \in \mathbb{Z}_2^3$ gives rise to the code word

$$(a_1, a_2, a_3, a_2 + a_3, a_1 + a_2).$$

A parity-check matrix is

$$\begin{bmatrix} 0 & 1 & 1 & 1 & 0 \\ 1 & 1 & 0 & 0 & 1 \end{bmatrix}.$$

PROBLEMS

57.1. List a set of code words for a (5, 2) linear binary code.
57.2. List a set of code words for a (5, 2) linear code over $\mathbb{Z}_3$.

57.3. Write a matrix whose rows are the code words of the linear binary code with generator matrix $\begin{bmatrix} 1 & 0 & 0 & 0 \\ 0 & 1 & 1 & 0 \end{bmatrix}$.

57.4. Determine a parity-check matrix for the code with the matrix in Problem 57.3 as generator matrix. [Remember that if a code has a $k \times n$ generator matrix then it has an $(n-k) \times n$ parity-check matrix.]

57.5. Write a matrix whose rows are the code words of the linear code over $\mathbb{Z}_3$ with generator matrix $\begin{bmatrix} 1 & 0 & 2 \\ 2 & 1 & 0 \end{bmatrix}$.

57.6. Determine a generator matrix for the linear binary code with parity-check matrix $\begin{bmatrix} 0 & 1 & 0 \\ 1 & 1 & 0 \end{bmatrix}$. (See the remark with Problem 57.4.)

57.7. (a) Write the standard generator matrix for the linear binary code with

$$\text{generator matrix } \begin{bmatrix} 0 & 0 & 1 & 1 & 0 \\ 0 & 1 & 0 & 1 & 0 \\ 1 & 1 & 1 & 0 & 0 \end{bmatrix}.$$

(b) Use Theorem 57.2 to write a parity-check matrix for the code in part (a).

57.8. Use the maximum-likelihood decoding rule to construct a decoding table for the linear binary code with standard generator matrix $\begin{bmatrix} 1 & 0 & 1 & 0 \\ 0 & 1 & 1 & 1 \end{bmatrix}$.

57.9. Define the weight, $w(x)$, of a word $x \in F^n$ to be the number of nonzero components of x. [Example: $w(01101) = 3$.] Prove that for a linear binary code either all of the code words have even weight or half have even weight and half have odd weight. (*Suggestion*: First explain why the code words of even weight form a subgroup.)

SECTION 58. STANDARD DECODING

Given a choice, it is clearly desirable to choose a code and a procedure for decoding that will minimize the probability of error. This section and the next will present some ways of satisfying this desire. We begin by describing a relevant decoding procedure. In Section 59 we shall see that for a large class of binary codes this procedure maximizes the average probability of correct decoding.

Let V be an (n,k) linear code over $\mathbb{Z}_p$. Then V is a subspace of $\mathbb{Z}_p^n$, and therefore, in particular, V is a subgroup of $\mathbb{Z}_p^n$ thought of as an Abelian group under vector addition. The cosets of V in $\mathbb{Z}_p^n$ form a partition of $\mathbb{Z}_p^n$, and so they can be used as the rows of a decoding table. Notice that $|\mathbb{Z}_p^n| = p^n$, $|V| = p^k$, and by Lagrange's Theorem $[\mathbb{Z}_p^n : V] = p^{n-k}$; therefore, there will be p^{n-k} rows. A *standard array* is a table constructed with cosets for rows with the elements arranged according to the following scheme:

Assume $V = \{v_1, v_2, \ldots, v_{p^k}\}$, with $v_1 = 0$.

Let $\{u_1, u_2, \ldots, u_{p^{n-k}}\}$ be a complete set of coset representatives of V in $\mathbb{Z}_p^n$, with $u_1 = 0$.

The vector in the ith row and jth column of the standard array is $u_i + v_j$ (Table 58.1).

Table 58.1 *Standard Array*

v_1	v_2	v_3		v_{p^k}
u_2	u_2+v_2	u_2+v_3	$\cdots$	$u_2+v_{p^k}$
u_3	u_3+v_2	u_3+v_3	$\cdots$	$u_3+v_{p^k}$
.	.	.		.
.	.	.		.
.	.	.		.
$u_{p^{n-k}}$	$u_{p^{n-k}}+v_2$	$u_{p^{n-k}}+v_3$	$\cdots$	$u_{p^{n-k}}+v_{p^k}$

In coding theory the coset representatives $u_1, u_2, \ldots, u_{p^{n-k}}$ are called *coset leaders*. The *weight* of a vector $v \in \mathbb{Z}_p^n$ is defined as the number of nonzero components of v. We shall see later that there is an advantage in choosing each coset leader to be an element of minimum weight in its coset. This can be done by constructing the array one row at a time, from the top, choosing each successive coset leader to be a vector of smallest weight from among all the remaining vectors.

Example 58.1. Table 58.2 gives a standard array for the (5, 3) code in Example 57.1.

If a code vector v is transmitted and a vector u is received, then $u-v$ is called the *error pattern*. For example, if the code vector 11010 is transmitted and 01110 is received, the error pattern is 10100. Notice that the weight of the error pattern $u-v$ is equal to the distance between u and v. Problem 58.6 asks you to show that for any linear code an error can be *detected* iff the error pattern is not a code vector. The following theorem tells which errors will be *corrected* if a standard array is used for decoding.

Theorem 58.1 *If a standard array is used as the decoding table for a linear code, then a received vector u will be decoded correctly into the transmitted code vector v iff the error pattern $u-v$ is a coset leader.*

Table 58.2

00000	10001	11100	01011	01101	11010	10111	00110
10000	00001	01100	11011	11101	01010	00111	10110
00010	10011	11110	01001	01111	11000	10101	00100
01000	11001	10100	00011	00101	10010	11111	01110

PROOF. Adopt the notation in Table 58.1, and assume that v_j is transmitted, u is received, and u is decoded as v_t. We must prove that $v_j = v_t$ iff $u - v_j$ is a coset leader.

If $v_j = v_t$, then u is in the column of v_j, so that $u = u_i + v_j$ for some coset leader u_i. Therefore, the error pattern $u - v_j = u_i$ is a coset leader. Conversely, if $u - v_j = u_i$ is a coset leader, then $u = u_i + v_j$; therefore, u is in the column of v_j, and hence $v_j = v_t$. □

Here is another way to look at this theorem. In using a standard array as a decoding table, we are in essence assuming that each received vector has its coset leader as an error pattern: if $u_i + v_j$ is received, it is decoded as v_j, which means that we are assuming an error pattern of $(u_i + v_j) - v_j = u_i$.

We shall now show that there is a more economical way of presenting the information given by a standard array. Let V be an (n,k) linear code over $\mathbb{Z}_p$, and let H denote the parity-check matrix for V. The *syndrome* of any received vector v is the $(n-k)$-dimensional vector vH'. By the very definition of parity-check matrix, a vector is a code word iff its syndrome is 0. This means that in a standard array the vectors of syndrome 0 are those in the first row. The next theorem shows that, more generally, two vectors have equal syndromes (not necessarily 0) iff they are in the same row of a standard array.

Theorem 58.2. *Two vectors are in the same coset of a linear code iff their syndromes are equal.*

PROOF. Two vectors v_1 and v_2 are in the same coset of a linear code V

$$\begin{aligned} &\text{iff } v_1 - v_2 = v \text{ for some } v \in V \\ &\text{iff } (v_1 - v_2)H' = 0 \\ &\text{iff } v_1H' - v_2H' = 0 \\ &\text{iff } v_1H' = v_2H'. \quad \square \end{aligned}$$

To capitalize on Theorem 58.2 and the remark following Theorem 58.1, we first form a *syndrome table*, which has two columns, one for coset leaders and the other for their syndromes (Table 58.3). Assume that a vector u is received. Compute the syndrome of u; by Theorem 58.2 this equals the syndrome of the coset leader of u, so that we can get this coset leader from Table 58.3. By the remark following Theorem 58.1, this coset leader (call it u_i) is the assumed error pattern in u. Therefore, u would be decoded by a standard array as $v_j = u - u_i$. To summarize, decoding by a standard array can be done by computing a syndrome and using a syndrome table.

Table 58.3

coset leaders	syndromes
u_1	u_1H'
u_2	u_2H'
u_3	u_3H'
.	.
.	.
.	.
$u_{p^{n-k}}$	$u_{p^{n-k}}H'$

Example 58.2. Table 58.4 is the syndrome table for the standard array in Table 58.2. The parity-check matrix was computed for this code in Example 57.2, and is

$$H=\begin{bmatrix} 0 & 1 & 1 & 1 & 0 \\ 1 & 1 & 0 & 0 & 1 \end{bmatrix}.$$

Therefore, the syndrome of 01000, for example, is

$$\begin{bmatrix} 0 & 1 & 0 & 0 & 0 \end{bmatrix}\begin{bmatrix} 0 & 1 \\ 1 & 1 \\ 1 & 0 \\ 1 & 0 \\ 0 & 1 \end{bmatrix}=\begin{bmatrix} 1 & 1 \end{bmatrix}.$$

To decode 11011, we first compute

$$\begin{bmatrix} 1 & 1 & 0 & 1 & 1 \end{bmatrix}\begin{bmatrix} 0 & 1 \\ 1 & 1 \\ 1 & 0 \\ 1 & 0 \\ 0 & 1 \end{bmatrix}=\begin{bmatrix} 0 & 1 \end{bmatrix}.$$

This is the syndrome of 10000, which is also the error pattern. Therefore, 11011 would be decoded as $11011-10000=01011$. This agrees with Table 58.2.

Table 58.4

coset leaders	syndromes
00000	0 0
10000	0 1
00010	1 0
01000	1 1

PROBLEMS

58.1. Construct a standard array for a (4, 2) linear binary code with generator matrix $\begin{bmatrix} 1 & 1 & 0 & 0 \\ 0 & 0 & 1 & 1 \end{bmatrix}$.

58.2. Construct the syndrome table for the standard array in Problem 58.1.

58.3. Construct a standard array for a (3, 2) linear code over $\mathbb{Z}_3$ with generator matrix $\begin{bmatrix} 1 & 2 & 0 \\ 0 & 1 & 2 \end{bmatrix}$.

58.4. Construct the syndrome table for the standard array in Problem 58.3.

58.5. Suppose that the vectors in the second row of a standard array for a linear binary code are 1000, 1100, 1111, and 1011 (in order). Complete the array. (There is more than one solution.)

58.6. Prove that for a linear code an error can be detected iff the error pattern is not a code vector.

58.7. How many error patterns can be detected by an (n,k) linear code over $\mathbb{Z}_p$? (See Problem 58.6.)

58.8. If a standard array is used for decoding, how many error patterns can be corrected by an (n,k) linear code over $\mathbb{Z}_p$?

58.9. If a standard array is used for a linear binary code, and if each vector of weight w or less appears as a coset leader, then each error pattern of weight w or less will be corrected. Why?

58.10. Construct a standard array, generator matrix, parity-check matrix, and syndrome table for a (3, 1) linear binary code that will correct all error patterns of weight 1. (See Problem 58.9.)

58.11. Construct a generator matrix and list of coset leaders for a (9, 2) linear binary code that will correct all error patterns of weight 2 or less.

SECTION 59. ERROR PROBABILITY

We shall now restrict our attention to linear binary codes. We shall assume, moreover, that there is a constant probability p that any symbol (binary digit) will be altered between transmission and reception; it is said in this case that the code is being transmitted over a *binary symmetric channel.* Theorem 59.1 adds the assumption that all code words are equally likely to be transmitted, and then gives a "best" decoding rule. Theorem 59.1 applies when a code (any code) has already been chosen; Example 59.2 goes further, and describes how to choose a "best" code as well as decoding procedure. To make these statements precise, we must use some elementary facts from probability theory. If you prefer to omit the proof of Theorem 59.1, simply move ahead now to the statement of the theorem, read that, and then pass on to Example 59.2.

If the probability is p that a digit will be altered, then the probability is $q=1-p$ that it will be transmitted without alteration. We assume that $p<\frac{1}{2}$. The probabilities of alteration are assumed to be independent for the different components of a code word; therefore, the probability that a word of length n will be altered in m specified positions is $p^m q^{n-m}$. In particular, the probability of a specified error pattern of weight w is $p^w q^{n-w}$. For instance, the probability is p^2q^3 that if 01110 is sent, then 10110 will be received, because the error pattern 11000 has weight 2.

Question: If we assume that the code word v_j is transmitted, and that decoding is done with the standard array in Table 58.1, what is the probability of correct decoding? Answer: There will be correct decoding iff the received word is in the column headed by v_j. Therefore, the probability is $p_1+p_2+\cdots+p_{2^{n-k}}$, where p_i denotes the probability of receiving u_i+v_j $(1 \le i \le 2^{n-k})$. But u_i is the error pattern, so that $p_i = p^{w_i}q^{n-w_i}$, where w_i denotes the weight of u_i. Thus the probability of correct decoding is

$$\sum_{i=1}^{2^{n-k}} p^{w_i}q^{n-w_i}.$$

Example 59.1. In Example 58.1, the code word 11100 will be decoded correctly if it is received as either 11100, 01100, 11110, or 10100. The error patterns here are 00000, 10000, 00010, and 01000, respectively; and these have weights 0, 1, 1, 1, respectively. Therefore, the probability of correct decoding is

$$q^5 + 3pq^4.$$

The probability of correct decoding will be the same for each of the other code words in Example 58.1, because the probability depends only on the weights of the error patterns, and these are the weights of the coset leaders (since Table 58.2 is a standard array).

If one error pattern has weight w_1, and another has weight w_2, and $w_1 < w_2$, then the first error pattern is more likely to occur, because

$$p^{w_1}q^{n-w_1} > p^{w_2}q^{n-w_2} \tag{59.1}$$

if $p < \frac{1}{2}$ and $w_1 < w_2$ (Problem 59.3). In particular, the error pattern $000\cdots 0$ is most likely of all; that is, if $p < \frac{1}{2}$, the most probable received word is the transmitted word. (This does not mean that the probability of no altered digit is always larger than the probability of an alteration. See Problem 59.4.) We are now ready for the following theorem.

Theorem 59.1. *If an (n,k) binary code is transmitted over a binary symmetric channel, and if all code words are equally likely to be transmitted, then the average probability of correct decoding will be maximum if the decoding table is chosen to be a standard array with each coset leader of minimum weight in its coset.*

PROOF. Consider a decoding table with ij-entry v_{ij} $(1 \le i \le 2^{n-k},\ 1 \le j \le 2^k)$, and with code words v_{1j} $(1 \le j \le 2^k)$. Let w_{ij} denote the weight of the assumed error pattern of v_{ij}, that is, the weight of $v_{ij} - v_{1j}$. Then the probability of correct decoding, if v_{1j} is transmitted, is

$$\sum_{i=1}^{2^{n-k}} p^{w_{ij}} q^{n-w_{ij}}. \tag{59.2}$$

Because the 2^k code words are equally likely to be transmitted, the average probability of correct decoding is

$$\left(\tfrac{1}{2}\right)^k \sum_{i,j} p^{w_{ij}} q^{n-w_{ij}}. \tag{59.3}$$

By (59.1), the individual terms $p^{w_{ij}} q^{n-w_{ij}}$ in the sum (59.3) will be maximum if the weights w_{ij} are minimum. Therefore, the average probability given in (59.3) will certainly be maximum if the table can be arranged so that the weights w_{ij} are simultaneously minimum. The theorem asserts that this is achieved by a standard array (Table 58.1) with each coset leader of minimum weight in its coset. To verify this, it suffices to show that if $u_i + v_j$ is a received vector, and v' is the code vector closest to it, then

$$w[(u_i + v_j) - v'] \ge w[(u_i + v_j) - v_j] = w[u_i].$$

But this is true because

$$(u_i + v_j) - v' = u_i + (v_j - v') \in u_i + V,$$

so that $(u_i + v_j) - v'$ is in the same coset as u_i, and u_i has minimum weight in its coset. □

Example 59.2. For each positive integer m, there is a special $(2^m - 1,\ 2^m - m - 1)$ binary code, known as the *Hamming code*, which yields the maximum probability of correct decoding among all $(2^m - 1,\ 2^m - m - 1)$ binary codes. This code can be defined in terms of its parity-check matrix H: the columns of H are all of the nonzero m-dimensional column vectors over $\mathbb{Z}_2$. A parity-check matrix for the Hamming code with $m = 2$ is

$$H = \begin{bmatrix} 0 & 1 & 1 \\ 1 & 0 & 1 \end{bmatrix}. \tag{59.4}$$

A parity-check matrix for the Hamming code with $m = 3$ is

$$H = \begin{bmatrix} 0 & 0 & 0 & 1 & 1 & 1 & 1 \\ 0 & 1 & 1 & 0 & 0 & 1 & 1 \\ 1 & 0 & 1 & 0 & 1 & 0 & 1 \end{bmatrix}. \tag{59.5}$$

Let us look more closely at the case $m=3$. The resulting code is a $(2^3-1, 2^3-3-1)=(7,4)$ code, and so there will be $2^4=16$ code words. By (57.5), a vector $v=(a_1,a_2,\ldots,a_7)$ is a code word iff $vH'=0$. This will be true iff

$$\begin{aligned} a_4+a_5+a_6+a_7 &= 0 \\ a_2+a_3 \qquad\quad a_6+a_7 &= 0 \\ a_1 \quad +a_3 \quad +a_5 \quad +a_7 &= 0 \end{aligned} \tag{59.6}$$

The symbols a_3, a_5, a_6, a_7 can be assigned arbitrarily, and then a_1, a_2, a_4 will be determined by the system (59.6). We can think of the 16 different 4-tuples $a_3a_5a_6a_7$ as the messages to be handled by this code, with $a_3a_5a_6a_7$ encoded as $a_1a_2\cdots a_7$. For example, 1010 would be encoded as $a_1a_21a_4010$, with a_1, a_2, a_4 determined by

$$\begin{aligned} a_4+1 &= 0 \\ a_2+0 &= 0 \\ a_1+1 &= 0. \end{aligned}$$

Therefore $a_1=1$, $a_2=0$, $a_4=1$; and 1010 would be encoded as 1011010.

Each Hamming code will correct all single errors. These codes were introduced by R. W. Hamming in 1950. The comparable double-error-correcting codes—known as BCH codes—were not developed until 10 years later. The properties and generalizations of Hamming and BCH codes are discussed in the references listed at the end of this chapter.

PROBLEMS

59.1. Assume that the code word 10101 is transmitted over a binary symmetric channel, with probability $\frac{1}{4}$ that a digit will be altered.
(a) What is the probability that 11011 will be received?
(b) What is the probability that the error pattern will have weight 2?

59.2. Assume that the code in Table 56.1 is being transmitted over a binary symmetric channel, with probability p that a digit will be altered. Compute the probability of correct decoding for each of the four different code words.

59.3. Verify inequality (59.1) for $p<\frac{1}{2}$, $q=1-p$, and $w_1<w_2$.

59.4. Consider Example 59.1. Prove that if $p>\frac{1}{4}$, then for each code word the probability of an altered digit exceeds the probability of no alteration.

59.5. For the Hamming code considered at the end of Example 59.2, how would 0111 be encoded?

59.6. Is 0100110 a code word for the Hamming code with parity-check matrix given by (59.5)?

59.7. Determine the generator matrix for the code with parity-check matrix given by (59.4).

59.8. Construct a standard array for the code with parity-check matrix given by (59.4). (See Problem 59.7.) Also construct the corresponding syndrome table.

59.9. Determine the standard generator matrix (57.7) for the code with parity-check matrix given by (59.5).

59.10. Determine a standard array for the code with parity-check matrix given by (59.5). (See Problem 59.9.) Also construct the corresponding syndrome table.

NOTES ON CHAPTER XVI

1. Berlekamp, E. R., *Algebraic Coding Theory*, McGraw-Hill, New York, 1968.
2. Blake, I. F., and R. C. Mullin, *An Introduction to Algebraic and Combinatorial Coding Theory*, Academic Press, New York, 1976.
3. Levinson, N., *Coding theory: a counterexample to G. H. Hardy's conception of applied mathematics*, American Mathematical Monthly **77** (1970) 249–258.
4. MacWilliams, F. J., and N. J. A. Sloane, *The Theory of Error-Correcting Codes, Parts I and II*, North-Holland, Amsterdam, 1977.
5. Peterson, W. W., and E. J. Weldon, Jr., *Error Correcting Codes*, 2nd ed., MIT Press, Cambridge, Mass., 1972.

CHAPTER XVII

Lattices and Boolean Algebras

Just as the axioms for groups and rings reflect properties of addition and multiplication in the familiar number systems, the axioms for lattices and Boolean algebras reflect properties of inclusion, union, and intersection in the theory of sets. But we shall see with lattices and Boolean algebras, as with groups and rings, that there are other important examples beyond those furnishing our original motivation. In particular, Boolean algebras provide appropriate algebraic settings for formal logic and the theory of computer design. Section 63 contains the most interesting theorem in this chapter, but it is not a prerequisite for the application in Section 64.

SECTION 60. PARTIALLY ORDERED SETS

The first of several fundamental ideas for this chapter is an abstraction from $\leq$ for numbers and $\subseteq$ for sets.

Definition. A *partially ordered set* is a set S together with a relation $\leq$ on S such that each of the following axioms is satisfied:

reflexive

if $a \in S$, then $a \leq a$,

antisymmetric

if $a, b \in S$, $a \leq b$, and $b \leq a$, then $a = b$,

transitive

if $a, b, c \in S$, $a \leq b$, and $b \leq c$, then $a \leq c$.

Example 60.1 The integers form a partially ordered set with respect to $\leq$. Here $\leq$ has its usual meaning; in other examples $\leq$ is replaced by whatever is appropriate for the relation involved.

Example 60.2. For each set S, let $\mathcal{P}(S)$ denote the set of all subsets of S. Then $\mathcal{P}(S)$ is a partially ordered set with $A \leq B$ defined to mean $A \subseteq B$. Notice that $\subseteq$ is a relation on $\mathcal{P}(S)$, not on S. [The set $\mathcal{P}(S)$ is called the *power set* of S; if S is finite, then $|\mathcal{P}(S)| = 2^{|S|}$, "two to the *power* $|S|$."]

Example 60.3. The set of all subgroups of any group is a partially ordered set with respect to $\subseteq$.

Example 60.4. The set $\mathbb{N}$ of all natural numbers (positive integers) is a partially order set with $a \leq b$ defined to mean $a|b$. The set of all positive divisors of a fixed positive integer n is also a partially order set with this relation.

The notation $a < b$ means that $a \leq b$ and $a \neq b$; $b \geq a$ means that $a \leq b$; and $b > a$ means that $a < b$. An element b in a partially ordered set S is said to *cover* an element a in S if $a < b$ and there is no x in S such that $a < x < b$.

It is often helpful to represent a finite partially ordered set by a diagram, in the following way. Each element of the set is represented by a small circle (or other appropriate symbol), and if b covers a, then the circle for b is placed above the circle for a and a line is drawn connecting the two circles. In this way $c < d$ iff d is above c and there is a sequence of segments connecting c to d.

Example 60.5. Figure 60.1 shows the diagram for the set of subsets of $\{x, y, z\}$ with the relation $\subseteq$.

Figure 60.1

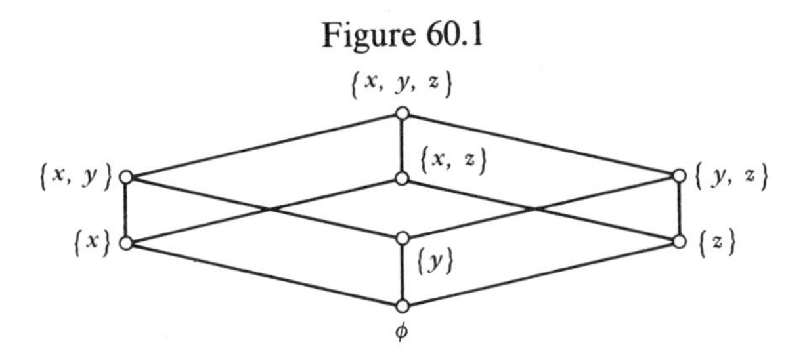

Figure 60.2

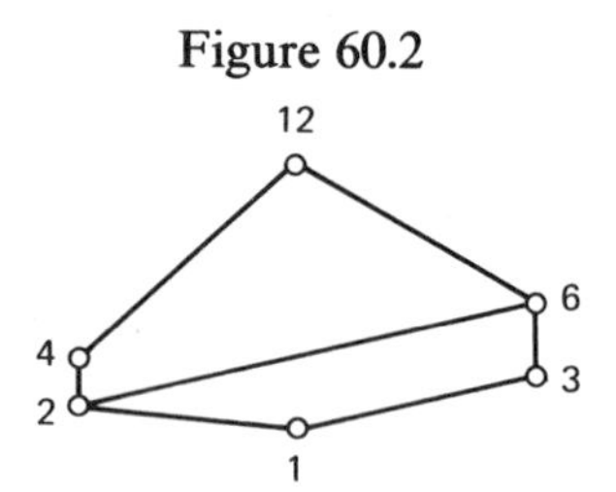

Example 60.6. Figure 60.2 shows the diagram for the set of divisors of 12, with the relation $a \leq b$ defined to mean that $a|b$.

An element u in a partially ordered set S is said to be an *upper bound* for a subset A of S if $x \leq u$ for each $x \in A$. The element u is a *least upper bound (l.u.b.)* for A if it is an upper bound and $u \leq v$ for each upper bound v of A. *Lower bound* and *greatest lower bound (g.l.b.)* are defined by replacing $\leq$ by $\geq$.

If a subset has a l.u.b., then it is unique, for if u_1 and u_2 are both l.u.b.'s for A, then $u_1 \leq u_2$ and $u_2 \leq u_1$, and therefore $u_1 = u_2$. Similarly, if there is a g.l.b. for A, then it is unique; simply replace l.u.b. by g.l.b. and $\leq$ by $\geq$ throughout the preceding sentence.

Example 60.7. Let S be a set, and consider the partially ordered set $\mathcal{P}(S)$ with $\subseteq$. If $A, B \in \mathcal{P}(S)$ (that is, if $A \subseteq S$ and $B \subseteq S$), then $A \cup B$ is a l.u.b. for $\{A, B\}$, and any set containing $A \cup B$ is an upper bound for $\{A, B\}$. Also, $A \cap B$ is a g.l.b. for $\{A, B\}$. More generally, if $\mathcal{C}$ is any subset of $\mathcal{P}(S)$, then the union of all the subsets in $\mathcal{C}$ is in $\mathcal{P}(S)$ and is the l.u.b. for $\mathcal{C}$; and the intersection of all the subsets in $\mathcal{C}$ is in $\mathcal{P}(S)$ and is the g.l.b. for $\mathcal{C}$.

Example 60.8. Consider Example 60.4. If A is any finite subset of $\mathbb{N}$, then the least common multiple of the integers in A is a l.u.b. for A, and the greatest common divisor of the integers in A is a g.l.b. for A.

Example 60.9. The diagram in Figure 60.3 represents a partially ordered set in which $\{a, b\}$ has no upper bound (and therefore no l.u.b.), and in which $\{c, d\}$ has no lower bound (and therefore no g.l.b.).

The question of the existence of least upper bounds and greatest lower bounds will be critical in the next section. Here is an another important example in which they always exist.

Figure 60.3

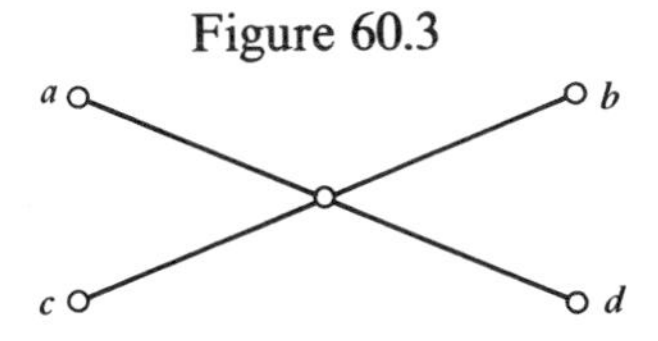

Example 60.10. Let G be a group, and consider the partially ordered set of subgroups of G with $\subseteq$ (Example 60.3). The subgroup generated by any collection of subgroups of G is a l.u.b. for that collection of subgroups. (Apply Theorem 7.3, with S being the set union of the subgroups in the given collection.) The point here is that the *set union* of subgroups will not in general be a subgroup, but there is a l.u.b. nonetheless. The intersection of any collection of subgroups *is* a subgroup (Theorem 7.2), and this intersection is a g.l.b. for the collection. Figure 60.4 shows the diagram for the subgroups of $\mathbb{Z}_6$; Figure 60.5 shows the diagram for the subgroups of S_3.

Figure 60.4

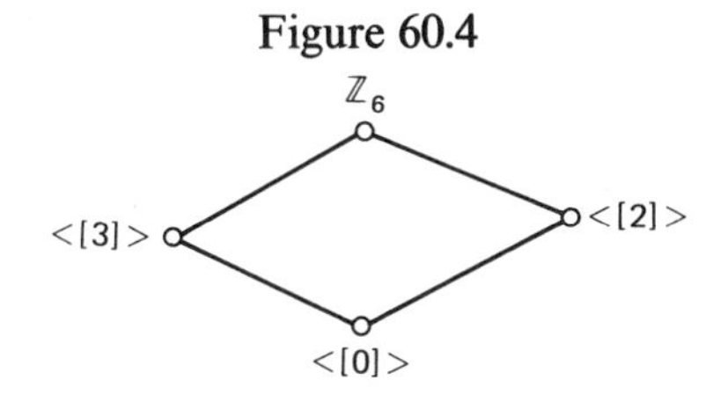

Figure 60.5

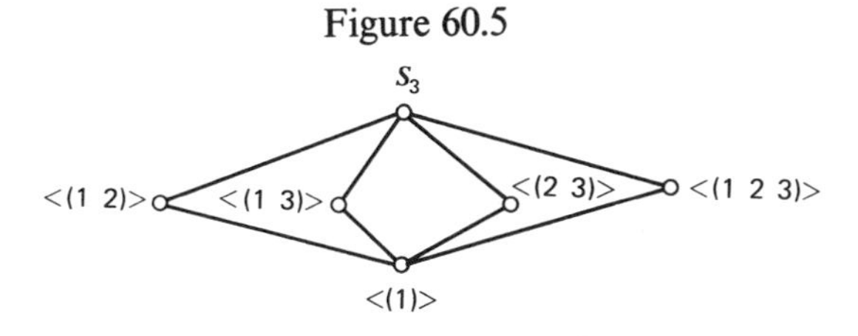

If either $a \leq b$ or $b \leq a$ for every pair of elements $\{a, b\}$ in a partially ordered set S, then S is said to be *linearly ordered* or to be a *chain*. With the usual meaning of $\leq$, each of $\mathbb{Z}$, $\mathbb{Q}$, and $\mathbb{R}$ is a chain. None of Figures 60.1 through 60.5 represents a chain.

Example 60.11. The subgroups of $\mathbb{Z}_8$ form a chain, as shown in Figure 60.6. In fact, the subgroups of any cyclic group of prime power order form a chain (Problem 60.13).

Figure 60.6

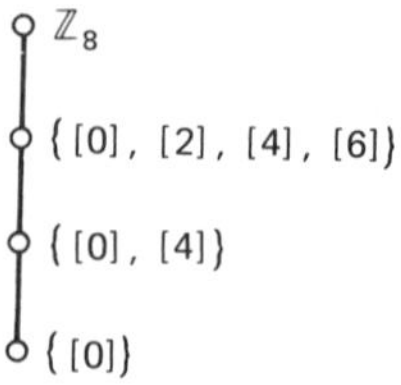

$\mathbb{Z}_8$

{[0], [2], [4], [6]}

{[0], [4]}

{[0]}

PROBLEMS

60.1. The set of all nonzero integers is *not* partially ordered with $a \le b$ defined to mean $a|b$. Why? (Compare Example 60.4.)

60.2. Draw the diagram for the set of subsets of $\{x,y\}$ with the relation $\subseteq$.

60.3. Draw the diagram for the set of subsets of $\{w,x,y,z\}$ with the relation $\subseteq$.

60.4. Draw the diagram for the set of positive divisors of 20, with $a \le b$ defined to mean $a|b$.

60.5. Assume that p and q are distinct primes, and let $n = p^2q$. Draw the diagram for the set of positive divisors of n, with $a \le b$ defined to mean $a|b$. (Compare Figure 60.2 and Problem 60.4.)

60.6. For which $n \in \mathbb{N}$ does the set of all positive divisors of n form a chain, with $a \le b$ defined to mean $a|b$?

60.7. For $\mathbb{N}$ partially ordered as in Example 60.4, determine the l.u.b. and the g.l.b. of the subset $\{12, 30, 126\}$.

60.8. Consider the partially ordered set $\mathbb{N}$ with $a \le b$ defined to mean $a|b$ (Example 60.4).

(a) Which integers are covered by 6?

(b) Which integers cover 6? (There are a lot.)

(c) For $m \in \mathbb{N}$, which integers are covered by m?

(d) For $m \in \mathbb{N}$, which integers cover m?

60.9. Consider the partially ordered set of all subgroups of the group of integers (Examples 60.3 and 60.10 with $G = \mathbb{Z}$).

(a) For which $m, n \in \mathbb{Z}$ is $\langle m \rangle \subseteq \langle n \rangle$?

(b) For $m, n \in \mathbb{Z}$, determine k so that $\langle k \rangle$ is the greatest lower bound of $\langle m \rangle$ and $\langle n \rangle$.

(c) For $m, n \in \mathbb{Z}$, determine k so that $\langle k \rangle$ is the least upper bound of $\langle m \rangle$ and $\langle n \rangle$.

60.10. Assume that A and B are subgroups of a group G. Prove that $A \cup B$ (set union) is a subgroup iff either $A \subseteq B$ or $B \subseteq A$. (This explains why $\langle A, B \rangle$ rather than $A \cup B$ is used in Example 60.10.)

60.11. A mapping $\theta: L_1 \to L_2$ of one partially ordered set onto another is called *order preserving* if $a \le b$ implies $\theta(a) \le \theta(b)$ for all a, $b \in L_1$. An invertible mapping $\theta: L_1 \to L_2$ is an *isomorphism* if both θ and θ^{-1} are order preserving. (Finite partially ordered sets are isomorphic iff their diagrams are the same except possibly for labeling.)

(a) Verify that isomorphism is an equivalence relation for partially ordered sets.

(b) There are two isomorphism classes for partially ordered sets with two elements. Draw a diagram corresponding to each class. (Remember that two elements need not be "comparable.")

(c) There are five isomorphism classes for partially ordered sets with three elements. Draw a diagram corresponding to each class.

(d) There are 16 isomorphism classes for partially ordered sets with four elements. Draw a diagram corresponding to each class.

(e) Formulate a necessary and sufficient condition on positive integers m and n for the partially ordered sets of positive divisors of m and n to be isomorphic, with $a \leq b$ defined to mean $a|b$. (*Suggestion*: See Problem 60.5.) Give several examples to support your condition, but do not write a formal proof.

60.12. Prove that $|\mathcal{P}(S)|=2^{|S|}$. (See Example 60.2. Assume $|S|=n$ and $x \in S$. The subsets of S can be put in two classes, those that do contain x and those that do not. The induction hypothesis will tell you how many subsets there are in each class.)

60.13. Prove that the subgroups of every cyclic group of prime power order form a chain, as claimed in Example 60.11.

SECTION 61. LATTICES

Definition. A *lattice* is a partially ordered set in which each pair of elements has a least upper bound and a greatest lower bound.

The l.u.b. of elements a and b in a lattice will be denoted $a \vee b$, and the g.l.b. will be denoted $a \wedge b$. The operations $\vee$ and $\wedge$ are called *join* and *meet*, respectively.

Example 61.1. The partially ordered sets in Examples 60.1 through 60.4 are all lattices. The proof for Example 60.1 is obvious, and proofs for the other cases follow from remarks in Section 60. We can now speak, for instance, of "the lattice of subgroups" of a group.

The definition of lattice demands that each *pair* of elements have a l.u.b. and a g.l.b. It follows from this that each *finite subset* has a l.u.b. and a g.l.b. For example, the l.u.b. of $\{a,b,c\}$ is $(a \vee b) \vee c$, which can be seen as follows. Let $u=(a \vee b) \vee c$. Then $a \vee b \leq u$, and therefore $a \leq u$ and $b \leq u$; also $c \leq u$. On the other hand, if $a \leq v$, $b \leq v$, and $c \leq v$, then $a \vee b \leq v$ and $c \leq v$; therefore $u=(a \vee b) \vee c \leq v$. Thus u is a l.u.b. for $\{a,b,c\}$, as claimed. A similar argument shows that $(a \wedge b) \wedge c$ is a g.l.b. for $\{a,b,c\}$ (replace l.u.b. by g.l.b. and $\leq$ by $\geq$ throughout). The l.u.b. and g.l.b. of a finite subset $a_1, a_2, \ldots, a_n$ will be denoted

$$a_1 \vee a_2 \vee \cdots \vee a_n \qquad \text{and} \qquad a_1 \wedge a_2 \wedge \cdots \wedge a_n,$$

respectively. The inductive proofs of the existence of these elements are left to Problem 61.16.

In the paragraph preceding Example 60.7 we proved that if a subset of a partially ordered set has a l.u.b., then that l.u.b. is unique; we then stated that the uniqueness of the g.l.b.'s could be proved similarly: replace l.u.b. by g.l.b. and $\leq$ by $\geq$. In the same way, the proof that $(a\vee b)\vee c$ is a l.u.b. for $\{a,b,c\}$ in a lattice, given above, can be transformed into a proof that $(a\wedge b)\wedge c$ is a g.l.b. for $\{a,b,c\}$. These are applications of a very useful principle that, for lattices, has the following form.

Principle of Duality. *Any statement that is true for every lattice remains true if $\leq$ and $\geq$ are interchanged throughout the statement and $\vee$ and $\wedge$ are interchanged throughout the statement.*

The Principle of Duality is valid because of three factors. First, $\geq$, as well as $\leq$, is reflexive, antisymmetric, and transitive. Second, g.l.b. is defined by replacing $\leq$ by $\geq$ in the definition of l.u.b. Third, a statement is true for every lattice only if it can be proved from the reflexive, antisymmetric, and transitive properties and the existence of l.u.b.'s and g.l.b.'s of finite sets.

If $\leq$ and $\geq$ are interchanged and l.u.b. and g.l.b. are interchanged in a statement, then the new statement obtained is called the *dual* of the original statement. For example, the dual of $a\vee a=a$ is $a\wedge a=a$. The Principle of Duality simply says that if a statement is true for every lattice, then so is its dual.

Although each finite subset of a lattice has both a l.u.b. and a g.l.b., an infinite subset of a lattice need not have a l.u.b. or a g.l.b. For example, in $\mathbb{Z}$, with the relation $\leq$, the set $\mathbb{Z}$ itself has neither a l.u.b. nor a g.l.b. In $\mathbb{N}$, with $a\leq b$ defined to mean $a|b$, no infinite subset has a l.u.b. A lattice is said to be *complete* if each subset (finite or infinite) has both a l.u.b. and a g.l.b. An element 1 in a lattice L is called a *unity* (or *identity*) for L if $a\leq 1$ for each $a\in L$. And an element 0 in L is called a *zero* for L if $0\leq a$ for each $a\in L$. Notice that if 1 and 0 exist in L, then

$$a\vee 0 = 0\vee a = a \qquad \text{and} \qquad a\wedge 1 = 1\wedge a = a \tag{61.1}$$

for each $a\in L$. Also, any finite lattice (one with a finite number of elements) has both a unity and a zero. We shall always require $0\neq 1$.

Example 61.2. (a) In $\mathbb{Z}$, with the relation $\leq$, there is neither a lattice unity nor a lattice zero.
(b) In $\mathcal{P}(S)$ (Example 60.2), S is a unity and $\varnothing$ (the empty set) is a zero.
(c) In the lattice of subgroups of a group G, G is a lattice unity and $\{e\}$ is a lattice zero.

(d) In $\mathbb{N}$, with $a \leq b$ defined to mean $a|b$, there is no lattice unity, but 1 is a lattice zero.

If L is a lattice, then both $\vee$ and $\wedge$ are operations on L, and they have properties much like those for $+$ and $\cdot$ in a ring. The next theorem lists the properties that are essential and shows that they give an alternative way to define a lattice.

Theorem 61.1. *The operations $\vee$ and $\wedge$ on a lattice L satisfy each of the following laws (for all $a, b, c \in L$):*

commutative laws

$$a \vee b = b \vee a, \qquad a \wedge b = b \wedge a,$$

associative laws

$$a \vee (b \vee c) = (a \vee b) \vee c, \qquad a \wedge (b \wedge c) = (a \wedge b) \wedge c,$$

idempotent laws

$$a \vee a = a, \qquad a \wedge a = a,$$

absorption laws

$$(a \vee b) \wedge a = a, \qquad (a \wedge b) \vee a = a.$$

Conversely, if operations $\vee$ and $\wedge$ on a set L satisfy each of these laws, and if a relation $\leq$ is defined on L by

$$a \leq b \quad \textit{iff} \quad \textit{either} \quad a \vee b = b \quad \textit{or} \quad a \wedge b = a, \tag{61.2}$$

then L is a lattice.

PROOF. The proof of the first half of the theorem is left to Problem 61.11. To prove the converse, we first show that in the presence of the assumptions being made on $\vee$ and $\wedge$, the conditions $a \vee b = b$ and $a \wedge b = a$ in (61.2) are equivalent (that is, either one implies the other). Assume that $a \vee b = b$. Then

$$\begin{aligned} a \wedge b &= a \wedge (a \vee b) && (a \vee b = b) \\ &= (a \vee b) \wedge a && \text{(commutative law)} \\ &= a && \text{(absorption law).} \end{aligned}$$

The proof of the other implication ($a \wedge b = a$ implies that $a \vee b = b$) is similar (Problem 61.12).

Now we shall prove that L is partially ordered relative to $\leq$. If $a \in L$, then $a \wedge a = a$ by one of the idempotent laws, and therefore $a \leq a$ by (61.2); thus $\leq$ is reflexive. If $a \leq b$ and $b \leq a$, then $a \wedge b = a$ and $b \wedge a = b$, so that $a = b$ by one of the commutative laws; thus $\leq$ is antisymmetric. If $a \leq b$

and $b \leq c$, then $a \wedge b = a$ and $b \wedge c = b$; therefore

$$a \wedge c = (a \wedge b) \wedge c = a \wedge (b \wedge c) = a \wedge b = a, \tag{61.3}$$

so that $a \leq c$. Thus $\leq$ is transitive.

To complete the proof that L is a lattice, we must verify that $a \vee b$ is a l.u.b. for a and b, and that $a \wedge b$ is a g.l.b. for a and b. Consider $a \vee b$. First, $a \leq a \vee b$ because $a \wedge (a \vee b) = (a \vee b) \wedge a = a$ by commutativity and absorption. Similarly, $b \leq a \vee b$. Now assume that $a \leq c$ and $b \leq c$. Then $a \vee c = c$ and $b \vee c = c$, so that $(a \vee b) \vee c = a \vee (b \vee c) = a \vee c = c$, which means $a \vee b \leq c$. This proves that $a \vee b$ is a l.u.b. for a and b. The proof that $a \wedge b$ is a g.l.b. for a and b now follows by using duality. □

Although $\vee$ and $\wedge$ satisfy the commutative and associative laws, they need not satisfy the *distributive law*

$$a \wedge (b \vee c) = (a \wedge b) \vee (a \wedge c) \tag{61.4}$$

for all $a, b, c \in L$. A lattice for which (61.4) is satisfied is called a *distributive lattice*. The lattice of subsets of a set is distributive. Each of the lattices in Figure 61.1 is nondistributive. In the one on the left, for instance, $a \wedge (b \vee c) = a \wedge 1 = a$ while $(a \wedge b) \vee (a \wedge c) = 0 \vee c = c$.

A subset M of a lattice L is called a *sublattice* of L if M is closed relative to the operations $\vee$ and $\wedge$ of L. It can be proved that any nondistributive lattice contains a sublattice whose diagram is like one of the two in Figure 61.1.

Another important property satisfied by the lattice of subsets of a set is that each of its elements has a complement: In a lattice with 0 and 1, an element a' is a *complement* of an element a if

$$a \wedge a' = 0 \qquad \text{and} \qquad a \vee a' = 1. \tag{61.5}$$

A lattice with 0 and 1 in which each element has a complement is called a *complemented lattice*. The divisors of 12 (Example 60.6) form a lattice that is not complemented (Problem 61.4). The diagrams in Figure 61.1 represent complemented lattices; they also show that the complement of an

Figure 61.1

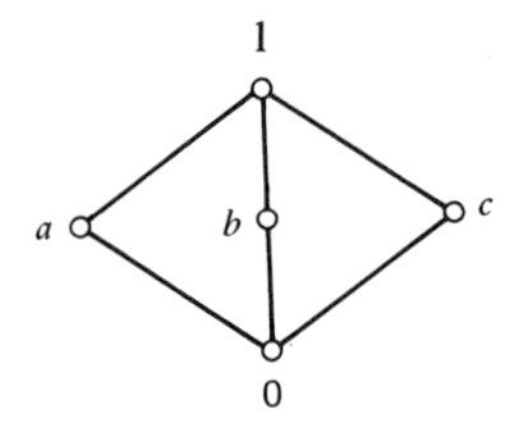

element need not be unique (in the example on the right, for instance, each of a, b, c is a complement of the other two).

All of these properties will be used together in the next section.

PROBLEMS

61.1. Show that the lattice on the right in Figure 61.1 is nondistributive.

61.2. Construct the diagram for the lattice of subgroups of $\langle(1\ \ 2), (3\ \ 4)\rangle$, and verify that it is nondistributive.

61.3. Prove: If a partially ordered set is a chain, then it is a distributive lattice.

61.4. Show that the lattice of divisors of 12 (Example 60.6) is not complemented. For which $n \in \mathbb{N}$ is the lattice of divisors of n complemented?

61.5. Is the lattice of subgroups of $\mathbb{Z}_6$ distributive? Is it complemented? (See Figure 60.4.)

61.6. Is the lattice of subgroups of S_3 distributive? Is it complemented? (See Figure 60.5.)

61.7. Construct the diagram for the lattice of subgroups of the group of rotations of a cube (Example 32.3).

61.8. Prove that if $n \in \mathbb{N}$, then the set of all positive divisors of n is a sublattice of the lattice in Example 61.2d. This sublattice has a lattice unity. What is it?

61.9. If S is a set, what is the complement of an element in the lattice $\mathcal{P}(S)$?

61.10. Construct the diagram for the lattice of subgroups of the group of symmetries of a square (Example 8.3).

61.11. Prove the first half of Theorem 61.1.

61.12. Supply the proof that $a \wedge b = a$ implies $a \vee b = b$, which was omitted from the proof of Theorem 61.1.

61.13. The lattice $\mathbb{N}$, with $a \leq b$ defined to mean $a|b$, is distributive. Determine and then prove the property of integers that makes this true. (*Suggestion*: See Problem 39.11.)

61.14. Explain why each distributive lattice satisfies the law $a \vee (b \wedge c) = (a \vee b) \wedge (a \vee c)$ for all a, b, c, as well as the law in Equation (61.4).

61.15. Lattices are *isomorphic* iff they are isomorphic as partially ordered sets (see Problem 60.11). Draw a diagram corresponding to each isomorphism class of lattices that have four or fewer elements. [See Problem 60.11 (b), (c), (d).]

61.16. Prove that each finite subset of a lattice has a g.l.b. and a l.u.b.

SECTION 62. BOOLEAN ALGEBRAS

In 1854 the British mathematician George Boole published a book entitled *An Investigation of the Laws of Thought, on Which Are Founded the Mathematical Theories of Logic and Probabilities*. This book amplified ideas Boole had introduced in a shorter work published in 1847, and brought the study of logic clearly into the domain of mathematics. Boolean algebra, which originated with this work, can now be seen as the proper tool for the study not only of algebraic logic, but also such things as the theory of telephone switching circuits and computer design.

Definition I. A *Boolean algebra* is a lattice with zero (0) and unity (1) that is distributive and complemented.

Lattices were presented in two forms in Section 61: first, in the definition, in terms of a partial ordering $\leq$; then, in Theorem 61.1, in terms of two operations $\bigvee$ and $\bigwedge$. Boolean algebras are most often discussed in the second of these forms. Because of this we shall give an alternative to Definition I. Theorem 62.1 will establish the equivalence of the two definitions. Hereafter, you may work only from Definition II, if you like. All that is required from Theorem 62.1 is the definition of $\leq$ given in (62.1), and the fact that this gives a partial ordering in a Boolean algebra as defined in Definition II.

Definition II. A *Boolean algebra* is a set B together with two operations $\bigvee$ and $\bigwedge$ on B such that each of the following axioms is satisfied (for all $a,b,c \in B$):

commutative laws

$$a \vee b = b \vee a, \qquad a \wedge b = b \wedge a,$$

associative laws

$$a \vee (b \vee c) = (a \vee b) \vee c, \qquad a \wedge (b \wedge c) = (a \wedge b) \wedge c,$$

distributive laws

$$a \wedge (b \vee c) = (a \wedge b) \vee (a \wedge c), \qquad a \vee (b \wedge c) = (a \vee b) \wedge (a \vee c),$$

existence of zero and unity

there are elements 0 and 1 in B such that

$$a \vee 0 = a, \qquad a \wedge 1 = a,$$

existence of complements

for each a in B there is an element a' in B such that

$$a \vee a' = 1 \quad \text{and} \quad a \wedge a' = 0.$$

Theorem 62.1. (*Definitions I and II are equivalent.*) *In any Boolean algebra as defined in Definition I, the lattice operations $\bigvee$ and $\bigwedge$ satisfy the axioms of Definition II. Conversely, any Boolean algebra as defined in Definition II is a Boolean algebra as defined in Definition I, with $\leq$ given by*

$$a \leq b \quad \text{iff} \quad a \vee b = b. \tag{62.1}$$

PROOF. Assume that B is a Boolean algebra according to Definition I. Then B is a lattice, and so its operations $\vee$ and $\wedge$ satisfy the commutative and associative laws by Theorem 61.1. The other axioms in Definition II are part of Definition I. (Also see Problem 61.14.) Therefore B satisfies all of the axioms of Definition II.

Assume that B is a Boolean algebra according to Definition II. To show that B is a Boolean algebra according to Definition I, it suffices, by Theorem 61.1, to show that the operations $\vee$ and $\wedge$ satisfy the idempotent and absorption laws. For the idempotent law $a \vee a = a$, write

$$a = a \vee 0 = a \vee (a' \wedge a) = (a \vee a') \wedge (a \vee a) = 1 \wedge (a \vee a) = a \vee a.$$

For the absorption law $(a \vee b) \wedge a = a$, write

$$(a \vee b) \wedge a = (a \vee b) \wedge (a \vee 0) = a \vee (b \wedge 0) = a \vee 0 = a.$$

The laws $a \wedge a = a$ and $(a \wedge b) \vee a = a$ can be proved by interchanging $\vee$and $\wedge$, and 0 and 1 (see the Principle of Duality below). □

The *dual* of a statement in a Boolean algebra is the statement that results by interchanging $\vee$ and $\wedge$, and 0 and 1. Because the dual of each axiom in Definition II is also an axiom, we have the following Principle of Duality for Boolean algebras.

Principle of Duality. *If a statement is true for every Boolean algebra, then so is its dual statement.*

For example, $(a \wedge b) \vee a = a$ is true in every Boolean algebra because $(a \vee b) \wedge a = a$ is true in every Boolean algebra, a fact that was used at the end of the proof of Theorem 62.1.

Example 62.1. If S is any nonempty set, then $\mathcal{P}(S)$, the power set of S, is a Boolean algebra relative to $\cup$ and $\cap$. The unity is S; the zero is $\varnothing$; and the complement of a subset is the ordinary set complement (Appendix A). We shall prove later that any finite Boolean algebra is essentially of this type (Theorem 63.1).

Example 62.2. A careful treatment of the connection between Boolean algebras and logic would require a separate chapter, but roughly the idea is as follows. By a *proposition* is meant a statement that is either true or false. For example, "Triangle ABC is equilateral" is a proposition. Propositions p and q are *logically equivalent* if either both are true or both are false.

Example: In Euclidean geometry, "ABC is equilateral" is logically equivalent to "ABC is equiangular." Logical equivalence is an equivalence relation on the set of propositions. For convenience we can treat equivalent propositions as being equal. Then by "the set of all propositions" is meant the set of equivalence classes of all propositions, and by a "proposition" is meant the equivalence class of that proposition. The logical connectives "or" and "and" are operations on the set of all propositions, and it can be verified that, used for $\vee$ and $\wedge$, respectively, they yield a Boolean algebra. If p and q denote propositions, then

$$p \vee q \text{ is true iff } p \text{ is true or } q \text{ is true or both are true}$$

and

$$p \wedge q \text{ is true iff } p \text{ is true and } q \text{ is true.}$$

This algebra is called the *algebra of propositions*.

The commutative, associative, and distributive laws are consequences of the usual meaning of "or" and "and." The negation of a proposition, "not p," is denoted by p'. Then $p \vee p'$ is true for every p, and $p \wedge p'$ is false for every p. (Example: "ABC is equilateral *or* ABC is not equilateral" is true. "ABC is equilateral *and* ABC is not equilateral" is false.) For a zero, we require a proposition 0 such that $q \vee 0 = q$ for every statement q. That is, $q \vee 0$ must be true iff q is true. This will be the case iff 0 is a false statement. Thus for 0 we can use (the equivalence class of) $p \wedge p'$ for any statement p ($p \wedge p'$ is called a *contradiction*).

For a unity, we require a proposition 1 such that $q \wedge 1 = q$ for every statement q. That is, $q \wedge 1$ must be true iff q is true. This will be the case iff 1 is a true statement. Thus for 1 we can use (the equivalence class of) $p \vee p'$ for any statement p ($p \vee p'$ is called a *tautology*). With these choices for 0 and 1, we also have p' ("not p") for the (Boolean algebra) complement of each statement p.

It is interesting to interpret $\leq$, as given by (62.1), in the algebra of propositions. For propositions p and q, $p \leq q$ iff $p \vee q = q$. Reflection shows, then, that $p \leq q$ iff p is true and q is true, or p is false (no matter what q). (Remember the meaning of $\vee$ and $=$.) Therefore $p \leq q$ is logically equivalent to "if p then q" or "p implies q." The property "$p \leq 1$ for every p" means that every statement implies a true statement. The property "$0 \leq p$ for every p" means that every statement is implied by a false statement.

The following theorem brings together some properties of Boolean algebras that are not listed in Definition II. The laws $(a \vee b)' = a' \wedge b'$ and $(a \wedge b)' = a' \vee b'$ in this theorem are called *DeMorgan's Laws*, after the British mathematician Augustus DeMorgan (1806–1871). DeMorgan helped lay the foundations for mathematical logic, and was one of the first to stress the purely symbolic nature of algebra.

Theorem 62.2. *If B is a Boolean algebra and $a,b,c\in B$, then*

$$a\vee a=a, \qquad a\wedge a=a,$$
$$(a\vee b)\wedge a=a, \qquad (a\wedge b)\vee a=a,$$
$$a\vee 1=1, \qquad a\wedge 0=0,$$
a has a unique complement, a', and $(a')'=a$,
$$(a\vee b)'=a'\wedge b', \qquad (a\wedge b)'=a'\vee b',$$
$$0'=1, \qquad 1'=0,$$

and

$$a\leq b \text{ iff } a\vee b=b \text{ iff } a\wedge b=a.$$

Proof. The first two pairs of properties (the idempotent and absorption laws) hold by Theorem 61.1 because B is a lattice. Certainly $a\vee 1=1$, because $a\leq 1$ for every a, and $a\wedge 0=0$ because $0\leq a$ for every a.

To prove that a has a unique complement, assume that $a\vee x=1$ and $a\wedge x=0$, and that $a\vee y=1$ and $a\wedge y=0$; we shall show that $x=y$:

$$\begin{aligned} x &= x\wedge 1 = x\wedge(a\vee y) = (x\wedge a)\vee(x\wedge y)\\ &= 0\vee(x\wedge y) = 0\vee(y\wedge x) = (y\wedge a)\vee(y\wedge x)\\ &= y\wedge(a\vee x) = y\wedge 1 = y. \end{aligned}$$

The law $(a')'=a$ follows from the definition and uniqueness of complements.

To prove $(a\vee b)'=a'\wedge b'$, it suffices by uniqueness of complements to show that $(a\vee b)\wedge(a'\wedge b')=0$ and $(a\vee b)\vee(a'\wedge b')=1$. By the distributive laws,

$$\begin{aligned} (a\vee b)\wedge(a'\wedge b') &= (a\wedge a'\wedge b')\vee(b\wedge a'\wedge b')\\ &= 0\vee 0\\ &= 0 \end{aligned}$$

and

$$\begin{aligned} (a\vee b)\vee(a'\wedge b') &= (a\vee b\vee a')\wedge(a\vee b\vee b')\\ &= 1\wedge 1\\ &= 1. \end{aligned}$$

The law $(a\wedge b)'=a'\vee b'$ now follows by the Principle of Duality.

The laws $0'=1$ and $1'=0$ are both consequences of $0\wedge 1=0$ and $0\vee 1=1$. The last part of the theorem is a consequence of the definition in Equation (62.1), and the first part of the proof of Theorem 61.1. □

PROBLEMS

62.1. (a) An axiom in Definition II requires that $a\vee 0=a$ for each $a\in B$. Why is $a\vee a=a$ also true? Why is $a\wedge 0=0$ also true? [*Suggestion:* To show that $a\wedge 0=0$, start with $a\wedge 0=(a\wedge 0)\vee(a\wedge a')$.]

(b) An axiom in Definition II requires that $a \wedge 1 = a$ for each $a \in B$. Why is $1 \wedge a = a$ also true? Why is $a \vee 1 = 1$ also true?

62.2. Justify each step in the proof of the idempotent law $a \vee a = a$ in the proof of Theorem 62.1. Also write a proof of the dual statement $a \wedge a = a$.

62.3. Justify each step in the proof of the absorption law $(a \vee b) \wedge a = a$ in the proof of Theorem 62.1. Also write a proof of the dual statement $(a \wedge b) \vee a = a$.

62.4. Justify each step in the proof of the uniqueness of a complement for a in the proof of Theorem 62.2.

62.5. Assume that B is a Boolean algebra and $a \in B$. Prove that $a = 0$ iff $b = (a \wedge b') \vee (a' \wedge b)$ for each $b \in B$.

62.6. Assume that B is a Boolean algebra and that $a, b \in B$.

(a) Prove that $a \leq b$ iff $b' \leq a'$.

(b) Prove that $a \leq b'$ iff $a \wedge b = 0$.

(c) Prove that $a \leq b$ iff $a' \vee b = 1$.

62.7. Construct Cayley tables for the operations $\vee$ and $\wedge$ on the lattice in Figure 62.1. Verify that this lattice is not a Boolean algebra by finding an element without a unique complement. (This will show that B is not a Boolean algebra because a condition in Theorem 62.2 is violated.) Is this lattice distributive?

Figure 62.1.

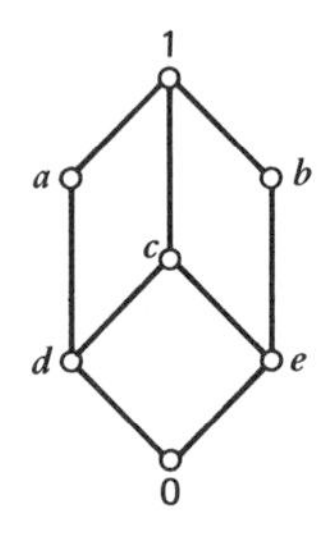

62.8. (a) Give an example to show that $a \vee c = b \vee c$ does not imply $a = b$ in Boolean algebras.

(b) Prove that if B is a Boolean algebra and a, b, $c \in B$, then $a \vee c = b \vee c$ and $a \vee c' = b \vee c'$ imply $a = b$.

62.9. Restate and do Problem 62.8 with $\vee$ replaced by $\wedge$ throughout.

62.10. Prove that if B is a Boolean algebra and $a, b, c \in B$, then $a \vee c = b \vee c$ and $a \wedge c = b \wedge c$ imply $a = b$. (Compare Problems 62.8 and 62.9.)

62.11. By considering the possible diagrams for partially ordered sets, show that there is no Boolean algebra containing exactly three elements. [(See Problem 60.11(c).]

62.12. For which $n \in \mathbb{N}$ will the lattice of all positive divisors of n be a Boolean algebra, with $a \leq b$ defined to mean $a|b$? (Problems 61.4, 61.8, and 61.13 will help. Begin by looking at specific examples.)

SECTION 63. FINITE BOOLEAN ALGEBRAS

The goal of this section is to prove Theorem 63.1, which characterizes all finite Boolean algebras. Boolean algebras, like groups and other algebraic structures, are classified according to isomorphism.

Definition. If A and B are Boolean algebras, an *isomorphism* of A onto B is a mapping $\theta: A \to B$ that is one-to-one and onto and satisfies

$$\theta(a \vee b) = \theta(a) \vee \theta(b)$$

and

$$\theta(a \wedge b) = \theta(a) \wedge \theta(b)$$

for all $a, b \in A$. If there is an isomorphism of A onto B, then A and B are said to be *isomorphic*, and we write $A \approx B$.

Theorem 63.1. *Every finite Boolean algebra is isomorphic to the Boolean algebra of all subsets of some finite set.*

Example 63.1. The divisors of 30 form a Boolean algebra with $a \leq b$ defined to mean $a|b$. Its diagram is shown in Figure 63.1. Here $a \vee b$ is the least common multiple of a and b, and $a \wedge b$ is the greatest common divisor of a and b.

A comparison of Figure 63.1 with Figure 60.1, the diagram for the Boolean algebra of subsets of $\{x, y, z\}$, suggests an isomorphism determined by $\theta(2) = \{x\}$, $\theta(3) = \{y\}$, and $\theta(5) = \{z\}$. The condition $\theta(a \vee b) = \theta(a) \vee \theta(b)$ forces $\theta(6) = \{x, y\}$, $\theta(10) = \{x, z\}$, $\theta(15) = \{y, z\}$, and $\theta(30) = \{x, y, z\}$. Also, the condition $\theta(a \wedge b) = \theta(a) \wedge \theta(b)$ forces $\theta(1) = \varnothing$. This mapping θ *is* an isomorphism. The idea here is to match the elements covering 1 (the prime divisors of 30) with the elements covering $\varnothing$ (the

Figure 63.1

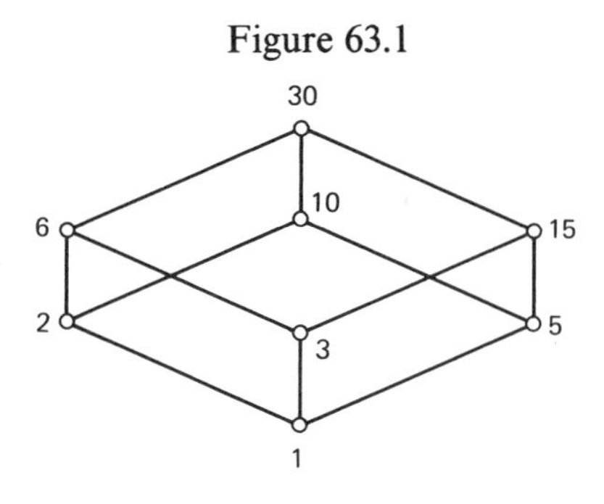

single-element subsets of $\{x,y,z\}$). This simple and important idea is the key to Theorem 63.1.

(Although the divisors of 30 form a Boolean algebra relative to $a|b$, the divisors of 12 (Figure 60.2) do not, because 6 has no complement among the divisors of 12. See Problem 63.7 for a more general statement.)

We shall lead up to the proof of Theorem 63.1 with a definition and several lemmas. This will amount to showing that every finite Boolean algebra has elements that play the same role as the single-element subsets in the Boolean algebra of subsets of a finite set. *In the remainder of this section B will denote a finite Boolean algebra.*

An element $a \in B$ is an *atom* of B if a covers 0 (the zero of B), that is, if $0 < a$ and there is no $x \in B$ such that $0 < x < a$ (Section 60). Equivalently, a is an atom iff

$$a \neq 0, \quad \text{and} \quad x \wedge a = a \quad \text{or} \quad x \wedge a = 0 \tag{63.1}$$

for each $x \in B$. The atoms in the Boolean algebra of all subsets of a set are the single-element subsets of that set. The atoms in the Boolean algebra of divisors of 30 are 2, 3, and 5, the prime divisors of 30 (Example 63.1).

Lemma 63.1. *If $b \in B$ and $b \neq 0$, then there is an atom $a \in B$ such that $a \leq b$.*

PROOF. If b is an atom, take $a = b$. Otherwise, choose an element $a_1 \in B$ such that $0 < a_1 < b$; there is such an element a_1 if b is not an atom. If a_1 is an atom, take $a = a_1$. Otherwise, choose an element $a_2 \in B$ such that $0 < a_2 < a_1 < b$. If a_2 is an atom, take $a = a_2$. Continue, if necessary, to get

$$0 < \cdots < a_3 < a_2 < a_1 < b.$$

This cannot continue indefinitely because B is finite. Therefore, a_k must be an atom for some k; take $a = a_k$. □

Lemma 63.2. *If a_1 and a_2 are atoms in B and $a_1 \wedge a_2 \neq 0$, then $a_1 = a_2$.*

PROOF. Using (63.1), first with $a = a_1$ and $x = a_2$, and then with $a = a_2$ and $x = a_1$, we conclude that $a_2 \wedge a_1 = a_1$ and $a_1 \wedge a_2 = a_2$. Therefore $a_1 = a_2$. □

Lemma 63.3. *For $b, c \in B$, the following conditions are equivalent.*

(*a*) $b \leq c$

(*b*) $b \wedge c' = 0$

(*c*) $b' \vee c = 1$.

PROOF. To prove the equivalence, it suffices to prove that (a) implies (b), (b) implies (c), and (c) implies (a).

(a) implies (b): If $b \leq c$, then $b \vee c = c$, so that (using substitution and one of DeMorgan's Laws)

$$\begin{aligned} b \wedge c' &= b \wedge (b \vee c)' \\ &= b \wedge (b' \wedge c') \\ &= (b \wedge b') \wedge c' \\ &= 0 \wedge c' \\ &= 0. \end{aligned}$$

(b) implies (c): If $b \wedge c' = 0$, then $(b \wedge c')' = 0'$. Therefore, by one of DeMorgan's Laws, $b' \vee c = 1$.

(c) implies (a): If $b' \vee c = 1$, then $b \wedge (b' \vee c) = b$, $(b \wedge b') \vee (b \wedge c) = b$, $0 \vee (b \wedge c) = b$, $b \wedge c = b$, and $b \leq c$. □

Lemma 63.4. *If $b, c \in B$ and $b \nleq c$, then there is an atom $a \in B$ such that $a \leq b$ and $a \nleq c$.*

PROOF. If $b \nleq c$, then $b \wedge c' \neq 0$ by Lemma 63.3. Therefore, by Lemma 63.1, there is an atom $a \in B$ such that $a \leq b \wedge c'$. For this a, $a \leq b$ and $a \nleq c$. □

Lemma 63.5. *If $b \in B$, and $a_1, a_2, \ldots, a_m$ are all the atoms $\leq b$, then $b = a_1 \vee a_2 \vee \cdots \vee a_m$.*

PROOF. Let $c = a_1 \vee a_2 \vee \cdots \vee a_m$. Then $c \leq b$ since $a_i \leq b$ for $1 \leq i \leq m$. Therefore, it suffices to show that $c \geq b$. Assume, to the contrary, that $c \ngeq b$. Then by Lemma 63.4 there is an atom a such that $a \leq b$ and $a \nleq c$. But $a \leq b$ implies $a = a_i$ for some i, by the definition of the a_i $(1 \leq i \leq m)$. The conditions $a = a_i$ and $a \wedge c = 0$ are contradictory, so that we must have $c \geq b$. □

The next lemma shows that if any of the atoms a_i $(1 \leq i \leq m)$ in Lemma 63.5 are omitted, then b is not the l.u.b. of the remaining atoms.

Lemma 63.6. *If $b \in B$, and $a, a_1, a_2, \ldots, a_m$ are atoms of B, with $a \leq b$ and $b = a_1 \vee a_2 \vee \cdots \vee a_m$, then $a = a_i$ for some i.*

PROOF. We have $a \wedge b = a$ because $a \leq b$. Therefore

$$a \wedge (a_1 \vee a_2 \vee \cdots \vee a_m) = (a \wedge a_1) \vee (a \wedge a_2) \vee \cdots \vee (a \wedge a_m) = a,$$

and so $a \wedge a_i \neq 0$ for some i. Lemma 63.2 gives $a = a_i$. □

PROOF OF THEOREM 63.1. Let B be a finite Boolean algebra, and let S be the set of all atoms in B. We shall prove that $B \approx \mathcal{P}(S)$.

If $b \in B$, then by Lemmas 63.5 and 63.6 b can be written uniquely in the form $b = a_1 \vee a_2 \vee \cdots \vee a_m$ with $a_1, a_2, \ldots, a_m \in S$; moreover, $\{a_1, a_2, \ldots, a_m\} = \{a \in S \mid a \leq b\}$. Therefore, we can define a mapping $\theta: B \to \mathcal{P}(S)$ by

$$\theta(a_1 \vee a_2 \vee \cdots \vee a_m) = \{a_1, a_2, \ldots, a_m\}.$$

The mapping θ is clearly onto.

To show that θ is one-to-one, it suffices to show that if $b, c \in B$ and $b \neq c$, then either there is $a \in S$ such that $a \leq b$ and $a \nleq c$, or else there is $a \in S$ such that $a \nleq b$ and $a \leq c$. But this follows from Lemma 63.4, since $b \neq c$ only if $b \nleq c$ or $b \ngeq c$.

Assume that $b, c \in B$. Then $\theta(b \vee c) = \theta(b) \cup \theta(c)$ because if a is an atom then

$$a \leq b \vee c \quad \text{iff} \quad a \leq b \quad \text{or} \quad a \leq c \tag{63.2}$$

(Problem 63.5). And $\theta(b \wedge c) = \theta(b) \cap \theta(c)$ because if a is an atom then

$$a \leq b \wedge c \quad \text{iff} \quad a \leq b \quad \text{and} \quad a \leq c \tag{63.3}$$

(Problem 63.6). □

Corollary. *If B is a finite Boolean algebra, then $|B| = 2^n$ for some positive integer n.*

Theorem 63.1 is not true if we remove the requirement that the Boolean algebra be finite. However, in 1936 the American mathematician M. H. Stone proved that if B is a Boolean algebra, then there is a set S such that B is isomorphic to a Boolean algebra formed by some collection $\mathcal{C}$ of subsets of S. The operations on $\mathcal{C}$ are union and intersection; the collection $\mathcal{C}$ is not necessarily all of $\mathcal{P}(S)$, but $\mathcal{C}$ is closed with respect to union, intersection, and complementation relative to S. (See [4].)

PROBLEMS

63.1. Prove that for Boolean algebras the definition of isomorphism in Problem 60.11 is equivalent to the definition in this section.

63.2. Prove: If $\theta: A \to B$ is a Boolean algebra isomorphism, then $\theta(1_A) = 1_B$ and $\theta(0_A) = 0_B$.

63.3. Prove: If $\theta: A \to B$ is a Boolean algebra isomorphism, then $\theta(a') = \theta(a)'$ for each $a \in A$.

63.4. The positive divisors of 1155 form a Boolean algebra when $a \leq b$ is taken to mean $a \mid b$. What are the atoms?

63.5. Assume that B is a Boolean algebra, $a, b, c \in B$, and a is an atom. Prove that $a \leq b \vee c$ iff $a \leq b$ or $a \leq c$. Also show that if a is not an atom then this is not necessarily true.

63.6. Assume that B is a Boolean algebra and $a,b,c\in B$. Prove that $a\leq b\wedge c$ iff $a\leq b$ and $a\leq c$. (Notice that in contrast to Problem 63.5, a need not be an atom here.)

63.7. Assume $n\in\mathbb{N}$. Prove that the lattice of divisors of n is a Boolean algebra (with $a\leq b$ iff $a|b$) iff n is a product of distinct primes. What is the order of the lattice? (See Problem 39.5.)

63.8. (This problem puts Boolean algebras in the context of rings, as discussed earlier in the book. This problem is longer than most of the problems in the book.)

(a) A ring R is called a *Boolean ring* if $x^2=x$ for each $x\in R$. Every Boolean ring is commutative and satisfies $2x=0$ for each $x\in R$ (Problem 21.15). Let R be a Boolean ring with unity 1, and define operations $\vee$ and $\wedge$ on R by

$$a\vee b = a+b-ab$$

and

$$a\wedge b = ab.$$

Prove that with these operations R is a Boolean algebra, with unity 1 and zero 0, and with $1-a$ for the complement a' of a for each $a\in R$.

(b) Let B be a Boolean algebra and define two operations on B by

$$a+b = (a\wedge b')\vee(a'\wedge b)$$

and

$$ab = a\wedge b.$$

(The first of these operations is called the *symmetry difference* of a and b.) Prove that with these operations B is a Boolean ring with unity.

SECTION 64. SWITCHING

Boolean algebras have drawn increasing interest in recent years because of their usefulness in such applications as computer design and the simplification of telephone switching circuits. This section will introduce some of the ideas underlying these applications. Although the terminology will be that of switches and ordinary electrical circuits, the ideas also apply to circuits with various kinds of magnetic or electronic two-state devices in place of conventional two-way switches.

Think of a *switch* as a device that can be located at a point of an electrical circuit and can be either *closed* or *open*. Current can pass through the point if the switch is closed, but not if the switch is open. If two switches in a circuit are always closed or open simultaneously, we denote them by the same letter. A switch that is always open when a is closed and always closed when a is open will be denoted by a'; a' is *opposite* to a.

Two switches such as a and b in Figure 64.1 are said to be in *series*—we denote this combination by $a\wedge b$. Current can pass from p to q (the *terminals*) in Figure 64.1 iff both a and b are closed. Two switches such as a and b in Figure 64.2 are said to be in *parallel*—we denote this combina-

Figure 64.1

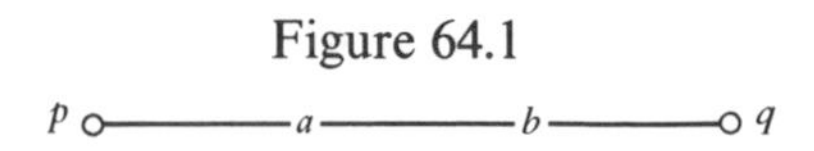

tion by $a \vee b$. Current can pass from p to q in Figure 64.2 iff either a is closed or b is closed. *Series-parallel circuits* are circuits that can be built up by beginning with a circuit with a single switch and repeatedly replacing switches by combinations of the type in either Figure 64.1 or Figure 64.2. Series-parallel circuits will always have two terminals, and the circuit will be called *closed* or *open* depending on whether current can or cannot pass between these terminals. All of the figures in this section represent series-parallel circuits. Two circuits are said to be *equivalent* if the switch positions that close one are the same as the switch positions that close the other.

Example 64.1. The two circuits in Figure 64.3 are equivalent. If a is open in either circuit, then the circuit will be open. If a is closed, then both circuits will be closed iff either b or c is closed.

If equivalent circuits are treated as being equal, then the series-parallel circuits form a Boolean algebra with $\vee$ and $\wedge$ as operations. The equivalence (equality) of the two circuits in Figure 64.3 amounts to one of the distributive laws: $a \wedge (b \vee c) = (a \wedge b) \vee (a \wedge c)$. The left-hand circuit in Figure 64.4 is $a \wedge a'$, the zero of the Boolean algebra: $a \wedge a'$ is always open; therefore $b \vee (a \wedge a') = b$ for every b. The right-hand circuit in Figure 64.4 is $a \vee a' = 1$, the unity of the Boolean algebra: $a \vee a'$ is always closed; therefore $b \wedge (a \vee a') = b$ for every b. Problem 64.1 asks you to draw diagrams for circuits representing most of the Boolean algebra laws from Definition II in Section 62.

Boolean algebras can help in the simplification of circuits and also in the design of circuits with specified properties. We shall have to be content with an illustration of the first of these two applications. References for full discussions of both applications can be found in the notes at the end of this chapter.

If we begin with a series-parallel circuit, then write the algebraic expression corresponding to that circuit, and then change this algebraic expres-

Figure 64.2

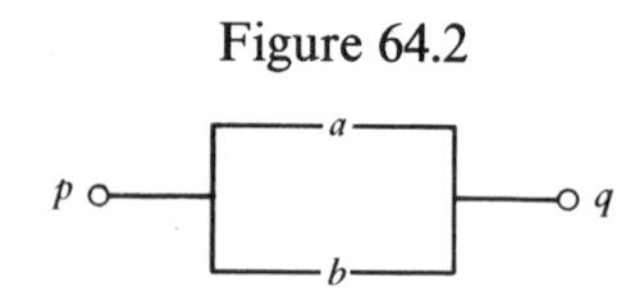

Figure 64.3

Figure 64.4

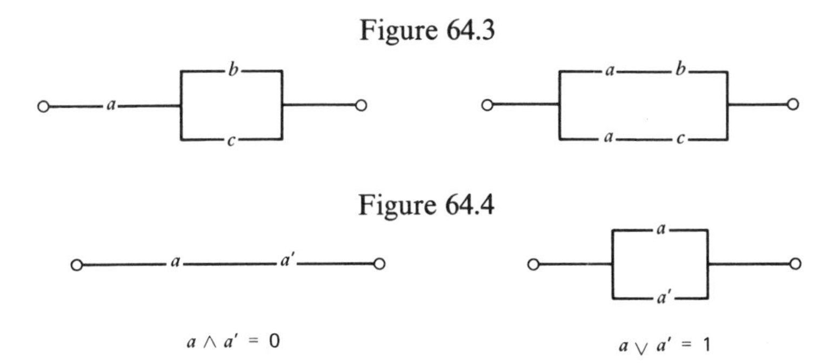

sion by using the laws of a Boolean algebra, the result will be an expression that corresponds to a circuit equivalent to the one with which we began. A circuit can often be simplified in this way.

Example 64.2. The circuit in Figure 64.5 is represented by

$$(a \vee b \vee c) \wedge (a' \vee b \vee c) \wedge (a' \vee b \vee c').$$

The laws of a Boolean algebra can be used to simplify this expression as follows:

$(a \vee b \vee c) \wedge (a' \vee b \vee c) \wedge (a' \vee b \vee c')$	associative law
$= [a \vee (b \vee c)] \wedge [a' \vee (b \vee c)] \wedge (a' \vee b \vee c')$	distributive law
$= [(a \wedge a') \vee (b \vee c)] \wedge (a' \vee b \vee c')$	complement
$= [0 \vee (b \vee c)] \wedge (a' \vee b \vee c')$	zero and commutative law
$= (b \vee c) \wedge [b \vee (a' \vee c')]$	distributive law
$= b \vee [c \wedge (a' \vee c')]$	distributive law
$= b \vee [(c \wedge a') \vee (c \wedge c')]$	zero and complement
$= b \vee (c \wedge a').$	

Therefore, the circuit in Figure 64.5 is equivalent to the circuit in Figure 64.6.

Figure 64.5

Figure 64.6

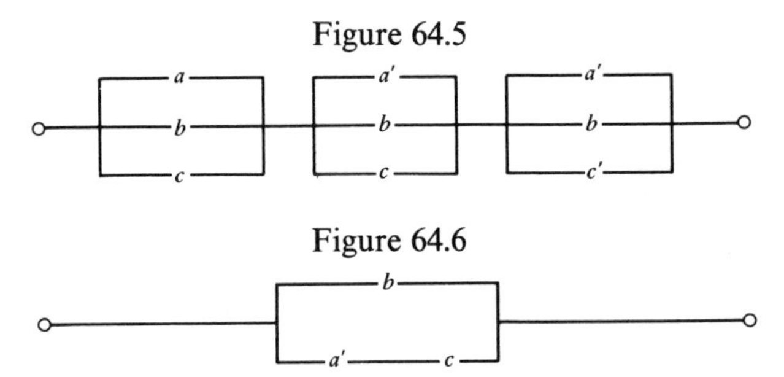

PROBLEMS

64.1. Draw diagrams for pairs of circuits representing the commutative laws, the associative laws, and the second distributive law from Definition II in Section 62.

64.2. Draw circuits equivalent to but simpler than each of the following circuits, by using Boolean algebra laws to simplify the algebraic expression corresponding to the circuit.

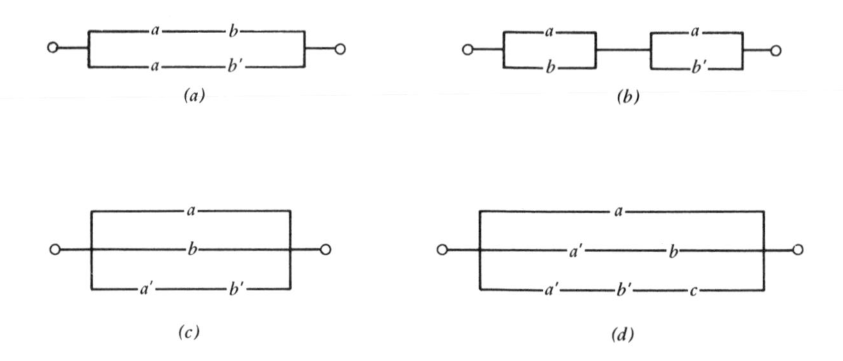

64.3. Draw a circuit equivalent to but simpler than the diagram

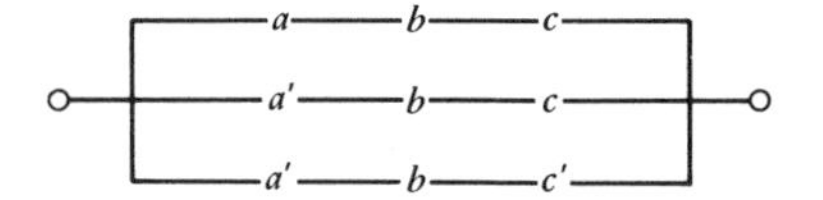

Compare Example 64.2, and use duality to state a general principle suggested by the two diagrams.

64.4. (a) Draw a diagram for a circuit that will be closed iff exactly one of a and b is closed.
(b) Draw a diagram for a circuit that will be closed iff at least one of a and b is closed.
(c) Draw a diagram for a circuit that will be closed iff at most one of a and b is closed.

64.5. (a) Draw a diagram for a circuit that will be closed iff exactly one of a, b, and c is closed.
(b) Draw a diagram for a circuit that will be closed iff at least one of a, b, and c is closed.
(c) Draw a diagram for a circuit that will be closed iff at most one of a, b, and c is closed.

NOTES ON CHAPTER XVII

1. Donnellan, T., *Lattice Theory*, Pergamon Press, Oxford, 1968.
2. Halmos, P. R., *The basic concepts of algebraic logic*, American Mathematical Monthly **53** (1956) 363–387.
3. Hohn, F. R., *Applied Boolean Algebra*, 2nd ed., Macmillan, New York, 1966.
4. Stone, M. H., *The theory of representations of Boolean algebras*, Transactions of the American Mathematical Society **40** (1936) 37–111.
5. Whitesitt, J. E., *Boolean Algebra and Its Applications*, Addison-Wesley, Reading, Mass., 1961.

APPENDIX A

Sets

This appendix contains a summary of basic facts and notation about sets.

A set is a collection of objects, called its *elements* or *members*. To indicate that x is an element of a set A, we write

$$x \in A.$$

To indicate that x is *not* an element of A, we write

$$x \notin A.$$

There are three commonly used methods of defining a set. First, simply by describing its elements:

The set of all positive integers.

Second, by listing its elements in braces:

$$\{a,b,c\} \qquad \text{or} \qquad \{1,2,3,\ldots\}.$$

(Not all elements in the second set can be listed, of course, but there should be no doubt that the set of all positive integers is intended.) Third, a set can be defined by *set-builder notation*:

$\{x|x \text{ has property } P\}$ means "the set of all x such that x has property P."

Thus $\{x|x \text{ is a positive integer}\}$ also denotes the set of all positive integers.

If A and B are sets, and each element of A is an element of B, then A is a *subset* of B; this is denoted by

$$A \subseteq B \qquad \text{or} \qquad B \supseteq A.$$

Notice that $A \subseteq B$ does not preclude the possibility that $A = B$. In fact,

$$A = B \qquad \text{iff*} \qquad A \subseteq B \quad \text{and} \quad A \supseteq B.$$

*We use "iff" to denote "if and only if."

The following three statements are equivalent:

(i) $A \subseteq B$.
(ii) If $x \in A$, then $x \in B$.
(iii) If $x \notin B$, then $x \notin A$.

Statement (iii) is the contrapositive of (ii) (see Appendix B); it is sometimes easier to prove that $A \subseteq B$ by using (iii) rather than (ii).

The *empty (null, vacuous) set* contains no elements, and is denoted by $\varnothing$. Thus

$$x \in \varnothing \text{ is always false.}$$

If A is not a subset of B, we write $A \nsubseteq B$. This is true, of course, iff A contains at least one element that is not in B. Set inclusion ($\subseteq$) has the following properties:

$\varnothing \subseteq A$ for every set A.
$A \subseteq A$ for every set A.
If $A \subseteq B$ and $B \subseteq C$, then $A \subseteq C$.

If A and B are sets, then $A \cap B$, the *intersection* of A and B, is defined by

$$A \cap B = \{x | x \in A \quad \text{and} \quad x \in B\}.$$

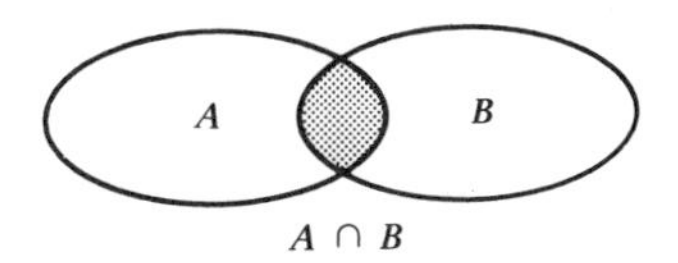

$A \cap B$

If A and B are sets, then $A \cup B$, the *union* of A and B, is defined by

$$A \cup B = \{x | x \in A \quad \text{or} \quad x \in B\}.$$

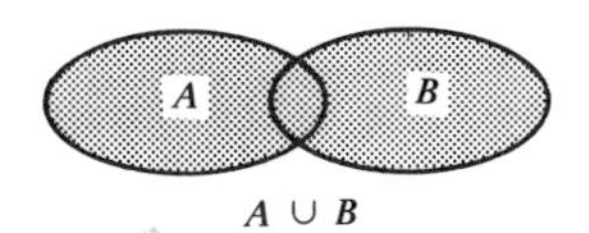

$A \cup B$

The diagrams above are called *Venn diagrams*.

Example A.1. Let $S = \{a,b,c\}$, $T = \{c,d,e\}$, $U = \{d,e\}$. Then

$a \in S$	$a \notin T$
$S \cap T = \{c\}$	$S \cap U = \varnothing$
$S \nsubseteq T$	$T \supseteq U$
$S \cup T = S \cup U = \{a,b,c,d,e\}$	

If A, B, and C denote any sets, then

$A \cap B = B \cap A$ $\qquad$ $A \cup B = B \cup A$

$A \cap B \subseteq A$ $\qquad$ $A \subseteq A \cup B$

$A \subseteq B$ implies $A \cap B = A$

$A \subseteq B$ implies $A \cup B = B$

$A \cap A = A$ $\qquad$ $A \cup A = A$

$$A \cap (B \cap C) = (A \cap B) \cap C$$
$$A \cup (B \cup C) = (A \cup B) \cup C$$
$$A \cap (B \cup C) = (A \cap B) \cup (A \cap C)$$
$$A \cup (B \cap C) = (A \cup B) \cap (A \cup C).$$

The *complement* of a set A in a set S that contains A is $A' = \{x | x \in S$ and $x \notin A\}$.

The *Cartesian product* of sets A and B is denoted by $A \times B$, and is defined by

$$A \times B = \{(x,y) | x \in A \quad \text{and} \quad y \in B\}.$$

Here each (x,y) is an *ordered pair*. The term *ordered* is used because (x,y) is to be distinguished from (y,x) if $x \neq y$. Ordered pairs (x,y) and (x',y') are *equal* iff $x = x'$ and $y = y'$.

Example A.2. If $A = \{1,2\}$ and $B = \{u,v,w\}$, then

$$A \times B = \{(1,u),(1,v),(1,w),(2,u),(2,v),(2,w)\}.$$

Notice that in this case $A \times B \neq B \times A$. In general, $A \times B = B \times A$ iff $A = B$.

Plane coordinate (analytic) geometry begins with a one-to-one correspondence between the points of the plane and the set of pairs of real numbers, that is, elements of the Cartesian product $\mathbb{R} \times \mathbb{R}$. This example explains the choice of the name "Cartesian product"—Descartes was one of the developers of analytic geometry.

The notion of Cartesian product can be used to give a definition of mapping that is preferred by some to that given in Section 1: A *mapping* from a set S to a set T is a subset of $S \times T$ such that each $x \in S$ is a first member of precisely one pair in the subset. The connection between this definition and that given in Section 1 is that if $\alpha : S \to T$ (in the sense of Section 1), then each $x \in S$ contributes the pair $(x, \alpha(x))$ to the subset of $S \times T$ in the definition of mapping as a subset of $S \times T$. For example, the mapping α in Example 1.1 corresponds to the subset $\{(x,2), (y,1), (z,3)\}$ of $\{x,y,z\} \times \{1,2,3\}$.

APPENDIX B

Proofs

In understanding or constructing proofs we are invariably concerned with *conditional* statements, that is, with statements of the form

if p then q.

Any such statement is logically equivalent to each of a number of other statements, where to say that two statements are *logically equivalent* means that they are either both true or both false. In particular, the conditional statement above is logically equivalent to each of the following statements:

p implies q.
p only if q.
q if p.
p is sufficient for q.
q is necessary for p.

Example B.1. Let A and B denote sets. The statement

A is a subset of B

is logically equivalent to the conditional statement

if $x \in A$ then $x \in B$.

This conditional statement, in turn, is logically equivalent to each of the following statements:

$x \in A$ implies $x \in B$.
$x \in A$ only if $x \in B$.
$x \in B$ if $x \in A$.
$x \in A$ is sufficient for $x \in B$.
$x \in B$ is necessary for $x \in A$.

Another statement logically equivalent to the conditional statement

$$\text{if } p \text{ then } q$$

is its *contrapositive*,

$$\text{if not } q \text{ then not } p.$$

This is the basis for the method of *indirect proof*, in which a conditional statement is proved by proving its contrapositive.

Example B.2. The statement

$$\text{if } x \in A \text{ then } x \in B$$

is logically equivalent to

$$\text{if } x \notin B \text{ then } x \notin A.$$

Therefore, to prove "$A \subseteq B$" is the same as to prove "if $x \notin B$ then $x \notin A$."

Another method, similar to indirect proof, is *proof by contradiction* (or *reductio ad absurdum*). With this method, a conditional statement "if p then q" is proved by showing that if p were true and q were not true, then some contradiction (absurdity) would result. For an example of a proof by contradiction see Theorem 25.1.

The statement

$$\text{if } q \text{ then } p$$

is the *converse* of

$$\text{if } p \text{ then } q.$$

A statement and its converse are *not* logically equivalent. For example, if we were to say that "if $x \in A$ then $x \in B$" is logically equivalent to "if $x \in B$ then $x \in A$," we would be saying that $A \subseteq B$ is the same as $B \subseteq A$, which is manifestly false.

Putting a conditional statement and its converse together, we get a *biconditional* statement:

$$p \quad \text{if and only if} \quad q,$$

which we are writing

$$p \text{ iff } q.$$

It is an immediate consequence of earlier remarks that this biconditional statement is logically equivalent to

$$p \quad \text{is necessary and sufficient for} \quad q.$$

Sometimes, "if p then q" or one of its equivalents is written when in fact "p iff q" is true. This happens especially with definitions. For example, the "if" in

> An integer p is a *prime* if $p > 1$ and p
> is divisible by no positive integer other than
> 1 and p itself,

really means "iff."

The connectives "or," "and," and "not" arise often in forming compound statements. Unless it is explicitly stated otherwise, the word "or" should always be taken in the *inclusive* sense:

$$p \quad \text{or} \quad q$$

means

$$p \quad \text{or} \quad q \quad \text{or both.}$$

The words "and" and "not" have their usual meanings.

The following rules are used frequently:

1. not (p and q) is logically equivalent to (not p) or (not q).
2. not (p or q) is logically equivalent to (not p) and (not q).

Example B.3. Let A and B denote sets, and let

$$p \quad \text{denote} \quad x \in A$$

and

$$q \quad \text{denote} \quad x \in B.$$

Then

$$p \text{ and } q \text{ means } x \in A \cap B,$$

and

$$p \text{ or } q \text{ means } x \in A \cup B.$$

This leads to the following two lists of logically equivalent statements:

$$\left.\begin{array}{l} \text{not } (p \text{ and } q) \\ x \notin A \cap B \\ x \notin A \text{ or } x \notin B \\ (\text{not } p) \text{ or } (\text{not } q) \end{array}\right\} \begin{array}{l}\text{logically} \\ \text{equivalent}\end{array}$$

$$\left.\begin{array}{l} \text{not } (p \text{ or } q) \\ x \notin A \cup B \\ x \notin A \text{ and } x \notin B \\ (\text{not } p) \text{ and } (\text{not } q) \end{array}\right\} \begin{array}{l}\text{logically} \\ \text{equivalent}\end{array}$$

In statements such as

$$x^2 - 4 = 0,$$

and

$$x \text{ is nonnegative,}$$

the letter "x" is called a *variable*. Variables are assumed to belong to some previously agreed-upon universal set. In the two examples, this might be the set of all real numbers, for instance. A statement with a variable can be combined with either a *universal quantifier* or an *existential quantifier*:

Universal quantifier: "for each x."

Existential quantifier: "there is an x such that."

Once a quantifier has been introduced it becomes meaningful to ask whether a statement with a variable is true or false.

Example B.4. Let the variable x represent a real number.

For each x, $x^2-4=0$.	False
There is an x such that $x^2-4=0$.	True
For each x, x^2 is nonnegative.	True

The terms "for every" and "for all" are often used in place of "for each." Also, "there exists an x such that" and "for some x" are often used in place of "there is an x such that." Moreover, statements that can be made using quantifiers are often made without their explicit use. For instance, "for each (real number) x, x^2 is nonnegative," is logically equivalent to "the square of every real number is nonnegative."

In a statement with more than one variable, each variable requires quantification. Example: For all real numbers a, b, and c, $a(b+c)=ab+ac$.

The negation of a statement with a universal quantifier is a statement with an existential quantifier, and the negation of a statement with an existential quantifier is a statement with a universal quantifier.

Example B.5.

Statement I: For each x, x^2 is nonnegative.

Negation of I: There is an x such that x^2 is negative.

Statement II: There is an x such that $x^2-4=0$.

Negation of II: For each x, $x^2-4\neq0$.

An example showing that a statement with a universal quantifier is false is called a *counterexample*. Thus $x=2$ is a counterexample to the statement "for each prime number x, x is odd."

The preceding remarks were concerned with the technical aspects of proofs. A slightly different problem is that of how to *discover* proofs; the following book is warmly recommended for its advice on this problem.

Pólya, G., *How to Solve It*, Princeton University Press, Princeton, 1945.

APPENDIX C

Mathematical Induction

Principle of Mathematical Induction

For each positive integer n, let $P(n)$ represent a statement depending on n. If

(a) $P(1)$ is true, and
(b) $P(k)$ implies $P(k+1)$ for each positive integer k,

then $P(n)$ is true for every positive integer n.

Intuitively, the principle is valid because $P(1)$ is true by (a), and then $P(2)$ is true by (b) with $k=1$, and then $P(3)$ is true by (b) with $k=2$, and so on. Logically, the principle is equivalent to the Least Integer Principle, stated in Section 10. (Also see Section 23, especially Problem 23.6.)

Example C.1. If $S(n)$ denotes the sum

$$a + ar + ar^2 + \cdots + ar^{n-1}$$

of the first n terms of a geometric progression with first term a and common ratio r, then

$$S(n) = \frac{a - ar^n}{1 - r}. \tag{C.1}$$

To prove this, for each positive integer n let $P(n)$ be the statement that the Formula (C.1) is correct.

(a) For $n=1$, (C.1) gives

$$S(1) = \frac{a-ar}{1-r} = a,$$

which is clearly correct.
(b) Assume that $P(k)$ is true. Then

$$S(k) = \frac{a-ar^k}{1-r}$$

is correct. Therefore

$$\begin{aligned} S(k+1) &= S(k) + ar^k \\ &= \frac{a-ar^k}{1-r} + ar^k \\ &= \frac{a-ar^k+(1-r)ar^k}{1-r} \\ &= \frac{a-ar^{k+1}}{1-r}, \end{aligned}$$

and the statement $P(k+1)$ is true. Therefore $P(k)$ does imply $P(k+1)$, as required.

Example C.2. Let a denote a real number. Define positive integral powers of a as follows: $a^1=a$, $a^2=aa$, $a^3=a^2a$, and, in general, $a^{n+1}=a^n a$. We shall prove that

$$a^m a^n = a^{m+n} \tag{C.2}$$

for all positive integers m and n. In doing this we shall assume the law

$$a(bc) = (ab)c \tag{C.3}$$

for all real numbers a, b, and c. [The law in Equation (C.2) will be familiar for real numbers from elementary algebra. The proof here will show that it is valid, more generally, in any system satisfying Equation (C.3). See Section 12.]

Here we let $P(n)$ be the statement that Equation (C.2) is correct for n and every positive integer m.
(a) With $n=1$, Equation (C.2) is $a^m a = a^{m+1}$, which is correct because $a^m a$ is the definition of a^{m+1}. Therefore $P(1)$ is true.
(b) Assume that $P(k)$ is true. Then $a^m a^k = a^{m+k}$ is correct. Therefore

$$\begin{aligned} a^m a^{k+1} &= a^m(a^k a^1) && \text{[by the definition of } a^{k+1}\text{]} \\ &= (a^m a^k)a && \text{[by (C.3)]} \\ &= a^{m+k}a && \text{[by } P(k)\text{]} \\ &= a^{m+k+1} && \text{[by the definition of } a^{m+k+1}\text{],} \end{aligned}$$

and the statement $P(k+1)$ is true.

Example C.3 (Binomial Theorem). If a and b denote any real numbers and n denotes a positive integer, then

$$(a+b)^n = C(n,n)a^n + C(n,n-1)a^{n-1}b + C(n,n-2)a^{n-2}b^2 + \cdots + C(n,r)a^r b^{n-r} + \cdots + C(n,0)b^n. \quad \text{(C.4)}$$

$C(n,r)$ is the *binomial coefficient*, defined by

$$C(n,r) = \frac{n!}{r!(n-r)!}$$

for $0 \leq r \leq n$ [with $r! = r(r-1)\cdots 1$ for $r > 0$, and $0! = 1$].

To prove this, let $P(n)$ be the statement that Equation (C.4) is correct.
(a) For $n = 1$, the right-hand side of Equation (C.4) gives

$$C(1,1)a + C(1,0)b = a + b,$$

which is equal to $(a+b)^1$, as required.
(b) Assume that $P(k)$ is true. If both sides of Equation (C.4) (with n replaced by k) are multiplied by $(a+b)$, then equality will result, and, moreover, the coefficient of $a^r b^{k+1-r}$ on the right will be

$$C(k,r-1) + C(k,r).$$

(It may take pencil and paper to verify that.) The coefficient of $a^r b^{k+1-r}$ on the right when n is replaced by $k+1$ in (C.4) will be $C(k+1,r)$. The truth of $P(k+1)$ is a consequence of the identity

$$C(k+1,r) = C(k,r-1) + C(k,r),$$

which can be verified by simple addition.

The statement $P(k)$, assumed in the second part of proofs by mathematical induction, is sometimes called the *induction hypothesis*. In the following alternative form of the Principle of Mathematical Induction an apparently stronger induction hypothesis is used: in place of assuming only $P(k)$, we assume $P(i)$ for all $i \leq k$. The proof that the two forms of the principle are logically equivalent will be left as an exercise; the appropriate context for this exercise is Section 23.

Second Principle of Mathematical Induction

For each positive integer n, let $P(n)$ represent a statement depending on n. If

(a) $P(1)$ is true, and
(b) the truth of $P(i)$ for all $i \leq k$ implies the truth of $P(k+1)$, for each positive integer k,

then $P(n)$ is true for every positive integer n.

APPENDIX D

Linear Algebra

This appendix contains a concise review of the facts about vector spaces, linear transformations, and matrices that are needed elsewhere in this book. It also presents important examples of groups and rings that arise from linear transformations and matrices. Ideas from Chapters I to VI are used freely, and proofs are omitted.

Definition. A *vector space* over a field F is a set V together with an operation $+$ on V and a mapping $F \times V \to V$ $[(a,v) \mapsto av]$ such that each of the following axioms is satisfied:

1. V is an Abelian group with respect to $+$,
2. $(ab)v = a(bv)$,
3. $(a+b)v = av + bv$,
4. $a(v+w) = av + aw$,
5. $ev = v$,

for all $a, b \in F$ and all $v, w \in V$, with e the unity of F.

The elements of V and F are called *vectors* and *scalars*, respectively. And av is called the *scalar multiple* of $v \in V$ by $a \in F$. Throughout this section V will denote a vector space over a field F.

Example D.1. Let F^n denote the set of n-tuples $(a_1, a_2, \ldots, a_n)$ with each $a_i \in F$. With the operations

$$(a_1, a_2, \ldots, a_n) + (b_1, b_2, \ldots, b_n) = (a_1 + b_1, a_2 + b_2, \ldots, a_n + b_n)$$

and

$$a(a_1,a_2,\ldots,a_n) = (aa_1,aa_2,\ldots,aa_n),$$

F^n is a vector space over F. The elements of $\mathbb{R}^2$ and $\mathbb{R}^3$ will sometimes be identified with geometric vectors in the usual way. If p is a prime, then $\mathbb{Z}_p^n$ is a vector space with p^n elements.

Example D.2. The complex field $\mathbb{C}$ can be thought of as a vector space over the real field $\mathbb{R}$. The elements of $\mathbb{C}$ are the vectors; the addition of vectors is the addition in $\mathbb{C}$; and $a(b+ci)=ab+aci$ ($a\in\mathbb{R}$, $b+ci\in\mathbb{C}$). That is, the scalar multiples are just multiples in $\mathbb{C}$ except that the first factors are restricted to $\mathbb{R}$. The same ideas can be extended to show that if E is any field and F is any subfield of E, then E is a vector space over F.

A subset W of V is a *subspace* of V if W is itself a vector space with respect to the addition of V and the scalar multiplication aw ($a\in F$, $w\in W$). It is useful to know that a nonempty subset W of V is a subspace of V iff

1. $v+w\in W$ for all $v,w\in W$, and
2. $av\in W$ for all $a\in F$, $v\in W$.

If $a_1, a_2,\ldots,a_n\in F$ and $v_1, v_2,\ldots,v_n\in V$, then

$$a_1v_1 + a_2v_2 + \cdots + a_nv_n$$

is called a *linear combination* of $v_1, v_2,\ldots,v_n$. If $\{v_1,v_2,\ldots,v_n\}$ is a subset of V, let

$$\langle v_1,v_2,\ldots,v_n\rangle = \{a_1v_1+a_2v_2+\cdots+a_nv_n|a_1,a_2,\ldots,a_n\in F\},$$

the set of all linear combinations of $v_1, v_2,\ldots,v_n$. Then $\langle v_1,v_2,\ldots,v_n\rangle$ is a subspace of V, called the subspace *generated* (or *spanned*) by $\{v_1,v_2,\ldots,v_n\}$. The vectors $v_1, v_2,\ldots,v_n$ are said to be *linearly independent* if

$$a_1v_1 + a_2v_2 + \cdots + a_nv_n = 0$$

implies that

$$a_1 = a_2 = \cdots = a_n = 0$$

for $a_1,a_2,\ldots,a_n\in F$; otherwise, $v_1,v_2,\ldots,v_n$ are *linearly dependent*.

If $v_1,v_2,\ldots,v_n$ are linearly independent and $\langle v_1,v_2,\ldots,v_n\rangle = V$, then $\{v_1,v_2,\ldots,v_n\}$ is said to be a *basis* for V. We shall be concerned only with vector spaces having finite bases. Any two bases for a vector space V have the same number of elements; this number is called the *dimension* of V and is denoted dim V.

Example D.3. If

$$\begin{aligned} e_1 &= (1,0,0,\ldots,0) \\ e_2 &= (0,1,0,\ldots,0) \\ &\vdots \\ e_n &= (0,0,0,\ldots,1), \end{aligned}$$

then $\{e_1, e_2, \ldots, e_n\}$ is a basis for $\mathbb{R}^n$. Thus dim $\mathbb{R}^n = n$. The vectors e_1, $e_2, \ldots, e_n$ will be called *standard unit vectors*.

Any set that generates V contains a basis for V. Any linearly independent subset of V is contained in a basis for V. The dimension of a subspace of V cannot exceed the dimension of V.

Example D.4. The set $\{1, i\}$ is a basis for $\mathbb{C}$ as a vector space over $\mathbb{R}$ (Example D.2). It generates $\mathbb{C}$ because each complex number can be written as $a \cdot 1 + b \cdot i = a + bi$ with $a, b \in \mathbb{R}$. It is linearly independent because if $a + bi = 0$ with $a, b \in \mathbb{R}$, then $a = 0$ and $b = 0$. Therefore, as a vector space over $\mathbb{R}$, dim $\mathbb{C} = 2$.

Let V denote a vector space over a field F. A mapping $\alpha: V \to V$ is a *linear transformation* if

$$\alpha(av + bw) = a\alpha(v) + b\alpha(w) \tag{D.1}$$

for all $a, b \in F$, $v, w \in V$. A linear transformation is *nonsingular* (or *invertible*) iff it is invertible as a mapping, that is, iff it is one-to-one and onto. The inverse of a nonsingular linear transformation is necessarily linear. Let

$L(V)$ denote the set of all linear transformations from V to V

and

$GL(V)$ denote the set of all nonsingular elements of $L(V)$.

If $\alpha, \beta \in L(V)$, then $\alpha + \beta$ and $\beta\alpha$ are defined as follows:

$$\begin{aligned} (\alpha + \beta)(v) &= \alpha(v) + \beta(v) \\ (\beta\alpha)(v) &= \beta(\alpha(v)) \end{aligned} \tag{D.2}$$

for all $v \in V$. It can be verified that both $\alpha + \beta$ and $\beta\alpha$ are linear; and so we have two operations on $L(V)$.

Theorem D.1. *(a) $L(V)$ is a ring with respect to the operations defined in* (D.2).
(b) $GL(V)$ is a group with respect to the product (composition) defined in (D.2).

If A and B are both $m \times n$ matrices over F (that is, with entries in F), then their sum, $A+B$, is defined to be the $m \times n$ matrix with ij-entry $a_{ij}+b_{ij}$, where a_{ij} and b_{ij} are the ij-entries of A and B, respectively. If A is an $m \times n$ matrix over F, and B is an $n \times p$ matrix over F, then their product, AB, is defined to be the $m \times p$ matrix with ij-entry

$$a_{i1}b_{1j} + a_{i2}b_{2j} + \cdots + a_{in}b_{nj} = \sum_{k=1}^{n} a_{ik}b_{kj}.$$

Let I_n denote the $n \times n$ *identity matrix*, which is defined by $a_{ii}=1$ for $1 \le i \le n$ and $a_{ij}=0$ for $i \ne j$. (Here 1 denotes the unity of F.) An $n \times n$ matrix A is *nonsingular* or (*invertible*) iff there is an $n \times n$ matrix A^{-1} such that $A^{-1}A = AA^{-1} = I_n$. Let

$M(n,F)$ denote the set of all $n \times n$ matrices over F

and

$GL(n,F)$ denote the set of all nonsingular elements of $M(n,F)$.

Theorem D.2. *(a) $M(n,F)$ is a ring with respect to matrix addition and multiplication.*
(b) $GL(n,F)$ is a group with respect to matrix multiplication.

Let $\{v_1, v_2, \ldots, v_n\}$ be a basis for V and let $\alpha \in L(V)$. Then each $\alpha(v_i)$ is in V $(1 \le i \le n)$, and can be written as a linear combination of $v_1, v_2, \ldots, v_n$:

$$\alpha(v_i) = \sum_{j=1}^{n} a_{ji}v_j = a_{1i}v_1 + a_{2i}v_2 + \cdots + a_{ni}v_n, \tag{D.3}$$

with all $a_{ij} \in F$. The matrix with ij-entry a_{ij} is the *matrix of α relative to the basis* $\{v_1, v_2, \ldots, v_n\}$. This matrix depends on the order of the vectors in the basis $\{v_1, v_2, \ldots, v_n\}$. Conversely, given an ordered basis for V, and an $n \times n$ matrix $A = [a_{ij}]$ over F, there is a unique $\alpha \in L(V)$ having the matrix A; this is defined by (D.3).

Example D.5. Assume that V is a vector space over $\mathbb{R}$, $\{v_1, v_2, v_3\}$ is a basis for V, $\alpha \in L(V)$, and

$$\begin{aligned}
\alpha(v_1) &= 5v_1 + v_2 - 2v_3 \\
\alpha(v_2) &= \quad -2v_2 + v_3 \\
\alpha(v_3) &= v_1 + 4v_2
\end{aligned}$$

Then the matrix for α relative to $\{v_1, v_2, v_3\}$ is

$$\begin{bmatrix} 5 & 0 & 1 \\ 1 & -2 & 4 \\ -2 & 1 & 0 \end{bmatrix}.$$

If $v = a_1v_1 + a_2v_2 + a_3v_3 \in V$, then the representation of $\alpha(v)$ as a linear combination of v_1, v_2, v_3 can be obtained by matrix multiplication, as follows:

$$\begin{bmatrix} 5 & 0 & 1 \\ 1 & -2 & 4 \\ -2 & 1 & 0 \end{bmatrix} \begin{bmatrix} a_1 \\ a_2 \\ a_3 \end{bmatrix} = \begin{bmatrix} 5a_1 + a_3 \\ a_1 - 2a_2 + 4a_3 \\ -2a_1 + a_2 \end{bmatrix}$$

and

$$\alpha(v) = (5a_1 + a_3)v_1 + (a_1 - 2a_2 + 4a_3)v_2 + (-2a_1 + a_2)v_3.$$

The correspondence $\alpha \leftrightarrow [a_{ij}]$ determined by (D.3) can be used to prove the following theorem.

Theorem D.3. (*a*) *If* $\dim V = n$, *then* $L(V) \approx M(n, F)$, *as rings.*
(*b*) *If* $\dim V = n$, *then* $GL(V) \approx GL(n, F)$, *as groups. Both* $GL(V)$ *and* $GL(n, F)$ *are called* general linear groups.

The following ideas are used in Chapter XVI, on algebraic coding. The *row space* of an $m \times n$ matrix A over F is the subspace of F^n generated by the rows of A (thought of as elements of F^n). The *column space* of A is the subspace of F^m generated by the columns of A (thought of as elements of F^m). The dimensions of the row space and column space of A are called the *row rank* and *column rank* of A, respectively. It can be proved that these two ranks are equal for each matrix A; their common value is called the *rank* of A.

The *transpose* of an $m \times n$ matrix A is the $n \times m$ matrix, denoted A', obtained by interchanging the rows and columns of A. Thus each ij-entry of A' is the ji-entry of A. The transpose of a $1 \times n$ (*row*) matrix is an $n \times 1$ (*column*) matrix. We shall identify each $1 \times n$ matrix with the corresponding vector in F^n.

Let A be an $m \times n$ matrix over F. The *null space* of A is the set of all $v \in F^n$ such that

$$Av' = 0.$$

The null space of A is a subspace of F^n, and its dimension is called the *nullity* of A. It can be proved that

$$\operatorname{rank} A + \operatorname{nullity} A = n.$$

The *elementary row operations* on a matrix are of three types:

I. Interchange two rows.
II. Multiply a row by a nonzero scalar.
III. Add a multiple of one row to another row.

If a matrix B can be obtained from a matrix A by a finite sequence of elementary row operations, then A and B are said to be *row equivalent*. Row equivalence is an equivalence relation on the set of $m \times n$ matrices over F. Each matrix is row equivalent to a unique matrix in *row-reduced echelon form*, that is, a matrix such that

1. the first nonzero entry (the *leading entry*) of each row is 1,
2. the other entries in any column containing such a leading entry are all 0,
3. the leading entry in each row is to the right of the leading entry in each preceding row, and
4. rows containing only 0's are below rows with nonzero entries.

PHOTO CREDIT LIST

FIGURE 1.
From *Historic Ornament, A Pictorial Archive* by C.B. Griesbach, Dover Publications, Inc.

FIGURE 2.
From *Historic Ornament, A Pictorial Archive* by C.B. Griesbach, Dover Publications, Inc.

FIGURE 3A and B.
From *Manual of Mineralogy* by C.S. Hurlbut and C. Klein, 19th ed., John Wiley, © 1977.

FIGURE 34.4.
R.B. Hoit/Photo Researchers.

FIGURE 35.8.
Courtesy Bernard Quaritch, Ltd., London. Reprinted with permission.

Index

Abel, N. H., 37, 251
Abelian group, 37
Absolute value, of complex number, 128
  in ordered integral domain, 111
Absorption laws, 287
Action of group, 149-150
  faithful, 150
Algebraic closure, 125
Algebraic coding, 263-279
Algebraic element, 125, 238
Algebraic extension, 125
Algebraic integer, 233
Algebraic number, 233
Algebraic number theory, 7, 213
Amplitude of complex number, 128
Archimedean property, 121
Argument of complex number, 128
Array, standard, 271
Associate, in integral domain, 213
  of polynomial, 210
Associative law, 22
  generalized, 64-65
Atom, 296
Aut(G), 152
Automorphism, of field, 252
  of group, 84
    inner, 152
  of ring, 103
Axis, m-fold, 168

Barlow, W., 183
Basis, 316
Bijection, 15
Binary operation, 22
Binary symmetric channel, 275
Binomial coefficient, 160, 314
Binomial theorem, 95, 314
Boole, George, 8, 289
Boolean algebra, 8, 290
  finite, 295-299
Boolean ring, 106, 299
Burnside's counting theorem, 155

$\mathbb{C}$, 15
C_n, 166
Cancellation, law, 65, 93
  property, 96-97
Cartesian product, 306
Cayley, Arthur, 22
Cayley's theorem, 85-86
  generalization of, 153
Cayley table, 22
Center, of group, 152
  of ring, 98, 223
Chain, 283
Characteristic of ring, 104
Circuits, 299-300
Closure, 21
Code, binary, 264
  block, 264
  Hamming, 277-278
  length of, 264
  linear, 267

minimum distance of, 265
systematic, 270
Codes, equivalent, 269
Codomain of mapping, 12
Coefficient, 199, 201
Common divisor, 190
Commutative law, 23
Complement, in lattice, 288
in set, 306
Complete set, of coset representatives, 71-72
of equivalence class representatives, 54
Complex numbers, 121-131
characterized, 125
Composition of mappings, 17
Composition series, 145
Congruence, class, 56
of integers, 55
Conjugate, class, 152
of complex number, 124
of group element, 152
of subgroup, 152, 161
Constructible circle, 256
Constructible line, 256
Constructible number, 256-257
Constructible point, 256
Contradiction, 292
Coset, left, 72
right, 69-70
Coset leader, 271-272
Counterexample, 311
Cover of element, 281
Crystallographic group, *see* Group, crystallographic
Crystallographic restriction, 184, 187
Cycle, 38

D_n, 166
Decoder, 264
Decoding, standard, 271-275
table, 265
Dedekind, Richard, 233
deg $f(x)$, 204
Degree of field extension, 241
DeMoivre's theorem, 129
DeMorgan, Augustus, 292
DeMorgan's laws, 292
Derivative of polynomial, 203
Descartes, René, 306
Dimension of vector space, 316
Diophantus, 6
Direct product of groups, 71, 84
Direct sum of rings, 92
Discriminant of quadratic polynomial, 249
Disjoint cycles, 38
Distance, between vectors, 186
between words, 265
Divisible, for integers, 55
in integral domain, 213
for polynomials, 207
Division algorithm, for integers, 57
for polynomials, 204
Division ring, 101
Domain of mapping, 12
Dual of statement, 286
Duplication of cube, 254-255, 261

$E(3)$, 185
Eisenstein irreducibility criterion, 212-213
Elementary row operations, 319
Embedded ring, 106
Encoder, 264
Endomorphism, 137
Epimorphism, 136
Equations, algebraic, 4-5
cubic, 250-251
quartic, 250-251
solvable by radicals, 5, 251-252
Equivalence class, 53
Error, correction, 266
detection, 266

pattern, 272
probability of, 275-279
Escher, M. C., 174
Euclidean algorithm, for integers, 191
for polynomials, 208
Euclidean domain, 215
Exponent of group, 75
Extension, field, 124
simple, 237
simple algebraic, 238
simple transcendental, 237
group, 144

Factor, of integer, 55
in integral domain, 213
of polynomial, 207
Factor group, 138
Factor theorem, 207
Fedorov, E. S., 183
Feit, Walter, 146
Fermat, Pierre de, 6-7
Fermat's last theorem, 7, 232
Field, 99
algebraically closed, 125
of algebraic numbers, 125
complete ordered, 119
finite, 240-244
Galois, 244
ordered, 118
of quotients, 113-117
root, 239
splitting, 239
Function, 11
Fundamental homomorphism theorem, for groups, 141-142
for rings, 225
Fundamental region, 178
Fundamental theorem, of algebra, 122
of arithmetic, 195
of finite Abelian groups, 83

Galois, Evariste, 5, 251
Galois, group, of polynomial, 252
of subfield, 252
theory, 250-253
Gauss, Carl Friedrich, 55
Gaussian integers, 215-216, 232
Geometric constructions, 6, 254-262
Glide-reflection, 165
$GL(n,F)$, 318
$GL(V)$, 317
"Greater than" concept, 110
Greatest common divisor, of integers, 190
in integral domain, 216
of polynomials, 208
Greatest lower bound, in ordered field, 120
in partially ordered set, 282
Greek alphabet, 10
Group, Abelian, 37
acting on set, 149-150
additive, of ring, 90
alternating, 145
automorphism, of group, 84
crystallographic, point, 178, 183
space, 178, 183-184
two-dimensional, 180-182
cyclic, 43
definition of, 31
dihedral, 166
discrete, 165-166
Euclidean, 185
finite, 34
frieze, 173
general linear, 319
icosahedral, 170
infinite, 34
of inner automorphisms, 152
non-Abelian, 37
octahedral, 170
order of, 34
permutation, 35
real orthogonal, 185

simple, 145
solvable, 251-252
special orthogonal, 186
of symmetries, 47, 164-184
tetrahedral, 170
of translations, 187

Hamilton, W. R., 127
Hamming, R. W., 278
Heesch, H., 184
Hilbert, David, 184
Hölder, Otto, 144
Homomorphic image, 134
Homomorphism, group, 133
natural, 140, 224
ring, 219

Ideal, 220
left, 223
prime, 225, 233
principal, 221
right, 223
Idempotent laws, 287
Identity, element, 23
uniqueness of, in group, 33
mapping, 12
Iff, 14*n*
Image, of element, 12
of mapping, 13
Index of subgroup, 74
Induction, mathematical, 312-314
Induction hypothesis, 314
Injection, 15
Inner product, 185
Inn(G), 152
Integer, square-free, 197
standard form for, 196
Integers, characterized, 112
modulo *n*, 61
relatively prime, 194
Integral domain, definitions of, 96-97
finite, 99
ordered, 108-109
well-ordered, 111
Invariant set, elementwise, 45
Inverse, of element, 23
uniqueness of, in group, 33
mapping, 18
Invertible element, in commutative ring, 102
Irreducible element, in integral domain, 214
Isometry, 46. *See also* Motion
Isomorphism, of Boolean algebras, 295
class, 81
of groups, 76-77
of lattices, 289
of partially ordered sets, 284
of rings, 102
Isomorphism theorem, for groups, first, 146-147
second, 147
for rings, first, 225
second, 226

Join, 285
Jordan-Hölder Theorem, 145

$\mathbb{K}$, 257
Kepler, Johannes, 167
Kernel, of group homomorphism, 134
of ring homomorphism, 220
Kummer, Ernst, 232

Lagrange, Joseph Louis, 75, 251
Lagrange's interpolation formula, 208
Lagrange's theorem, 73-74
Lattice, complemented, 288
complete, 286
definition of, 285
distributive, 288
Lattice group, 182
lattice associated with, 182
Law of quadratic reciprocity, 231

Least common multiple, of integers, 197
  of polynomials, 212
Least element, 111
Least integer principle, 56, 111
Least upper bound, in ordered field, 119
  in partially ordered set, 282
Legendre, A. M., 231
Length of vector, 185
Leonardo da Vinci, 167
"Less than" concept, 110
Linear algebra, 315-320
Linear combination, of integers, 192
  of vectors, 316
Linear transformation, 317
  invertible, 317
  non-singular, 317
Lindemann, C. L. F., 262
Lower bound, in ordered field, 120
  in partially ordered set, 282
$L(V)$, 317

Mapping, 12, 306
  codomain of, 12
  domain of, 12
  image of, 13
  invertible, 18
  one-to-one, 13
  onto, 13
  order preserving, 284
  range of, 15
Mappings, equal, 15
Matrix, addition, 90
  column, 319
  column rank of, 319
  column space of, 319
  generator, 267
  invertible, 318
  of linear transformation, 318
  multiplication, 22
  non-singular, 318
  nullity of, 319
  null space of, 319
  orthogonal, 186
  parity-check, 269
  rank of, 319
  row rank of, 319
  row space of, 319
  standard generator, 269
  transpose of, 319
Maximum-likelihood decoding rule, 266
Meet, 285
Miller, G. A., 50
$M(n,R)$, 318
Modulus of complex number, 128
Monomorphism, 136
Motion, 46
  proper, 165
$M(S)$, 21
Multiple of integer, 55

$\mathbb{N}$, 15
Negative element, 109
Noether, Emmy, 233
Norm, 214
Normalizer, 153
Number, algebraic, 262
  constructible, 256-257
  imaginary, 124
  irrational, 117
  see also Complex numbers; Integers; *and* Real numbers

$O(3)$, 185
Operation, 20
  associative, 22
  commutative, 23
  well-defined, 60
Orbit, 150
Order, of element, 67
  of group, 34
Ordered pair, 12, 306

Partition, 52
Peano postulates, 132
Perfect square, 197
Permutation, 35
Polar form of complex number, 128
Polyhedra, regular convex, 169
Polynomial, 199, 201
  cyclotomic, 213
  degree of, 199, 201
  irreducible, 210
  monic, 199
  prime, 210
  reducible, 210
Polynomials, equal, 199, 201
Positive element, 109
Power set, 281
  as ring, 95
Prime integer, 55
Principal ideal domain, 230-234
Principle of duality, for Boolean algebras, 291
  for lattices, 286
Prism, *n*-, 169
Progression geometric, 312
Proofs, 307-311
Propositions, algebra of, 291-292
Pyramid, *n*-, 168
Pythagoreans, 117
Pythagorean triple, 6, 232, 235

$\mathbb{Q}$, 15
Quadrature of circle, 254-255, 262
Quantifier, existential, 310
  universal, 310
Quaternions, 126-127
Quintilian, 10
Quotient, for integers, 57
  for polynomials, 204
Quotient, group, 138
  ring, 223-224
    of polynomial ring, 226-230

$\mathbb{R}$, 15
Rational numbers, characterized, 117
Real numbers, characterized, 119
Reductio ad absurdum, 308
Reflection through line, 46
Reinhardt, K., 184
Relation, antisymmetric, 280
  equivalence, 51
  reflexive, 51, 280
  symmetric, 52
  transitive, 52, 280
Remainder, for integers, 57
  for polynomials, 204
Remainder theorem, 206
Residue, biquadratic, 231
  quadratic, 231
Residue class, 56. *See also* Congruence, class
Ring, commutative, 94
  definition of, 89-90
  of endomorphisms, 137
  noncommutative, 94
  of polynomials, 199-202
Root, of polynomial, 207
    multiplicity of, 246
  of unity, 129
    primitive, 131
Roots, conjugate, 249
  rational, 248
Rotation, 26
Row-reduced echelon form, 320
Ruffini, Paulo, 251

S_n, 35
Scalar, 315
Schönflies, A., 183
Schreier, Otto, 144
Set, empty (null, vacuous), 305
  infinite, 14
  linearly ordered, 283
  partially ordered, 280
Sets, 304-306

Statement, biconditional, 308
  conditional, 307
  contrapositive of, 308
  converse of, 308
Stone, M. H., 298
Subfield, 100
  prime, 239
Subgroup, definition of, 40
  diagonal, 73
  generated by set, 43
  normal, 135
  p-, 161
  Sylow p-, 161
Subgroups, intersection of, 42
Sublattice, 288
Subring, 97
Subspace, 316
Surjection, 15
Switch, closed, 299
  open, 299
  opposite, 299
Sylow, Ludwig, 75
Sylow's theorem, 159
  extended version of, 161
Symbols, check, 270
  information, 270
Symmetric difference, 299
Symmetric group, 35
Symmetry, 1-4, 164-189
Symmetry groups, 47, 164-184
Sym(S), 35
Syndrome of code vector, 273
Syndrome table, 273-274

$T(3)$, 187
Tautology, 292
Thompson, John, 146
Transcendental element, 237
Translation, 46
Trigonometric form of complex number, 128
Trisection of arbitrary angle, 254-255, 262

Unique factorization domain, 214
Unique factorization theorem, 211
Unit, in integral domain, 213
  element in ring, 94
Unity, of lattice, 286
  of ring, 94
Upper bound, in ordered field, 119
  in partially ordered set, 282

Vector space, 315
Vectors, linearly dependent, 316
  linearly independent, 316
  standard unit, 317
Venn diagrams, 305

Weight of code vector, 272
Well-ordering principle, 111
Weyl, Hermann, 167
Wielandt, Helmut, 160
Word, code, 264
  received, 265

$\mathbb{Z}$, 15
$\mathbb{Z}_n$, 59
Zero, of lattice, 286
  of polynomial, 207
  of ring, 90
Zero divisor, 96
$Z(G)$, 152